Studies in Inorganic Chemistry 21

Luminescence and the Solid State

2nd Edition

Studies in Inorganic Chemistry 21

Other titles in this series

Studies in Inorganic Chemistry 21

Luminescence and the Solid State

2nd Edition

R. C. Ropp

138 Mountain Avenue,
Warren, NJ 07059, U.S.A.

2004

ELSEVIER

Amsterdam – Boston – Heidelberg – London – New York – Oxford
Paris – San Diego – San Francisco – Singapore – Sydney – Tokyo

ELSEVIER B.V.   ELSEVIER Inc.   ELSEVIER Ltd   ELSEVIER Ltd
Sara Burgerhartstraat 25 525 B Street, Suite 1900 The Boulevard, Langford Lane 84 Theobalds Road
P.O. Box 211, 1000 AE San Diego, CA 92101-4495 Kidlington, Oxford OX5 1GB London WC1X 8RR
Amsterdam,   USA   UK   UK
The Netherlands

Second edition 2004

First edition 1991, ISBN 0-444-88940-X

Library of Congress Cataloging in Publication Data
A catalog record is available from the Library of Congress.

British Library Cataloguing in Publication Data
A catalogue record is available from the British Library.

ISBN: 0-444-51661-1

∞ The paper used in this publication meets the requirements of ANSI/NISO Z39.48-1992 (Permanence of Paper).
Printed in The Netherlands.

PREFACE

Since the first date of publication of this book in 1991, the subject of phosphors and luminescence has assumed even more importance in the overall scheme of technological development. Many new types of displays have appeared which depend upon phosphors in their operation. Some of these were pure conjecture in 1991 but are a reality in 2003. The computer has continued to mature and manipulation of bit-rates by several billion per second are common in desktop applications (compared to kilobits per second in 1991). Many, if not all, displays now use computers (or programmable controllers) to operate properly. I have included descriptions of the newer (as well as the older) types of displays in this edition along with an annotated portrait of the phosphors used in each category. Many of these new light sources promise to displace and make obsolete our current light sources, i.e.- incandescent lamps, fluorescent lamps and the ubiquitous color Cathode Ray Tube now used in TV and computer monitors.

The original volume of this book was written to introduce the reader to the science and art of preparing inorganic luminescent materials, namely "phosphors". In order to understand how and why luminescent materials exhibit such specific intrinsic properties, one needs to be thoroughly versed in the science of the solid state. Since that time, I have published a separate volume entitled:

"The Chemistry of Artificial Lighting Devices" - Lamps, Phosphors and Cathode Ray Tubes" - published by Elsevier in 1993

In this volume, I presented the exact formulas and conditions required to make all of the phosphors known at that time. Each formula was the result of many hours of experimentation to optimize the final phoshor composition. You may wish to consult this volume from time to time in order to get an idea of how a given phosphor was made.

Because I believe that one needs to know what the history of any given subject entails, I have retained much of the material presented in the initial edition, while adding information to update the current state of affairs concerning phosphor usage. I have tried to be as comprehensive as possible.

I trust that you will be able to completely understand the complexities of phosphors once you have studied this current volume. It is clear to me that the use of phosphors is changing and will continue to change as new types of electronic devices and displays continue to mature. If you are working in the field, I hope that this presentation is of value in your project.

I have enjoyed preparing this manuscript and hope that you find reading it both profitable and enlightening.

R. C. Ropp,
October, 2003

ACKNOWLEDGEMENT

This book is dedicated to my wife, Francisca Margarita, who has staunchly supported me over the number of years that it has taken to compose the material in this book.

INTRODUCTION

As many of you may not have had the opportunity to investigate the body of knowledge concerning phosphors, I have continued the survey and summarization of the important parts of solid state science in the introductory chapters of this 2nd edition. Many of these chapters have been completely rewritten or modified from the original descriptions. Each chapter has a special contribution to make in your overall understanding of the solid state science of phosphors and luminescence.

For example, Chapter 1 begins with a survey of factors needed to define the solid state. It also contains a survey of x-ray techniques which one needs to know concerning solid state lattice structures of phosphors. Chapter 2 deals with the point-defect aspects of solids in general, which are shown to be thermal in origin. Examples taken from phosphors are presented. The thermodynamics and equilibria of point defects is presented, as well as practical ways to evaluate the effects of point defects in solids. Chapter 3 is new in that mechanisms of nucleation, solid state diffusion and sequences in particle growth are described together. Chapter 4 presents methods of measuring particle shapes and sizes, particle size distributions and growing single crystals. Chapter 5 is very important since the factors associated with energy conversion, electronic transition intensities, energy transfer mechanisms and phonon processes (vibronic coupling) concerning phosphors are presented. Chapter 6 presents the factors identified with design of phosphors and those required to obtain efficient luminescence. The factors affecting phosphor efficiencies are delineated as well. Chapter 7, the final one, describes first the recent advances made in both CRT's and fluorescent lamps, then vacuum fluorescent displays, field emission displays, light emitting diodes and lasers, x-ray intensifying screens and plasma display tubes. All of these utilize phosphors in their construction in some way. I have divided these new displays and light sources into 2 categories: those that use electrons to excite the phosphors and those that depend upon generated high energy photons to excite the phosphors to operate as a device. I have enjoyed the preparation of this manuscript and hope that you find reading it both profitable and enlightening.

CHAPTER ONE

Introduction to The Solid State

This treatise is written to elucidate and explain the characteristics of phosphors and the phenomenon of luminescence. That is, the nature of materials which absorb energy and reĕmit it as light. We encounter phosphors every day when we watch television. The television tube (or plasma panel as it may be) has dots (or lines) composed of red, green and blue-emitting phosphors. These produce the active pictures that we view from Cable or TV-Network broadcasts such as ABC, CBS and NBC. We also use phosphors in fluorescent lamps as inside (and outside) lighting in our homes and offices. However, in order to understand how phosphors perform, we need to know something about the nature of solids in order to comprehend the luminescent state. This chapter will summarize the nature of solids as related to phosphors in general. Of necessity, we cannot be completely comprehensive. Many scientific texts have been written on each of these individual subjects and you can refer to them to get more specific answers to certain questions that cannot be covered here. This chapter will cover the basics of:

1. A comparison of the three states of matter
2. How one determines the structure of solids
3. An introduction to the defect solid state

Because of the importance of the defect solid (all phosphors fall into this category), we will devote a chapter to the defect-state. Since we need to know how phosphors are made, a chapter covering solid state reactions is also mandatory. Finally, we will cover both single crystal formation and particle size determinations. Some phosphors are single crystal. A good example is that of the laser. The first laser invented by Maiman used a single crystal of ruby, i.e.- aluminum oxide or sapphire (corundum) activated by trivalent chromium (Cr^{3+}). A great variety of laser crystals are now in use including YAG:Nd^{3+}, i.e.- $Y_3Al_5O_{11}$: Nd^{3+}. Since most phosphors are used as powders, we will also include a chapter showing how one determines particle size and particle size distributions. We will also include a description of the nature of light and photons in general since phosphors emit photons once they are excited by absorption of energy. Color is also an important component of photons as we perceive them.

Because I believe that one needs to be aware of the history of science in order to thoroughly understand any part of it, I will try to include as much background and results of investigations as they pertain to phosphors. For example, the methods of determining structures of solids did not occur overnight but were developed slowly over a period of time once Röntgen discovered x-rays. Nowadays, if you are in this field, you probably use a computer-controlled program to collect diffraction data automatically and to do the calculations for you. Yet, you must to know how the calculations need to be accomplished in order to ascertain whether you have obtained an accurate depiction of the structure of any material, or not.

When a detailed study of solids is undertaken, one quickly determines that solids have properties which differ profoundly from those of the other states of matter. We will be concerned mainly with inorganic solids, although the same principles apply to organic solids as well. It is well to note here that OLEDs, or "Organic Light Emitting Diodes" are the promise of the future to replace inorganic phosphors as light-emitting displays in certain applications. Inorganic solids are the primary materials of construction for use in electronics, lighting, communications, information storage and display, superconductors, and others at this time, but have been supplanted by organic based materials in some applications.

In comparing the three states of matter, our approach will be as follows. First, we will compare the three (3) basic states of matter, namely gases, liquids and solids. We will then contrast these states energetically and atomistically. Next, we will discuss structures of solids and the factors involved in determination of crystal structure. Finally, we will introduce the concept of the defect solid and how such defects affect the macroscopic properties of the solid state.

We can summarize differences between the three states of matter as follows:

a. The solid is the most condensed phase of the three normally possible, and contains the least energy.

b. The solid has but three (3) degrees of freedom in contrast to those usually present in the gaseous state. (The number actually present in a gas depends upon the number of atoms in the molecule).

c. All solids contain defects. There is no such thing as a perfect solid.

1.1. - CHANGES OF STATE

Of the three states of matter, gaseous, liquid and solid, **the** major difference involves their atomistic freedom of movement. For example, a gas molecule can move in three (3) dimensions without restriction and has full vibrational and rotational modes within the molecule. We know that it is possible to condense gases to liquids to solids by removal of energy. Each state represents a succeedingly lower degree of condensation, and has a characteristic temperature range at which it exists, the gas being the "hottest", and solids the "coldest". **Thus, solids contain the least energy.** The actual temperature range for the gaseous, liquid and solid state of a given material depends, of course, upon the nature of the atoms involved. Even elements differ quite widely. Gallium, for example, is a metal that is liquid slightly above room temperature, but does not vaporize until above 1100 °C. Tungsten, also a metal, does not melt until the temperature is above 3850 °C.

An elementary way to understand how the solid differs from the gaseous or liquid states involves the following example. Take the MX_2 molecule. In the gaseous state, there will be a maximum of 3 x 3, or nine degrees of freedom, for the molecule composed of these three atoms (there are three dimensions, x, y, & z, in which to operate). These can be divided into 3-translational, 3-vibrational and 3-rotational degrees of freedom. But as we change the state of matter, we know that the translational degrees of freedom present in gases disappear in solids (and are restricted in liquids). For gaseous molecule, the 3 vibrational degrees of freedom will have ($2J+1 = 7$) rotational states superimposed upon them (J here is the

number of atoms in the molecule, assuming quantized vibrational states).
If we measure the absorption spectra of our molecule, MX_2 , in the infra-
red region of the spectrum, we obtain results similar to those shown as
follows:

1.1.1.-

Energy States of a Triatomic Molecule as a Function of States of Matter

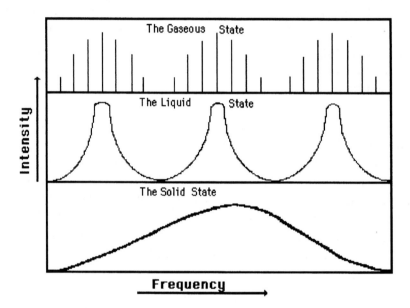

The gaseous phase shows seven well separated rotational and vibrational
states, the liquid phase broadened vibrational plus rotational states (they
are no longer distinct), while the solid exhibits only one broad featureless
band in this region of the spectrum. We conclude that we are limited to
three degrees of freedom in the solid state. The 3-translational and 3-
rotational degrees of freedom have disappeared because of the imposition
of **long range ordering** of MX_2 molecules in the solid. This point cannot
be over-emphasized since long range ordering is the **major difference**
between solids and the other states of matter.

1.2.- ENERGETICS OF CHANGES OF STATE

When heat energy is added to a solid, one of two changes will occur. Either the internal heat content (temperature) of the material will change or a ***change of state*** will occur. On an atomistic level, this involves an increase in vibrational energy which manifests itself either as an increase in internal temperature, or the breaking of bonds in the solid to form either a liquid or gas. The most common example used to illustrate energy involved in changes of state is that of water. We know that water, a liquid, will change to a solid (ice) if its internal temperature falls below a certain temperature. Likewise, if its temperature rises above a certain point, water changes to a gas (steam). Because water is so plentiful on the Earth, it was used in the past to define Changes of State and even to define **Temperature Scales**.

The invention of the thermometer is generally credited to Galileo. In his instrument, built about 1592, the changing temperature of an inverted glass vessel produced the expansion or contraction of the air within it, which in turn changed the level of the liquid with which the vessel's long, open-mouthed neck was partially filled. This general principle was perfected in succeeding years by experimenting with liquids such as alcohol or mercury and by providing a scale to measure the expansion and contraction brought about in such liquids by rising and falling temperatures.

By the early 18th century as many as 35 different temperature scales had been devised. The German physicist, Daniel Gabriel Fahrenheit, in the period of 1700-30, produced accurate mercury thermometers calibrated to a standard scale that ranged from 32, the melting point of ice, to 96 for body temperature. The unit of temperature (degree) on the Fahrenheit scale is 1/180 of the difference between the boiling (212) and freezing points of water. The first centigrade scale (made up of 100 degrees) is attributed to the Swedish astronomer Anders Celsius, who developed it in 1742. Celsius used 0 for the boiling point of water and 100 for the melting point of snow. This was later inverted to put 0 on the cold end and 100 on the hot end, and in that form it gained widespread

use. It was known simply as the centigrade scale until in 1948 the name
was changed to honor Celsius. In 1848 the British physicist William
Thompson (later Lord Kelvin) proposed a system that used the degrees
that Celsius used, but was keyed to absolute zero (-273.15 °C). The unit of
this scale is now known as the Kelvin, i.e.- °K. The Rankine scale employs
the Fahrenheit degree keyed to absolute zero (-459.67 °F) , i.e.- °R.
These are the four temperature scales that we employ today.

The factors involved in energy change, i.e.- temperature change, include:
"heat capacity", i.e.- C_p or C_v, and "heat of transformation", **H**. The former
is connected with internal temperature change whereas the latter is
involved in changes of state. The actual names we use to describe **H**
depend upon the direction in which the temperature change occurs, vis:

1.2.1.-	CHANGE OF STATE	H, HEAT OF:	TEMPERATURE	
			°C.	°F
	ice to water	Fusion	0	32
	water to ice	Solidification	0	32
	water to steam	Vaporization	100	212
	steam to water	Condensation	100	212

Note that when a change of state, i.e.- water to ice or ice to water, occurs,
there is no change in temperature while this is occurring (this is a
concept that is sometimes difficult for beginning scholars to grasp). Heat
capacity was originally defined in terms of water. That is, heat capacity
was defined as the amount of heat required to raise the temperature of
one *cubic centimeter* of water (whose density is defined as 1.0000 @ 4
°C.) by one(1) degree. "Heat" itself is defined as the amount of energy
required to raise 1 cc. of water by one degree, i.e.- one (1.0) calorie. The
calorie in turn was originally defined as the amount of heat required to
raise the temperature of **one gram of water** from 14.5 °C to 15.5 °C at a
constant pressure of one (1) atmosphere. Heat capacity was also known as
thermal capacity. We thus label heat capacity as C_p, meaning the thermal
capacity at **constant pressure.** Originally, C_v, the thermal capacity at
constant volume was also used, but its use is rare nowadays. This is due to
the fact that some materials do not have a linear temperature expansion.

These are two types of "heat" involved in the thermal changes of any given material. We specify the "heat" of a material, i.e.- its internal energy, by $H_{S,L,G}$. where S, L, or G refer to solid, liquid or gas. Thus, the relation between $H_{S,L,G}$ and C_p is:

1.2.2.- $\{H = C_p \cdot T\}_{S,L,G}$,   or $\Delta H_{S,L,G} = C_{p(S,L,G)} \; \Delta T_{S,L,G}$

where the **actual** state of matter is either S, L, **or** G.

> Note that: **The internal temperature of a material does not change as the material undergoes a change of state. All of the energy goes into forming a new state of matter.**

For a change of state between solid and liquid, we would have:

1.2.3- $\{H_S - H_L\} = C_p (T_{(S)} - T_{(L)})$ or $\Delta H = C_p \; \Delta T$

ΔH here is a heat of transformation involved in a **change of state** whereas $\Delta H_{S,L,G}$ (sometimes written $\Delta E_{S,L,G}$) refers to **change in internal heat** for a given state of matter.

We can now calculate the amount of energy required to raise **one gram of ice** at - 10 °C. to form **one gram of steam** at 110 °C:

1.2.4.- <u>CALORIES REQUIRED TO CHANGE 1 Gm OF ICE TO STEAM</u>

FORM	ΔT	C_p	$C_p \Delta T$	ΔH_X	TOTAL
Ice	-10 to 0	0.5	5.0 cal.	---	5.0 cal.
(Ice	0	---	---	80 cal.	80.0 cal.
to water)				(fusion)	
Water	100	1.0	100 cal.	---	100.0 cal.
(Water	0	---	---	540 cal.	540.0 cal.
To steam)				(vaporization)	
Steam	10	0.5	5.0 cal.	-----	5.0 cal.

$$=======$$
$$730.0 \text{ cal.}$$

Note that in this case ΔH_x is specified in terms of the type of change of state occurring, while $C_p \, \Delta T$ (= ΔH) is the **change in internal heat** which occurs as the temperature rises. At a given change of state, all of the energy goes to the change of state and the temperature does not change until the transformation is complete.

1.3. - PROPAGATION MODELS AND THE CLOSE-PACKED SOLID

We have already said that the solid differs from the other states of matter in that long range ordering of atoms or molecules has appeared. To achieve long range order in any solid, one must stack atoms in a symmetrical way **to completely fill space**. If the atoms are all of one kind, i.e.- one of the elements, the problem is straight forward. Sets of eight atoms, each set arranged as a cube, will generate a cubic structure. Two sets of three atoms, each set of three arranged in a triangle, will propagate a hexagonal pattern with three dimensional symmetry. The following diagram illustrates this point:

1.3.1.

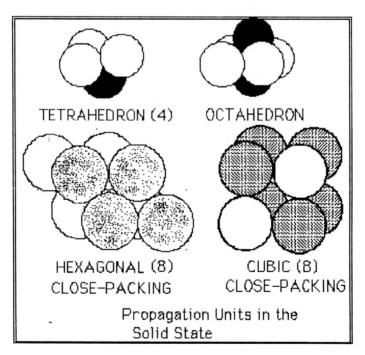

TETRAHEDRON (4)   OCTAHEDRON

HEXAGONAL (8)   CUBIC (8)
CLOSE-PACKING   CLOSE-PACKING

Propagation Units in the
Solid State

This diagram shows a perspective of cubic and hexagonal close-packing propagation units in their space-filling aspects. What we mean by propagation units are solid state building blocks that we can stack in a symmetrical form to infinity. We do not consider 1-, 2- or 3-atom units since they are trivial. That is, they are only 1- or 2-dimensional at best. But, 4-atoms will form a 3-dimensional **tetrahedron** (half a cube is only 2-dimensional) which is a valid propagation unit. This means is that we can take tetrahedrons and fit them together 3-dimensionally to form a symmetrical structure which extends to infinity. However, the same is not true for 5-atoms, which forms a four-sided pyramid. This shape cannot be completely fitted together in a symmetrical and space-filling manner. Even if we stack these pyramids, we find that their translational properties preclude formation of a symmetrical structure (There is too much lost space!) However, if one more atom is added to the pyramid, we then have an **octahedron** which is space-filling with translational properties.

Going further, combinations of seven atoms are asymmetrical, but eight atoms form a cube which can be propagated to infinity. Note that by turning the top layer of four atoms (see 1.3.1.) by 45° , we have a hexagonal unit which is related to the hexagonal unit composed of 6 atoms, two triangles atop of each other. By taking Ping-Pong balls and gluing them together to form the propagation units as shown in 1.3.1., one can get a better picture of these units.

Although we can continue with more atoms per propagation unit, it is easy to show that all of those are related to the four basic propagation units found in the solid state (and depicted in 1.3.1.), to wit:

1.3.2.-	Tetrahedron (4)	Octahedron	(6)
	Hexagon (6 or 8)	Cube	(8)

We thus conclude that structures of solids are based, in general, upon four **basic** propagation units, which are stacked in a symmetrical and space-filling form to near infinity. Variation of structure in solids depends upon whether the other atoms forming the structure are larger or smaller than the basic propagation units composing the structure. In many cases they are smaller and will fit into the **interstice** of the propagation unit. A good example is the phosphate tetrahedron, PO_4. The P^{4+} atom is small enough to fit into the center (interstice) of the tetrahedron formed by the four oxygen atoms. If we combine it with Er^{3+} (which is slightly smaller than PO_4), we obtain a tetragonal structure, a sort of elongated cube of high symmetry. But if we combine it with La^{3+} , which is larger than PO_4^{3-} , a monoclinic structure with low symmetry results.

Now, you may think that this discussion is mainly conjecture since the case in nature may differ considerably. Actually, it is easy to ascertain that most of the elements (which are homogeneous, i.e.- having the all of the same type of atoms) form structures that are either cubic or hexagonal. This is shown in the following diagram:

1.3.3.-

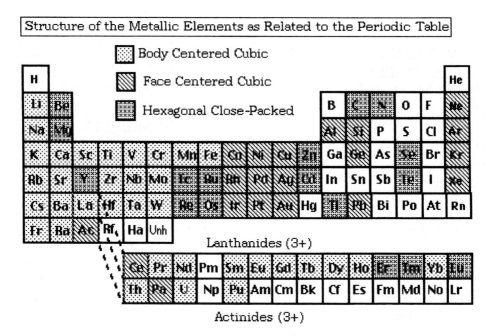

Structure of the Metallic Elements as Related to the Periodic Table

It should be clear that the **metal** elements have either a face-centered, body-centered or hexagonal close-packed structure. These three structural units are shown in the following diagram:

1.3.4.-

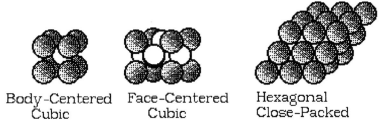

Body-Centered   Face-Centered   Hexagonal
    Cubic   Cubic   Close-Packed

A better perspective is given in the 1.4.14 diagram, given below. Therein, we show the seven types of close-packed structures, i.e.- Bravais lattices, that appear in the solid state. Of those elements whose structures are not indicated in the above diagram, the following table shows their structure:

Table 1-1
Crystal Structures of Some Elements

Gases at Room Temp	Liquids at Room Temp.	Other Solid Metals
H, He = hexagonal		
Ne, Ar, Kr, Xe, Rn (Solid = fcc)	Ga = orthorhombic Hg = rhombohedral	Rhombohedral B, As, Sb, Bi
F = cubic Cl, Br = orthorhombic		Orthorhombic I, S, Np
N = hexagonal O = cubic		P, Po - Monoclinic In, Sn- Tetragonal
fcc= face-centered cubic		Co, Ni, Cu, etc.

You will note that even though some elements are gases or liquids at room temperature, the structure that they form in the solid has been determined. Some of the temperatures required to condense them into the solid phase are close to absolute zero.

But, if the atoms are not all of the same kind, i.e.- heterogeneous, we have a different problem. Part of the problem lies in the fact that 6.023 x 10^{23} molecules comprise one mole and we must stack these in a symmetrical manner to form a close-packed solid. For $CaCO_3$, this represents 100. 1 grams of powder. We find in general that solid structures are based on the largest atom present, as well as how it stacks together (its valence) in space filling-form. For most inorganic compounds, this is the **oxygen atom**, e.g.- oxides, silicates, phosphates, sulfates, borates, tungstates, vanadates, etc. The few exceptions involve chalcogenides, halides, hydrides, etc., but even in those compounds, the structure is based upon aggregation of the largest atom, i.e.- the sulfur atom in ZnS. Zinc sulfide exhibits two structures, sphalerite- a cubic arrangement of the sulfide atoms, and wurtzite- a hexagonal arrangement of the sulfide atoms. Divalent zinc atoms have essentially the same coordination in both structures.

This brings us to a discussion of the structure of solids and how structure is determined.

1.4.- THE STRUCTURE OF SOLIDS

We have indicated that solids can have several symmetries. Let us now examine the structure of solids in more detail. In 1895, Röntgen experimentally discovered "x-rays" and produced the first picture of the bones of the human hand. This was followed by work by von Laue in 1912 who showed that solid crystals could act as diffraction gratings to form symmetrical patterns of "dots" on a photographic film whose arrangement depended upon how the atoms were arranged in the solid. It was soon realized that the atoms formed "planes" within the solid. In 1913, Sir William Henry Bragg and his son, William Lawrence Bragg, analyzed the manner in which such x-rays were reflected by planes of atoms in the solid. They showed that these reflections would be most intense at certain angles, and that the values would depend upon the distance between the planes of atoms in the crystal and upon the wavelength of the x-ray. This resulted in the Bragg equation:

1.4.1.- $n \lambda = 2d \sin \theta$

where d, the distance, is in angstroms (Å = 10^{-8} cm) between planes and
θ is the angle in degrees of the reflection. In Bragg's x-ray diffraction
equation, i.e.- the angle , θ , is actually the angle between a given **plane** of
atoms in the structure and the **path** of the x-ray beam. The unit, "d", is
defined as the distance between planes of the lattice and λ is the
wavelength of the radiation. It was Georges Friedel who in 1913
determined that the intensities of reflections of x-rays from these planes
in the solid could be used to determine the symmetry of the solid. Thus,
by convention, we usually define planes, not points, in the lattice. We can
define the structure of any given solid in terms of its **lattice points.** What
this means is that if we substitute a point for each atom (ion) composing
the structure, we find that these points constitute a lattice, i.e.- an
internal configuration, having certain symmetries.

A lattice is not a structure per se. **A lattice is defined as a set of three-
dimensional points.** These points may, or may not, be totally occupied by
the atoms composing the structure. Consider a cubic structure such as
that shown in the following diagram:

1.4.2.-

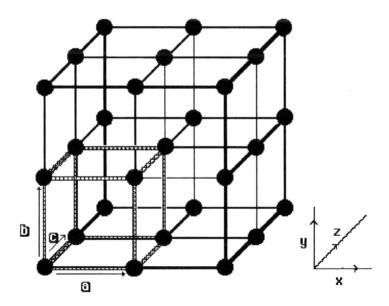

Here we have a set of atoms (ions) arranged in a simple three-dimensional cubic pattern. The lattice directions are defined, **by convention**, as x , y & z. Again, by convention, "x" is defined as the right hand direction from the origin, "y" is in the vertical direction and "z" is at an angle to the plane of x & y (In our cube, the angle is 90°) . Note that there are eight (8) cubes in our example. The **unit-cell** is the smallest cube. The unit-cell directions are defined as the "lattice-unit-vectors". That is, the x, y, & z directions of the unit cell are vectors having directions corresponding to x @ a ; y @ b ; z @ c with lengths of each unit- vector being equal to 1.0. (Our notation for a **vector** henceforth is a letter which is "outlined", i.e.- the a unit-cell translation vector). T, the **translation vector** , is then:

1.4.3.- $$T = n_1 a + n_2 b + n_3 c$$

where n_1 , n_2 , and n_3 are intercepts of the unit-vectors, a , b , c , on the x , y , z - directions in the lattice, respectively. The **unit cell volume** is then:

1.4.4.- $V = \{ a \cdot b \times c \}$ (This is a "dot - cross" vector product).

Thus, it is easy to see that in Bragg's x-ray diffraction equation, the angle, θ , is actually the angle between a given **plane** of atoms in the structure and the **path** of the x-ray beam. The unit, "d", is defined as the distance between planes of the lattice and λ is the wavelength of the radiation.

As we said, we usually define planes, not points, in the lattice. The reason that we do this is that the waves of electromagnetic radiation are constructively diffracted by planes of atoms in the solid rather than points in the lattice. One system that has come into general use is that of "MILLER INDICES" which is represented by: { **h** , **k** , **l** }.

In this system, we use **a , b, & c** (not unit-cell vectors) to represent the lengths of intercepts which define the planes **within the unit cell**. Miller Indices are the reciprocals of the intercepts, a , b & c, of the chosen plane on the x , y , z - directions in the lattice.

To illustrate this concept, examine the symmetry elements of our cube, shown as follows:

1.4.5.-

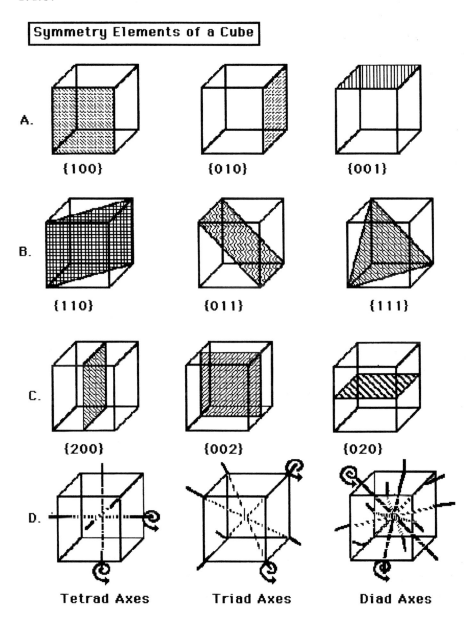

Note that in all cases, we have the intercepts of the a, b & c specified as Miller Indices. This simply means that given these indices, we can determine where the atoms lie within the unit cell of the lattice. This is indicated in the following:

1.4.6.-   a, b, c   MILLER INDICES
                  (1/2, 0 , 0)   (200)
                  (0 , 1/2 , 0)   (020)
                  (0, 0, 1/2)   (002)

Note also the (100), (110) and (111) planes are illustrated. Planes are important in solids because, as we will see, they are used to locate atom positions within the lattice structure. The TETRAD, TRIAD , AND DIAD AXES ARE ALSO SHOWN IN **Part D.** These are **rotational symmetry** axes. That is, the triad axis must be rotated 3-times in order to bring a given corner back to its original position.

The final **factor** to consider is that of the **angle** between the x , y , and z directions in the lattice. In our examples so far, angles were 90° in all directions. If the angles are not 90°, then we have additional lattices to define. For a given unit-cell defined by the axes a, b and c, the corresponding angles are defined as: α , β , γ , where α is the angle in the x-direction, etc.

In 1921, Ewald developed a method of calculating the sums of diffraction intensities from different planes in the lattice by considering what is called the "Reciprocal Lattice". The reciprocal lattice is obtained by drawing perpendiculars to each plane in the lattice, so that the axes of the reciprocal lattice are perpendicular to those of the crystal lattice. This has the result that the planes of the reciprocal lattice are at right angles (90°) to the real planes in the unit-cell. Ewald used a sphere to represent how the x-rays interact with any given lattice plane in three dimensional space. He employed what is now called the *Ewald Sphere* to show how reciprocal space could be utilized to represent diffraction of x-rays by lattice planes. Ewald originally **rewrote** the Bragg equation as:

1.4.7.- $\quad \sin q \quad = \quad \dfrac{n\,l}{2d\,\{hkl\}} \quad = \quad \dfrac{1/d\,\{hkl\}}{2/l}$

Using this equation, Ewald applied it to the case of the diffraction sphere which we show in the following diagram:

1.4.8.-

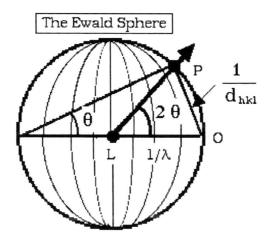

In this case, the x-ray beam enters the sphere enters from the left and encounters a lattice plane, L. It is then diffracted by the angle 2θ to the point on the sphere, P, where it is registered as a diffraction point on the reciprocal lattice. The distance between planes in the reciprocal lattice is given as $1/d_{hkl}$ which is readily obtained from the diagram. It is for these reasons, we can use the Miller Indices to indicate planes in the real lattice, based upon the reciprocal lattice.

The reciprocal lattice is useful in defining some of the electronic properties of solids. That is, when we have a semi-conductor (or even a conductor like a metal), we find that the electrons are confined in a band, defined by the reciprocal lattice. This has important effects upon the conductivity of any solid and is known as the "band theory" of solids. It turns out that the reciprocal lattice is also the site of the Brillouin zones, i.e.- the "allowed" electron energy bands in the solid. How this originates is explained as follows.

The free electron resides in a quantized energy well, defined by k (in wave-numbers). This result can be derived from the Schrödinger wave-equation. However, in the presence of a periodic array of electromagnetic potentials arising from the atoms confined in a crystalline lattice, the energies of the electrons from all of the atoms are severely limited in orbit and are restricted to specific allowed energy bands. This potential originates from attraction and repulsion of the electron clouds from the periodic array of atoms in the structure. Solutions to this problem were made by Bloch in 1930 who showed they had the form (for a one-dimensional lattice):

1.4.9.- $\Psi = e^{ikx} u(x)$ - one dimensional

               $\Psi_k (a) = e^{ika} u_k(x)$ - three dimensional

where k is the wave number of the allowed band as modified by the lattice, a may be x, y or z, and $u_k(x)$ is a periodical function with the same periodicity as the potential. One representation is shown in the following diagram, given as 1.4.10. on the next page.

We have shown the least complicated one which turns out to be the simple cubic lattice. Such bands are called "Brilluoin" zones and , as we have said, are the allowed energy bands of electrons in any given crystalline lattice. A number of metals and simple compounds have been studied and their Brilluoin structures determined.

However, when one gives a representation of the energy bands in a solid, a "band-model" is usually presented. The following diagram, presented as 1.4.11. on a succeeding page, shows three band models as used to depict energy states of any given solid. They include insulators, semi-conductors and metals (conductive). This is. In the solid, electrons reside in the valence band, (as defined by the Brilloin zones in the reciprocal lattice) but can be excited into the conduction band by absorption of energy. The energy gap of various solids depends upon the nature of the atoms comprising the solid. Semi-conductors have a rather narrow energy gap (forbidden zone) whereas that of insulators is wide (metals have little or no gap). That is, the allowed bands in metals overlap with the valence

1.4.10.- The Allowed Energy Bands (Brillouin Zones) in a Crystal

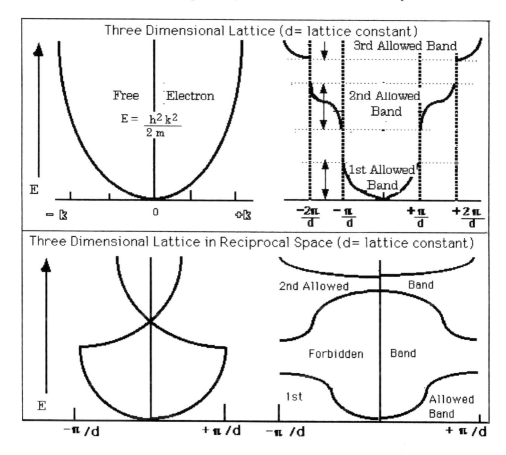

band. Note that energy levels of the atoms "A" in 1.4.11. are shown in the valence band. These will vary depending upon the nature atoms present. Electrons can be excited into these low-lying states, depending upon the temperature of the material.

Thus, we find that a total of three (3) factors are needed to define a given lattice and its structure. This is shown as follows:

1.4.12.-

      I - unit-cell axes , intercepts and angles

      II - rotational symmetry

      III - localized space group symmetry

1.4.11.- Energy Band Models Used to Depict Energy States of Solids

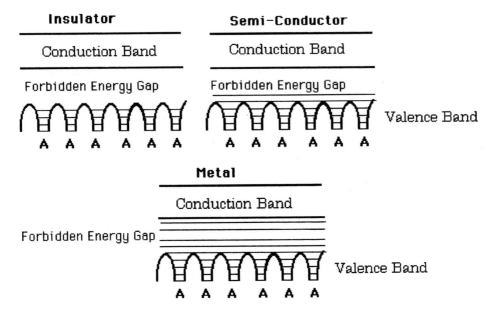

It is these three factors (see 1.4.12.) which give rise to the different symmetries of solids:

> Factor I gives rise to the 14 Bravais lattices
> Factor II generates the 32 point- groups
> Factor III creates the 232 space- groups

In these three factors, each contributes to the total number of symmetrical lattices that can appear in the solid state. This will become evident in the following discussion: In Factor I (which we have already considered), if a ≠ b ≠ c and α ≠ β ≠ γ, then we have a different lattice than that defined by a = b ≠ c ; α = β ≠ γ The number of combinations that we can make from these 3-lengths and 3-angles is seven (7) and these define the 7 unique lattice structures, called BRAVAIS LATTICES. These have been given names, as shown in 1.4.13. on the next page.

Each of these seven lattices may have sublattices, the total being **14**. If we arrange the crystal systems in terms of symmetry, the cube has the

1.4.13.-   CUBIC   TETRAGONAL   HEXAGONAL
           ORTHORHOMBIC   TRIGONAL   MONOCLINIC
           TRICLINIC

highest symmetry and the triclinic lattice, the lowest symmetry. We can therefore arrange the seven (7) systems into a hierarchy as shown in the following diagram:

1.4.14.-

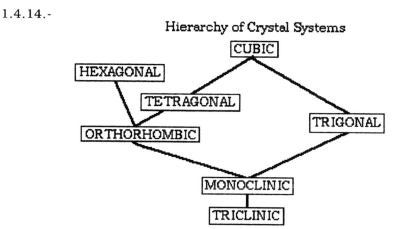

Hierarchy of Crystal Systems

The highest symmetry lattice is at the top, while the lowest is at the bottom. These 14 Bravais lattices are shown in 1.4.15. given on the next page.

If we now apply **rotational symmetry** (Factor II) to the 14 Bravais lattices, we obtain the 32 Point-Groups with the factor of symmetry imposed upon them. The symmetry elements that have been used are shown as follows:

1.4.16.-   Rotation axcs
                Plane symmetry   <   horizontal
                                          vertical
                Inversion symmetry (mirror)

Table 1-2, lists these on the following page with corresponding symbols, and the relation between axes and angles associated with each structure. These are the 14 Bravais Lattices which are unique in themselves.

1.4.15.

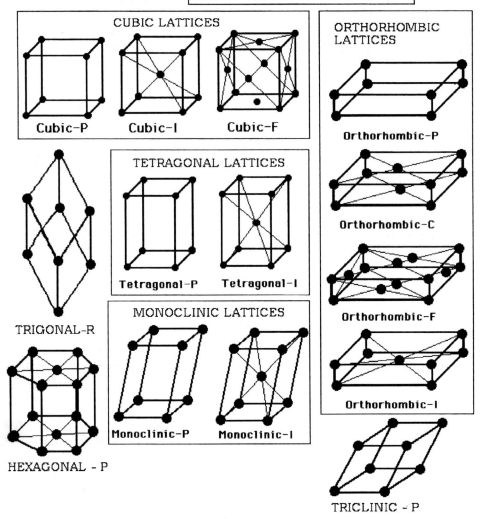

THE 14 BRAVAIS SPACE LATTICES

CUBIC LATTICES

Cubic-P   Cubic-I   Cubic-F

TETRAGONAL LATTICES

Tetragonal-P   Tetragonal-I

MONOCLINIC LATTICES

Monoclinic-P   Monoclinic-I

TRIGONAL-R

HEXAGONAL - P

ORTHORHOMBIC LATTICES

Orthorhombic-P

Orthorhombic-C

Orthorhombic-F

Orthorhombic-I

TRICLINIC - P

Note that only cubic, tetragonal and orthorhombic lattices have 90° angles in all lattices directions. Monoclinic and hexagonal lattices have two 90° directions while trigonal and triclinic lattices have none. Table 1.2. lists the angles of all of the 14 Bravais lattices, with lattice types, i.e.- body-centered, etc., symbols and angles within the lattices defined. Note that in trigonal- and hexagonal- P symmetries, $\gamma = 120°$.

AMERICAN SOCIETY FOR TESTING AND MATERIALS 1916 Race St., Philadelphia, Penna. 19103, we can look up the most probable composition. This turns out to be $LaAl_{11}O_{18}$. Usually, we will know how the material was made and the components used to make it. If not, we can analyze for constituents. In this case, we would find La and Al, and would surmise that we have an oxidic compound. We find that there are two compounds possible, viz.- $LaAlO_3$ and $LaAl_{11}O_{18}$. However, the 3 **most intense lines** of $LaAlO_3$ are:

1.5.4.-   d = 2.66 Å(100), 3.80 Å(80) and 2.19 Å(80).

This does not fit our pattern. We do find that the pattern for $LaAl_{11}O_{18}$ is identical to the x-ray pattern we obtained.

Using 1.5.3., d-values can then be calculated from the 2θ values using the Bragg equation, for the hexagonal composition, $LaAl_{11}O_{18}$. The {h,k,l} values can be calculated from special formulas developed for this purpose. These are given in Table 1-3, presented on the next page. These values are obtained by trying certain values in the hexagonal formula, and seeing if the results conform. What this means is that we substitute Miller Indices into the formula and see if the calculated value matches $1/d^2$. Once this is done, we have characterized our material. Because of the physical geometry of the x-ray diffraction goniometer (angle-measuring device), one obtains values of 2θ directly.

Note that the equations in Table 1-3 for the high symmetry lattices are rather simple while those for low symmetry lattices are complicated. Since we know that the compound, $LaAl_{11}O_{18}$ is hexagonal, we use the equation:

1.5.5.- $1/d^2 = 4/3\ [(h^2 + k^2 + l^2)/\ a^2]$

If we do this, we obtain the data in Table 1-4 also shown on the next page. This allows us to determine the unit cell lengths for our compound as:

Hexagonal: $a_0 = 5.56.$ Å ; $b_0 = 22.04$ Å

Table 1- 3
Plane Spacings for Various Lattice Geometries

CUBIC

$1/d^2 = h^2 + k^2 + l^2/a^2$

TETRAGONAL

$1/d^2 = h^2 + k^2/a^2 + l^2/c^2$

HEXAGONAL

$1/d^2 = 4/3\,[(h^2 + k^2 + l^2)/a^2]$

ORTHORHOMBIC

$1/d^2 = (h^2/a^2) + (k^2/b^2) + (l^2/c^2) + l^2/c^2$

RHOMBOHEDRAL

$$1/d^2 = \frac{(h^2 + k^2 + l^2)\sin^2 a + 2(hk + kl + hl)(\cos^2 a - \cos a)}{a^2(1 - 3\cos^2 a + 2\cos^3 a)}$$

MONOCLINIC

$1/d^2 = \{1/\sin^2 b\}\{ h^2/a^2 + (k^2 \sin^2 b)/b^2 + l^2/c^2 - 2hl \cos b / ac \}$

TRICLINIC

$1/d^2 = 1/V^2\{ S_{11} h^2 + S_{22} k^2 + S_{33} l^2 + 2 S_{12} hk + 2 S_{23} kl + 2 S_{13} hl$

where: V = volume of unit cell ; $S_{11} = b^2 c^2 \sin^2 a$; $S_{22} = a^2 c^2$
$\sin^2 b$; $S_{33} = a^2 b^2 \sin^2 g$; $S_{12} = abc^2 (\cos a \cos b - \cos g)$;
$S_{23} = a^2 bc (\cos b \cos g - \cos a)$; $S_{13} = ab^2 c (\cos g \cos a - \cos b)$

TABLE 1-4
Conversion of 2θ Values of the Diffraction Pattern to {hkl} Values

2θ	I/I_0	d	{hkl}	2q	I/I_0	d	{hkl}
8.02	16	11.02	002	36.18	74	2.48	114
16.1	6	4.81	004	39.39	27	2.29	023
18.86	32	4.71	001	40.94	29	2.20	0010
20.12	28	4.41	012	42.79	66	2.11	025
22.07	10	4.03	013	45.01	46	2.01	026
24.23	21	3.67	001	53.36	15	1.72	029
24.55	11	3.63	014	58.57	36	1.58	127
32.19	44	2.78	110	60.07	60	1.54	0211
32.50	15	2.76	008	67.35	48	1.39	220
33.23	15	2.70	112	71.61	18	1.32	0214
34.02	100	2.64	017	95.42	14	1.04	2214

Additionally, we can list the x-ray parameters and convert them to structural factors as shown. These values are averaged over all of the reflections used for calculation. Note that this pattern has several planes where the "d" value is more than ten.

Let us consider one other example. Suppose we obtained the following set of 2θ values and intensities for a compound:

TABLE 1-5

DIFFRACTION LINES AND INTENSITIES OBTAINED

Intensity	2θ in degrees
96	29.09
100	33.70
56	48.40
50	57.48
14	60.28
18	78.38
10	90.55
14	80.78
17	97.80
19	118.26
11	121.00
7	144.34
9	148.49

The first three values are the strongest diffraction lines. After calculating "d" values and looking up the set of strong lines which correspond to our set, we find that the probable compound is CdO_2 , or cadmium peroxide. This compound turns out to be cubic in structure, with a_0 = 5.313 Å . When we calculate the {h,k,l} values of the diffracting planes, the strongest line is found to be {200}. We can then make the determination that since Cd^{2+} is a strongly diffracting atom (it has high atomic weight, which is one way of stating that it has many electron shells, i.e.- $1s^2 \ 2s^2 \ 2p^6 \ 3s^2 \ 3p^6 \ 3d^{10} \ 4s^2 \ 4p^6 \ 4d^{10}$, the structure is probably face-centered cubic. Indeed, this turns out to be the case. In the unit cell, Cd atoms are in the special positions of : $\{0,0,0\}$, $\{1/2,1/2,1/2\}$; $0,1/2.1/2\}$; $\{1/2,1/2,0\}$. There are four molecules per unit cell. We could continue further and calculate intensities, I_C , using atomic scattering factors already present in prior literature. We would scale calculated intensities to our observed intensities by $S\ I_C = S\ I_0$. We then calculate a reliability factor, called R, from $R = S\ (I_0 - I_c\ )\ /S\ I_0$. A low value indicates that our selection of lattice parameter was correct. If not, we choose a slightly different value and apply it. The details of the procedure for determining exact structure and atomic positions in the lattice are well known, but are beyond the scope of this Chapter. However, for your own edification, I request that you **index** the lines (calculate the {hkl} values) given above, similar to that presented in Table 1-4.

Summarizing to this point, we have shown that only certain propagation units can be stacked to infinity to form **close-packed solids**. We have also shown how the units fit together to form specific solids with specific symmetries. The nature of the structure of solids has also been reviewed in some detail. Now let us look at the solid from the standpoint of stacking and stacking defects.

1.6. - THE DEFECT SOLID

We have shown that by stacking atoms or propagation units together, a solid with certain symmetry aspects results. If we have done this properly, a perfect solid should have resulted with no defects in it. Yet, defects are related to the **entropy of the solid**, and a perfect solid would violate the second law of thermodynamics. This law states that **zero entropy** is only possible at absolute zero temperature. Thus, most of the solids that we encounter **are** defect-solids. It is natural to ask concerning the nature of these defects. It should be obvious that most of them will be stacking defects or faults. **Nature** (and indeed **we** in our propagation unit example as well) finds it impossible to stack atoms (molecules) in perfect order to infinity. Moreover, even if we could obtain a perfect solid, it would likely be non-reactive and would be **singularly** stable. The reason for this is that it is the defects in solids which give them special properties.

Consider the surface of a solid. In the interior, we see a certain symmetry which depends upon the structure of the solid. As we approach the surface from the interior, the symmetry begins to change. At the very surface, the surface atoms see only half the symmetry that the interior atoms do (and half of the bonding as well). Reactions between solids take place at the surface. If there were some way to complete the symmetry of the surface atoms, then they too would likely be nearly non-reactive.

In a three-dimensional solid, we can conceive of three major types of defects, one-, two- and three- dimensional in nature. These are called: point, line (edge) and volume (plane), respectively. Point defects are changes at atomistic levels, while line and volume defects are changes in stacking of groups of atoms (molecules). An easy way to visualize point defects is shown in the following:

1.6.1.-

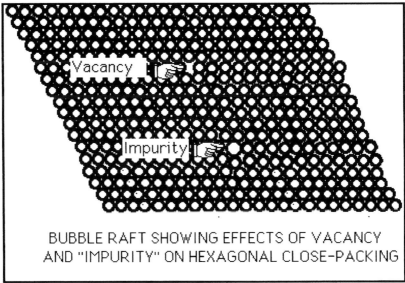

BUBBLE RAFT SHOWING EFFECTS OF VACANCY AND "IMPURITY" ON HEXAGONAL CLOSE-PACKING

A bubble raft is made by creating bubbles in a soap solution which float to the surface of the liquid (in this case, water) to create a raft. The trick is to get the size of the bubbles all the same. If not, one does not get a way to compare the effects of close packing using bubbles to simulate atom positions in a lattice structure. This problem was worked on until the investigator was successful.

The first thing one notices is that packing defects have a significant effect on the close packing of a bubble raft. Two types of defects can be seen. One is a "vacancy", that is, the bubble supposed to be there, is missing. The other is an "impurity", here as a larger bubble. Note its effect on the degree of ordering. Again, lattice compensation is the norm and the close packing is compensated by an adjustment in the lattice. Note that most of the bubbles are all the same and are hexagonally-close packed. You can imagine the effect on stacking in the lattice of a very impure solid where the compound is only 95% pure.

In the following is another view of defects:

1.6.2.-

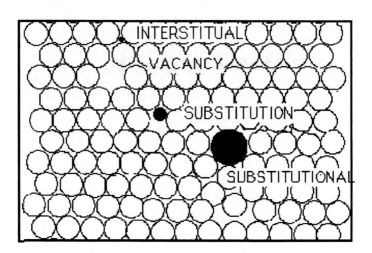

In this diagram, we see three types of point defects. In addition to the vacancy, we also see two types of substitutional defects. Both are direct substitutions in the "lattice", or arrangement of the atoms. One is a smaller atom, while the other is larger than the atoms comprising the lattice. Note the difference, due to size of the impurity, upon the ordering in the hexagonally close-packed lattice. In addition, we have also included the "interstitial" atom, that is, one that is able to insinuate itself into the interstices of the lattice. This completes the list of possible point defects in a mono-atomic lattice. In the next Chapter, we will show other point defects possible when both cation and anion atom-species are present in the lattice. In the following diagram, we see the effect of an edge dislocation upon hexagonal close packing:

1.6.3.-

LINE DEFECTS DUE TO PACKING (1) AND DEFECTS
DUE TO INHOMOGENEITIES (2).

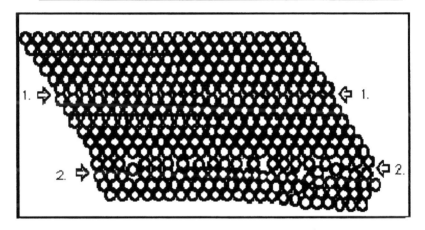

At "1" in 1.6.3., the hexagonal packing has changed to that of cubic close-packing, causing a line defect. At "2", the line defect is caused by lack of ordering along the plane. This may also be regarded as a series of point defects affecting ordering. What has happened is that many defect-vacancies have congregated and have caused a boundary between the "grains". Within the grain-boundary, the ordering is uniform but differs from the next grain. It is actually a line imperfection (2-dimensional), or an edge imperfection.

A better perspective of an edge dislocation is shown in 1.6.4., presented on the next page.

1.6.4 -

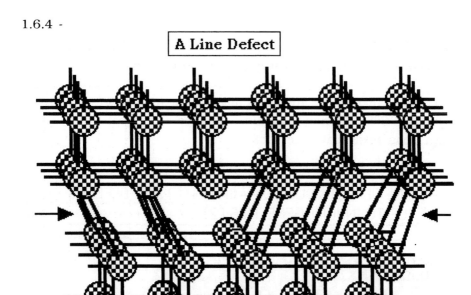

A Line Defect

Note that in its simplest form, an edge dislocation is an omission of a line of atoms composing the lattice. Another perspective is presented in the following, shown as 1.6.5. on the next page.

In this diagram, the area within the lattice around the line defect is under both compression and tension due to the difference in atom-density as one passes through it in a direction perpendicular to the line defect. Thus, there is excess energy in the lattice due to the compressive-tensile forces present. This has the effect of causing the edge dislocations to propagate through the solid by a hopping motion until they reach the surface of the solid. There have been several cases where it has been possible to directly observe line imperfections by suitable preparation and microscopic examination of the surface in reflected light. One example is the MgO crystal. MgO is a cubic crystal and it is possible to etch it along the {100} direction (this is the direction along the x-axis). What is observed is a series of surface lines of specific length. The reason that

CHAPTER 2

The Point Defect

There are two types of defects associated with phosphors. One involves controlled point defects in which a foreign activator cation is incorporated in the solid in defined amounts. The other involves line and point defects inadvertently formed in the solid structure because of impurity and entropy effects. This chapter will define and characterize the nature of all of these point defects in the solid, their thermodynamics and equilibria. It will become apparent that the type of defect present will depend upon the nature of the solid in which they are incorporated. That is, the characteristics of the point defects in a given phosphor will depend upon its chemical composition. Of necessity, this chapter is not intended to be exhaustive, and the reader is referred to the many treatises concerned with the point defect.

2.1. - TYPES OF POINT DEFECTS

Let us now consider the defect solid from a general perspective. Consider the case of semi-conductors, where most of the atoms are the same, but the total of the charges is not zero. In that case, the excess charge (n- or p- type) is spread over the whole lattice so that no single atom, or group of atoms, has a charge different from its neighbors. However, most inorganic solids are composed of charged moieties, half of which are positive (cations) and half negative (anions). The total charge of the cations equals, in general, that of the anions. If an atom is missing, the lattice readjusts to compensate for this loss of charge. If there is an extra atom present, the charge-compensation mechanism again manifests itself. Another possibility is the presence of an atom with a charge **larger or smaller** than that of its neighbors. In a given structure, cations are usually surrounded by anions, and vice-versa (Remember what we said in Chapter 1 wherein it was stated that most structures are oxygen-dominated). Thus, a cation with an extra charge needs to be compensated by a like anion, or by a nearest neighbor cation with a lesser charge. An example of charge-compensation for a divalent cation sub-lattice would be the following **defect equation:** :

2.1.1.- $2 M^{2+} \leftrightharpoons M^+ + M^{3+}$

where the M^{3+} and M^+ are situated on nearest neighbor cation sites, *which were originally divalent.*

> **Thus, the charge compensation mechanism represents the single most important mechanism which operates within the defect solid.**

Because of this, the number and types of defects, which can appear in the solid, are limited. This restricts the number of defect types we need to consider, in both elemental (all the same kind of atom) and ionic lattices (having both cations and anions present). We have shown that by stacking atoms or propagation units together, a solid with specific symmetry results. If we have done this properly, a perfect solid should result with no holes or defects in it. Yet, the 2nd law of thermodynamics demands that a certain number of point defects (vacancies) appear in the lattice. It is impossible to obtain a solid without some sort of defects. A perfect solid would violate this law. The 2nd law states that **zero entropy** is only possible at absolute zero temperature. Since most solids exist at temperatures far from absolute zero, those that we encounter **are** defect-solids. It is natural to ask what the nature of these defects might be, particularly when we add a foreign cation (activator) to a solid to form a phosphor.

Consider the surface of a solid. In the interior, we see a certain symmetry which depends upon the structure of the solid. As we approach the surface from the interior, the symmetry begins to change. At the very surface, the surface atoms see only half the symmetry that the interior atoms do. Reactions between solids take place at the surface. Thus, the surface of a solid represents a defect in itself since it is not like the interior of the solid.

In a three-dimensional solid, we can postulate that there ought to be

three major types of defects, having either one-, two- or three-dimensions. Indeed, this is exactly the case found for defects in solids, as we briefly described in the preceding chapter. We have already given names to each of these three types of defects. Thus a one-dimensional defect of the lattice is called a "point" defect, a two-dimensional defect a "line" or "edge" defect and a three-dimensional defect is called a "plane" or "volume" defect. We have already described, in an elementary way, line and volume defects and will not address them further except to point out how they may arise when certain point defects are present. It is sufficient to realize that they exist and are important for anyone who studies homogeneous materials such as metals.

Point defects are changes at atomistic levels, while line and volume defects are changes in stacking of planes or groups of atoms (molecules) in the structure. The former affect the **chemical** properties of the solid whereas the latter affect the **physical** properties of the solid. Note that the arrangement (structure) of the individual atoms (ions) are not affected, only the method in which the structure units are assembled. That is, the structure of the solid remains intact in spite of the presence of defects. Let us now examine each of these defects in more detail, starting with the one-dimensional lattice defect and then with the multi-dimensional defects. We will find that specific types have been found to be associated with each type of dimensional defect which have specific effects upon the stability of the solid structure. It should be clear that the type of point defect prevalent in any given solid will depend upon whether it is homogenous (same atoms) or heterogeneous (composed of differing atoms).

1. The Point Defect in Homogeneous Solids

We begin by identifying the various defects which can arise in solids and later will show how they can be manipulated to obtain desirable properties not found in naturally formed solids. Let us look first at the homogeneous type of solid. We will first restrict our discussion to solids which are stoichiometric, and later will examine solids which can be classified as "non-stoichiometric", or having an excess of one or another of one of the building blocks of the solid. These occur in semi-conductors as well as other types of electronically or optically active solids.

Suppose you were given the problem of identifying defects in a homogeneous solid. Since all of the atoms in this type of solid are the same, the problem is somewhat simplified over that of the heterogeneous solid (that is- a solid containing more than one type of atom or ion). After some introspection, you could speculate that the homogeneous solid could have the following types of point defects:

2.1.2.- Types of Point Defects Expected in a Homogeneous Solid

      *   Vacancies   *****   Substitutional Impurities
      *   Self-interstitial   *****   Interstitial Impurities

On the left are the two types of point defects which involve the lattice itself, while the others involve impurity atoms (Note that interstitial atoms can involve either an impurity atom or the same atom that makes up the lattice structure itself). Indeed, there do not seem to any more than these four, and indubitably, no others have been observed. Note that we are limiting our defect family to point defects in the lattice and are ignoring line and volume defects of the lattice. These four point defects, given above, are illustrated in the following diagram, given as 2.1.3. on the next page.

Note that what we mean by an "interstitial" is an atom that can fit into the spaces between the main atoms in the crystalline array. In this case, we have shown a hexagonal lattice and have labeled each type of point defect. Observe that we have shown a vacancy in our hexagonal lattice, as well as a foreign interstitial atom which is small enough to fit into the **interstice** between the atoms of the structure. Also shown are two types of substitutional atoms, one larger and the other smaller than the atoms composing the principal hexagonal lattice. In both cases, the hexagonal packing is disrupted due to a "non-fit" of these atoms in the structure. Additionally, we have illustrated another type of defect that can arise

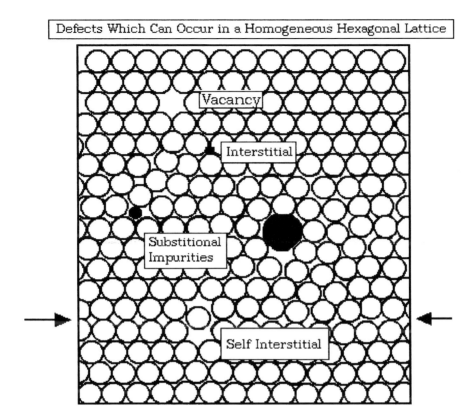

within the homogeneous lattice (in addition to the vacancy and substitutional impurities that are bound to arise). This is called the "self-interstitial". Note that it has a decisive effect on the structure at the defect. Since the atoms are all the same size, the self-interstitial introduces a **line-defect** in the overall structure. It should be evident that the line-defect introduces a difference in packing order since the close packing **at the arrows** has changed to cubic and then reverts to hexagonal in both lower and upper rows of atoms. It may be that this type of defect is a major cause of the line or edge type of defects that appear in most homogeneous solids. In contrast, the other defects produce only a disruption in the **localized** packing order of the hexagonal lattice, i.e.- the defect does not extend throughout the lattice, but only close to the specific defect. It should be evident that metals or solid solutions of metals (alloys) show such behavior in contrast to heterogeneous lattices which involve compounds such as ZnS. This accounts for the tremendous discrepancy between theoretical and actual strength of certain alloys in practical applications due to "fatigue" failure when the object is being used.

Now, suppose that we have a **solid solution** of two (2) elemental solids. Would the point defects be the same, or not? An easy way to visualize such point defects is shown in the following diagram:

2.1.4.- Defects in the Homogeneous Solid Containing 2 Solids in Solution

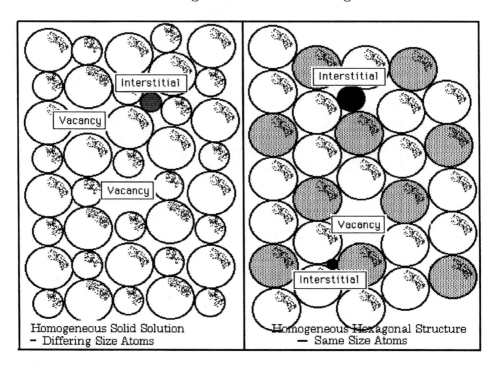

Here, we use a hexagonally-packed representation of atoms to depict the close-packed solid. Both types of homogeneous solids are shown, where one solid is composed of the same sized atoms while the other is composed of two different sized atoms. On the right are the types of point

defects that could occur for the same sized atoms in the lattice. That is, given an array of atoms in a three dimensional lattice, only these two types of lattice point defects could occur where the size of the atoms are the same. The term "vacancy" is self-explanatory but "self-interstitial" means that one atom has slipped into a space between the rows of atoms. In a lattice where the atoms are all of the same size, such behavior is energetically very difficult unless a severe disruption of the lattice occurs (usually a "line-defect" results). This behavior is quite common in certain types of homogeneous solids. In a like manner, if the metal-atom were to have become misplaced in the lattice and were to have occupied one of the interstitial positions, as shown in the different sized atom solid (see 2.1.4.) then the lattice is disrupted by its presence at the interstitial position. This type of defect has also been observed. Note that the atoms are usually not charged in the homogeneous lattice.

Summarizing, three types of point defects are evident in a homogeneous lattice. In addition to the Vacancy, two types of substitutional defects can also be delineated. Both are direct substitutions in the "lattice", or arrangement of the atoms. One is a smaller atom, while the other is larger than the atoms comprising the lattice. In both cases, the lattice arrangement affects the hexagonal ordering of the lattice atoms around it. The lattice packing is seen to be affected for many lattice distances.

It is for this reason that compounds containing impurities sometimes have quite different chemical reactivities than the purest ones. However, the interstitial impurity does not affect the lattice ordering at all. Now, let us look at the heterogeneous lattice

II. The Point Defect in Heterogeneous Solids

The situation concerning defects in heterogeneous inorganic solids is similar to that given above, except for one very important factor, that of **charge** on the atoms. Covalent inorganic solids are a rarity while ionicity or partial ionicity seems to be the norm. Thus, heterogeneous solids are usually composed of charged moieties, half of which are positive (cations) and half negative (anions). In general, the total charge of the cations will equal that of the anions (Even in the case of semi-conductors, where the total of the charges is not zero, the excess charge (n- or p- type) is spread over the whole lattice so that no single atom, or group of atoms, ever has a charge different from its neighbors. Note also that most of the semi-conductors that we use are homogeneous in nature, modified by homogeneous additions to form p- or n-type electrically charged areas). In a given structure, cations are usually surrounded by anions, and vice-versa. Because of this, we can regard the lattice as being composed of a **cation sub-lattice** and an **anion sub-lattice**. (Remember what was stated in Chapter 1 concerning the fact that most structures are oxygen-dominated). What we mean by a "sub-lattice" is illustrated in the following diagram:

2.1.5.-

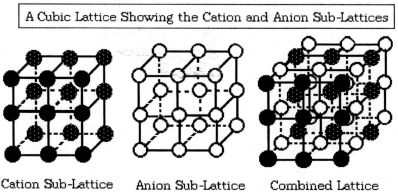

A Cubic Lattice Showing the Cation and Anion Sub-Lattices

Cation Sub-Lattice   Anion Sub-Lattice   Combined Lattice

In this case, we have shown both the cubic cation and anion "sub-lattices separately, and then the combination. It should be clear that all positive charges in the cation sub-lattice will be balanced by a like number of negative charges in the anion sub-lattice, even if excess charge exists in one or the other of the sub-lattices. If an atom is missing, then the overall lattice readjusts to compensate for this loss of charge. If there is a different atom present, having a differing charge, the charge-compensation mechanism again manifests itself. Thus, a cation with an extra charge needs to be compensated by a like anion, or by a nearest neighbor cation with a lesser charge. An example of this type of charge-compensation mechanism for a divalent cation sub-lattice would be the following **defect equation:**

2.1.6.- $2\ Ca^{2+} \leftrightharpoons Li^+ + Sb^{3+}$

where the Sb^{3+} and Li^+ are situated on nearest neighbor cation sub-lattice sites, in the divalent Ca^{2+} sub-lattice. Note that a total charge of 4+ exists on both sides of the above equation.

Thus, the charge compensation mechanism represents the single most important mechanism which operates within the *defect ionic* solid.

We find that the number and types of defects, which can appear in the heterogeneous solid, are limited because of two factors:

1) The charge-compensation factor
2) The presence of two sub-lattices in the ionic solid.

These factors restrict the number of point defect types we need to consider in ionic heterogeneous lattices (having both cations and anions present). For ionic solids, the following types of defects have been found to exist:

*   Schottky defects (absence of both cation and anion)
*   Cation vacancies
*   Anion vacancies
*   Frenkel defects (Cation vacancy plus same cation as interstitial)
*   Interstitial impurity atoms (both cation and anion)
*   Substitutional impurity atoms(both cation and anion)

These defects are illustrated in the following diagram, given as 2.1.7. on the next page.

Note that, in general, anions are larger in size than cations due to the extra electrons present in the former. A hexagonal lattice is shown in

2.1.7.-

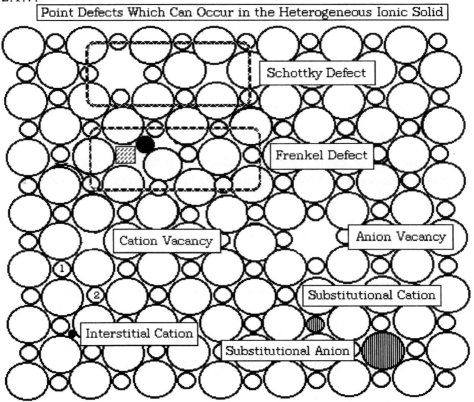

Point Defects Which Can Occur in the Heterogeneous Ionic Solid

2.1.7. with both Frenkel and Schottky defects, as well as substitutional defects. Thus, if a cation is missing (cation vacancy) in the cation sub-lattice, a like anion will be missing in the anion sub-lattice. This is known as a Schottky defect (after the first investigator (1935) to note its existence).

In the case of the Frenkel defect, the "square" represents where the cation was supposed to reside in the lattice before it moved to its interstitial position in the cation sub-lattice. Additionally, "Anti-Frenkel" defects can exist in the anion sub-lattice. The substitutional defects are shown as the same size as the cation or anion it displaced. Note that if they were not, the lattice structure would be disrupted from regularity at the points of insertion of the foreign ion.

To summarize, the categories of point defects possible for these two types of lattices are illustrated are:

    1). In an elemental solid, we may have:

        ✳   Vacancies
        ✳   Self-interstitial
        ✳   Interstitial Impurities
        ✳   Substitutional Impurities

    2). In the ionic solid, i.e.- heterogeneous solid, we may have:

        ✳   Schottky defects (absence of both cation and anion)
        ✳   Cation or anion vacancies
        ✳   Frenkel defects (Cation vacancy plus the same cation as interstitial)
        ✳   Interstitial impurity atoms (both cation and anion)
        ✳   Substitutional impurity atoms(both cation and anion)

All of these point defects are intrinsic to the solid. The factors responsible for their formation are entropy effects (point defect faults) and impurity effects. At the present time, the highest-purity materials available still contain about 1.0 part per billion of various impurities, yet are 99.9999999 % pure. Such a solid will contain about 10^{14} impurity atoms per mole. So it is safe to say that all solids contain impurity atoms, and that it is unlikely that we shall ever be able to obtain a solid which is **completely pure and does not contain defects.**

2.2. - THE PLANE NET

Now, let us consider how such defects arise in any given solid. The easiest way to visualize how intrinsic defects occur in the solid is to study the **PLANE NET**. Imagine that we have M as cations and X as anions (we shall ignore formal charge for the moment). This is shown in the following diagram:

2.2.1.-

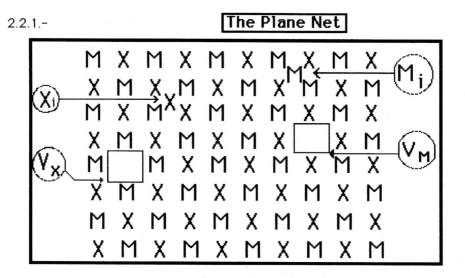

It is easy to see that we can stack a series of these "NETS" to form a three-dimensional solid. Note that we have used the labeling: V = vacancy; i = interstitial ; m = cation site; x= anion site and s = surface site. One might think that perhaps V^-_m and V^+_x ought to be included in our list of vacancies. However, a negatively-charged cation vacancy **alone,** particularly when it is surrounded by negative anions, would not be very stable. Neither should a positively-charged anion vacancy be any more stable. Either arrangement would require high energy stabilization to exist. Therefore, we do not include them in our listing.

However, a V_m could capture a positive charge to become V^+_m and likewise for the V_x which then becomes a V^-_x Both of these are stable when surrounded by oppositely charged sites. We have already stated that surface sites are special. Hence, they are included in our listing of intrinsic defects. The same criteria apply concerning charge on the defect.

For the plane net, we can expect the following types of intrinsic defects:

termed the "M-Center" as shown above in the two diagrams, 2.2.8. and 2.2.9.

The optical properties of these two types of centers are given in the following diagram:

2.2.10.-

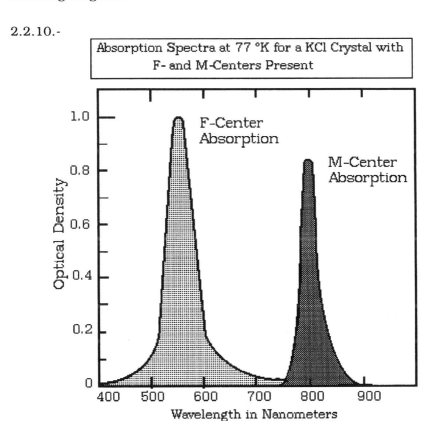

Absorption Spectra at 77 °K for a KCl Crystal with F- and M-Centers Present

The defect equation for formation of the M-center is also given as follows:

2.2.11.-   M-center: $KCl = KCl_{1-\delta} + \delta/2\, Cl_2 \uparrow + \delta/4\, [\, V^{\cdot}_{Cl} \mid V^{-}_{Cl}\,]$

Note that each vacancy in 2.2.9. has captured an electron, in response to the charge-compensation mechanism which is operative for the defect reactions. These associated, negatively-charged, vacancies have quite different absorption properties than that of the F-center.

There are other impurity systems to which this notation can be applied. For the case of an AgCl crystal containing the Cd^{2+} cation as an impurity, we have:

2.2.12.- $2\ Ag^+ \leftrightarrows [\ Cd^{2+}, V_{Ag}\ ]$

This is an example of a **heterotype** system. Another such system is $CdCl_2$ containing Sb^{3+}.

Here, we can write at least three different equations involving defect equilibria:

2.2.13.- $2\ Cd^{2+} \leftrightarrows Sb^{3+} + p^+ + V_{Cd}$

$2\ Cd^{2+} \leftrightarrows Sb^{3+} + V^+{}_{Cd}$

$2\ Cd^{2+} \leftrightarrows Sb^{3+} + Li^+$

In the last equation, charge-compensation has occurred due to inclusion of a monovalent cation. All of these equations are cases of impurity substitutions.

Another type is the so-called homotype impurity system. The substance, nickelous oxide, is a pale-green insulator, when prepared in an inert atmosphere. If it is reheated in air, or if a mixture of NiO and Li^+ is reheated in an inert atmosphere, the NiO becomes a black semi-conductor. This is a classical example of the effect of defect reactions upon the intrinsic properties of a solid:

2.2.14.- $2\ Ni^{2+} \leftrightarrows [\ Ni^{3+} / V_{Ni}\ ] + p^+$

$2\ Ni^{2+} \leftrightarrows Ni^{3+} + Li^+$

This behavior is typical for transition metals which easily undergo changes in valence in the solid state.

Up to this point, we have only investigated stoichiometric lattices. Let us now examine non-stoichiometric lattices in light of our symbolism. Consider the semi-conductor, Ge. The defect reactions associated with the formation of p-type and n-type lattices are:

2.2.15. -   **n-type:** $Ge + \delta\, As = [\,Ge/As_\delta\,] + \delta\, e^-$

          p-type: $Ge + \delta\, Ga = [Ge/Ga_\delta\,] + \delta\, p^+$

The excess charges shown are spread over the entire lattice, as stated before.

2.3. - DEFECT EQUATION SYMBOLISM

Whether you realize it or not, we have already developed our own symbolism for defects and defect reactions based on the Plane Net. It might be well to compare our system to those of other authors, who have also considered the same problem in the past. It was probably Rees (1930) who wrote the first monograph on defects in solids. Rees used $\square_M$ to represent the cation vacancy, as did Libowitz (1974). This has certain advantages since we can write the first equation in 2.2.12. as:

2.3.1- $2\ Cd^{2+} \rightleftharpoons \left\{ \boxed{Sb}^{+}_{Cd} + \square_{Cd} \right\} + p^+$

Likewise, the other equations become:

2.3.2.- $2\ Cd^{2+} \rightleftharpoons \boxed{Sb}^{+}_{Cd} + \boxed{p}^{+}_{Cd}$

and:

2.3.3.- $2\ Cd^{2+} \rightleftharpoons \boxed{Sb}^{+}_{Cd} + \boxed{Li}_{Cd} + p^+$

Although the results are equal as far as utility is concerned, we shall continue to use our symbolism, for reasons which will become clear later.

The following compares defect symbolism, as used by prior Authors. Note that our symbolism most resembles that of Krœger, but not in all aspects.

2.3.4-	Rees(1930)	Krœger (1954)	Libowitz (1974)
Cation Site Vacancy:	□☐$_M$	V_M	□☐$_M$
Anion Site Vacancy	□☐$_X$	V_X	□$_X$
Cation Interstitial	Δ$_M$	M_i , M^+_i	M_i
Anion Interstitial	Δ$_X$	X_i , X^-_i	X_i
Negative Free Charge	e	e^-	e^-
Positive Free Charge	p	h+	h+
Interstices	---	$\alpha\, V_i$	$\alpha\, V_i$
Unoccupied Interstitial	---	V_i	Δ
Anti-structure Occupation	---	---	M_M , X_X , M_X , X_M

These prior authors have considered some intrinsic defects that we have not touched, namely interstices and the so-called "anti-structure" occupation. The latter deals with an impurity anion on a cation site coupled with an impurity cation on an anion site, both with the proper charge.

We have mentioned interstices but not in detail. They appear as a function of structure. There is one site in a tetrahedron, four in a body-centered cube, and six in a simple cube. Thus, α in {$\alpha\, V_i$} is 1, 4 or 6, respectively. We shall need this symbol later, as well as V_i , the unoccupied interstitial.

2.4. - SOME APPLICATIONS FOR DEFECT CHEMISTRY

Before we proceed to analyze defect reactions by a mathematical approach, let us consider two applications of solid state chemistry. We begin with a description of some phosphor defect chemistry.

I. - <u>Phosphors</u>

In the prior literature, it was found (Kinney- 1955) that $Ca_2 P_2 O_7$ could be activated by Sb^{3+} to form the phosphor: $Ca_2P_2O_7$: $Sb_{.02}$ (this formalism actually means a solid-solution of two pyrophosphate compounds, i.e.- $[(Ca_{.99},Sb_{.01})_2P_2O_7]$). The brightness response of this phosphor was moderate when excited by ultraviolet radiation but was improved four times by the addition of Li^+ . The optimum amount proved to be that exactly equal to the amount of Sb^{3+} present in the phosphor. The defect reactions occurring were:

2.4.1.- Defect Reactions Occurring in Calcium Pyrophosphate Phosphor

PHOSPHOR <u>BRIGHTNESS</u>

$$2\ Ca^{2+} \leftrightharpoons Sb^{3+}{}_{Ca} + V^+{}_{Ca} \qquad\qquad 25\ \%$$
$$\text{or } (2\ Ca^{2+} \leftrightharpoons Sb^{3+}{}_{Ca} + V_{Ca} + p^+)$$

$$2\ Ca^{2+} \leftrightharpoons Sb^{3+}{}_{Ca} + Li^+{}_{Ca} \qquad\qquad 100$$

It is well known that phosphor brightness in a phosphor is proportional to the numbers of activator ions, i.e.- Sb^{3+} ions, actually incorporated into the pyrophosphate structure. Phosphors are prepared by heating the ingredients at high temperature (> 1000° C.) to obtain a compound having high crystallinity. The sintering process decreases entropy and is **counterproductive** to the formation of vacancies in the pyrophosphate lattice. In the absence of Li^+ , lack of vacancy-formation actually decreases the amount of Sb^{3+} incorporated into activator sites. Apparently, four times as many activator ions were incorporated into the lattice when the charge-compensating Li^+ ions were present on nearest neighbor sites.

Note that we have written two defect reactions for the case of vacancy formation in 2.4.1. Pyrophosphate is an insulator and the formation of a positively-charged vacancy is much more likely than the vacancy plus a free positive charge.

THUS, ALTHOUGH MORE THAN ONE DEFECT REACTION MAY BE
APPLICABLE TO A GIVEN SITUATION, ONLY ONE IS USUALLY FAVORED
BY THE PREVAILING THERMODYNAMIC AND ELECTRICAL
CONDITIONS.

II.- Lithium Niobate

Lithium niobate, $LiNbO_3$, is a photorefractive material, discovered in 1966
(Ashkin et al). That is, it is an electroöptic material in which the indices
of refraction can be changed by an applied electric field. It is used in
optical devices which employ its nonlinear optical and electroöptic
properties. As a single crystal, $LiNbO_3$ has high transparency to
electromagnetic radiation. If a laser beam is directed down the length of
such a crystal, its frequency is doubled. $YAG:Nd^{3+}$, i.e.- $Y_3Al_5O_{12}:Nd^{3+}$, is a
commonly used laser crystal, whose emission lies at 10,600 Å. (near-infrared radiation). The combination of a $LiNbO_3$ crystal and $YAG:Nd^{3+}$,
produces a laser beam at 5,300 Å (green light). If a second crystal is also
incorporated in the optical setup, radiation at 2,650 Å (ultraviolet
radiation) is obtained. Although the radiation frequency changes, so does
the intensity of light produced. Losses of 100 times or more are common.
However, this detrimental factor is overcome by increasing the power of
the laser beam. Fortunately, $LiNbO_3$ has a high resistance to damage by a
laser beam, unlike many other similar crystals. It gains its unique
characteristics because the crystal structure has "built-in" defects.

$LiNbO_3$ is a ferroelectric structure related to that of the cubic perovskite
($CaTiO_3$) structure. That is, electric polarization of the lattice electrons
occurs upon application of an electric field. Although a voltage will induce
such polarization, so will the electric vectors of a beam of light,
particularly that of a laser beam. However, $LiNbO_3$ consists of distorted
oxygen-octahedra sharing faces so that a planar hexagonal arrangement
results. The pile-up of the octahedra along the perpendicular direction, c-
axis, follows the cation sequence "Li, Nb and **vacant site**". The point
symmetry group is C_3 (see 1.4.14. and 1.4.16 of the first chapter) with
the trigonal axis along the cation rows. It is very close to a C_{3v} (3m)
configuration.

The following diagram shows the arrangement of the three ions in the LiNbO$_3$ structure.

2.4.2.-

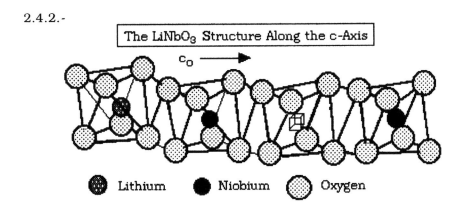

Note that we have a structure with a "built-in" crystal defect, a vacancy. Both the lithium and niobium cations are in an octahedral coordination. In fact, the two ions, Li$^+$ and Nb^{5+}, have nearly the same radius and occupy octahedral sites with the same C$_{3v}$ symmetry. The lithium deficiency in congruent crystals is accommodated by means of Nb$_{Li}$ anti-sites and Nb^{5+} vacancies in a relative concentration that guaranties overall electrical neutrality. Note that many physical properties depend upon stoichiometry, e.g.- Curie temperature, absorption spectra, lattice parameters and photorefractive yield.

However, it has been found that impurities play a major role in the operation of a frequency-doubling crystal like LiNbO$_3$. Many impurities have ionic radii similar to that of lithium. They substitute at Li$^+$ sites rather than Nb^{5+} sites (possibly because the NbO$_4$ coordination is stronger at the niobium site). Among these are: Mn^{2+}, Fe^{3+} and Ni^{2+}. As we have already seen, substitution of such multivalent cations on a monovalent site results in lattice compensation such as oxygen vacancies and the formation of color-centers at the oxygen vacancies. These interfere with the photorefractive properties of such defect crystals since the ease of electric polarization of the lattice is impaired during use. Although , as we will see, the crystal-growing process is also a purification process, it is not able to exclude all of the impurities as the crystal grows.

III. Bubble Memories

"Bubble memory" is the term applied to the device which uses a "soft" magnetic material to carry information. If a ferromagnetic film such as europium gallium garnet is grown epitaxially upon a suitable substrate such as gadolinium gallium garnet, i.e.- $Gd_3 Ga_5 O_{11}$ (= GGG), it forms magnetic domains in which the electron spins of the cations are aligned in the same direction in the same domain. This is shown in the following:

2.4.3.- A Ferromagnetic Film Grown Epitaxially on GGG

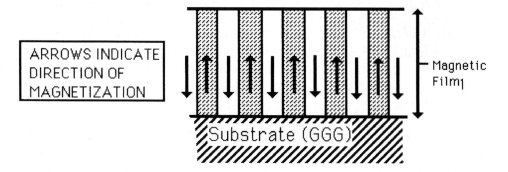

The following diagram, given as 2.4.4. on the next page, illustrates how these would look under polarized light (the Faraday effect) using crossed Nicol polarizers. (The black and white parts are domains of opposite polarity).

When a magnetic field is applied, with the field vector horizontal to the film, the domains collapse to form separated cylinders within the film, as shown. These appear to be "bubbles" when viewed from the top, hence the name. The bubbles then become mobile under the influence of a separate electric field and will move. Actually, the electric field causes the domain-wall to collapse by a spin-flip mechanism, while the cylinder **volume** is maintained by the magnetic field.

2.4.4.-

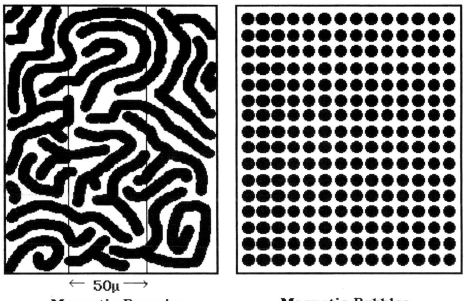

| Magnetic Bubbles as Viewed from the Top of the Film in Polarized Light |

← 50μ →

Magnetic Domains   **Magnetic Bubbles**

This causes the apparent movement of the wall. Bubbles densities as high as 10^7 per cm^2, and bubble velocities of up to $10^3 - 10^4$ cm./sec. have been reported. Obviously, bubble velocity depends upon E, the electric field strength, as well as the composition of the epitaxially grown film. The use of a vapor-phase-deposited metal grid upon the surface of the film serves to switch bubbles from site to site. The presence (1) or absence (0) of a bubble is detected by polarized light beams. Thus, information can be stored, and retrieved, on a "chip" in binary language.

However, the epitaxial film must be defect-free. GGG is difficult to grow as a defect-free single crystal. Frenkel defects appear, and give rise to line dislocations. When the crystal is cut into wafers, the line dislocations remain and can be revealed by chemical polishing in hot phosphoric acid. There is a strong chance that these dislocations will be propagated into the epitaxial film when it is deposited. The resulting film-defects then "pin" one or more bubbles to one location on the film, making it (them) unusable. In addition, a lattice mismatch tends to accentuate the defects of the GGG substrate onto the film grown on it. The a_0 lattice parameter for GGG is 12.53 Å (this is the length of the side of the cube enclosing the 19 atoms) while that of $Eu_3 Ga_5 O_{11}$ is 12.48 Å. This mismatch of lattice parameter is about 50x greater than that desired so that a composition such as $Y_{2.45} Eu_{5.5} Fe_{3.8} Ga_{1.2} O_{12}$ is generally more suitable to correctly match the lattice parameter of GGG. In this way, defects present in the substrate can be avoided in the epitaxially grown film, if proper growth conditions are maintained. Nevertheless, although the "bubble-memory" approach to higher density memories in computers was thought to hold much promise, the problems associated with intrinsic defects in the GGG film and those involving the substrate proved to be too daunting and today the bubble-memory approach has been abandoned in favor of more advantageous methods.

IV. Calcium Sulfide Phosphor

Lenard in 1928 reported on a calcium sulfide phosphor, i.e.- $CaS:Bi^{3+}$, having a blue emission. Actually, he probably prepared: $CaS:Bi^{3+}:V^-$ initially. As you can see, the divalent cation site would have to be occupied by a negative vacancy since charge compensation dictates that the vacancy will appear. Lenard found that the emission intensity of $CaS:Bi^{3+}:V^-$ was very low. It is likely that the V^- site is a color center which dissipated most of the excitation energy. He found that the use of chloride fluxes greatly improved the emission intensity and that the use of KCl produced the brightest phosphor. He concluded that K^+_{Ca} was a coactivator, i.e.- as $CaS:Bi^{3+}:K^+$, since he used first used NaCl as a flux during its preparation and then KCl. Even today, the phosphor is referred to in that manner although it is obvious that the charge compensation mechanism is most likely the correct mechanism in the formation of the phosphor.

This concludes our consideration of defect applications. Let us now consider a more mathematical approach to the description of point defects.

It has been said: "If you cannot calculate the properties of any given theory, you really do not understand it".

2.5. - THERMODYNAMICS OF THE POINT DEFECT

We shall use two approaches to derive some working values for the point defect in solids, namely that of Statistical Mechanics and that of the Thermodynamics of Defects. There are those who have some familiarity with statistical mechanics. For others, some explanation is due. Statistical Mechanics as a discipline was originally derived in the early 1920's when it was realized that one had to deal with large populations of atoms or molecules in various energy states (particularly gases at that time). These states arise because each molecule, for example, is vibrating in a manner slightly different than its neighbor.

What we describe as the energy state for a given set of conditions is actually the average of that of a Boltzmann population. The discipline best suited for handling such a system is statistics, hence the name. The approach used for manipulating molecular populations in Statistical Mechanics is quite involved, and we shall touch very briefly on the mathematics involved. We will first describe each of these approaches separately and then a combined version. Hopefully, this will aid in your understanding of the two methods of determining the effect of the point defect upon the properties of the solid.

I. Statistical Mechanics Approach

The language of Statistical Mechanics evolved over a considerable period of time. For example, the term "ensemble" is used to denote a statistical population of molecules; "partition function" is the integral, over phase-space of a system, of the exponential of {-E/kT} [where E is the energy of the system, k is Boltzmann's constant, and T is the temperature in °K]. From this "function", all of the thermodynamic functions can be derived. The definitions that we shall need are given as follows in 2.5.1. on the next page.

Here, Nj is the sum of the individual atoms times an entropy factor and **N** is the sum of all of the NJ 's.

2.5.1.- Statistical Mechanics Definitions Needed

$$W \quad \equiv \quad \text{thermodynamic probability}$$
$$G \quad \equiv \quad \text{ensemble of related atoms(molecules)}$$
$$N_j \quad \equiv \quad \textstyle\sum n_i$$
$$\mathbf{N} \quad \equiv \quad \textstyle\sum N_j$$
$$Q \quad \equiv \quad \text{Partition Function}$$

Using these, we can derive the following equation for the total energy of any given system:

2.5.2.- $E_T = \text{total energy of system} = \mathbf{N}\ E(\text{ave.}) = \sum N_j\ E_j$

where N_J is the total number of atoms (molecules) involved. Using the methods of Statistical Mechanics, we can derive by probability relations:

2.5.3.-
$$W \quad = \quad \Delta \Gamma / h^3\ N_j\ !$$
$$Q \quad = \quad \sum \exp^{-\ E_j\ /\ k\,T} \quad \text{(for states)}$$
$$Q \quad = \quad \omega \sum \exp^{-\ E_j\ /\ k\,T} \quad \text{(for levels)}$$

In these equations, Γ is a so-called "partition coefficient", ω is a degeneracy, k is Boltzsmann's constant, and T is in degrees Kelvin. The first equation in 2.5.3. is a statistical mechanical definition of work, whereas the last two describe total energy states. Having these definitions and equations allows us to define point defects from a Statistical Mechanical viewpoint.

II. Schottky and Frenkel Defects

Consider a plane net having N sites, of which N_L are lattice sites, N_V are vacancies, and N_i are Interstitials. (Note that we do not consider charge at the sites for the moment). Using these. we have the following two equations:

equations for specific defects in the solid. To show how this is accomplished, we let N_L be the total number of lattice sites and N_V the number of vacancies. The way that N_V vacancies can be arranged upon N_L lattice sites is given by combinatorial statistics as:

2.5.18.- $\quad W = \dfrac{N_L \; !}{N_V \; ! \; (N_V - N_V) \; !}$

The configurational or mixing entropy will be defined by:

2.5.19.- $\qquad \Delta S_M = k \ln W$

Using Stirling's approximation for large numbers (this method is used extensively in Statistical Mechanics), we can get:

2.5.20.- $\quad \Delta S_M = k \; [N_L \ln N_L - N_V \ln N_V - (N_L - N_V) \ln (N_L - N_V) \;]$

You will note that we have the change in entropy as a function of the differences between normal lattice sites and vacancies. Since we know that $N_L \gg N_V$, we can write for the entropy of mixing:

2.5.21.- $\quad \Delta S_M \approx [N_L \ln N_L - N_V \ln N_V \;]$

$\qquad$ and $\quad \Delta S_M = - \; k \; [N_V \ln N_V \; / \; N_L \;] = - \; R \; X_V \; \ln X_V$

where X_V is the fraction of defects actually present. As expected, for non-interacting defects, the entropy of mixing is ideal so that the free energy of the system can be written as:

2.5.22.- $\quad \Delta F_V = N_V \; (\Delta E_V - T \; \Delta S^v{}_V) + RT \; [N_L - N_V / \; N_L \; (\ln N_L - N_V / \; N_L \;)$
$\qquad\qquad\qquad + N_V \; / N_L \ln - N_V \; / \; N_L)]$

where $\Delta S^v{}_V$ is the change in vibrational energy of the lattice arising from the change in vibrational frequency around the vacant lattice site. If we minimize the free energy, i.e.-

2.5.23.- $\quad \partial F_V / \partial N_V = 0 = \Delta E_V - T \; \Delta S^v{}_V) + RT \ln (N_V \; / \; N_L - N_V)$
$\qquad$ and: $\quad \ln (N_V \; / \; N_L - N_V) = \mathbf{\ln X_v} = \Delta E_V - T \; \Delta S^v{}_V) \; / \; RT$

Thus, we can write for the atomic fraction of vacancies present:

2.5.24.- $X_V = \exp - \Delta F^0_V / RT = e^{-\Delta E_V / RT}$

You may wonder why we have examined this method for specifying vacancies in general terms. The reason is that we can apply the same method to the problem of point-defect pairs. This is intended to help you understand how these various types of defects in the solid arise.

a. Interstitial Atoms

The process of creating an interstitial is just the opposite of creating a vacancy. If there are N_I possible interstitial positions in the lattice, and if ΔF_i is the free energy needed to move the atom into its interstitial position, then the interstitial concentration at equilibrium will be:

2.5.25.- $N_i / N_I = X_i = \exp(- \Delta F_i / kT)$

The same approach applies to Frenkel pairs.

b. Frenkel Pairs

We have, in this case, both a vacancy associated with an interstitial atom. Using the approach shown in 2.5.18., we have:

2.5.26.- $W = \dfrac{N_L!}{N_V! (N_L - N_V)!} \quad X \quad \dfrac{N_I!}{N_i! (N_I - N_i)!}$

Now, we have two combinatorial functions, one for the vacancies and one for the interstitials. Since the numbers of vacancies and interstials are equal, i.e.- $N_F = N_L = N_i$, we can use the same steps given in 2.5.19., ,20, 21., and 2.5.22. to get:

2.5.27.- $\Delta S_M = RT [(N_L \ln N_L) - (2N_F \ln N_F) + (N_I \ln N_I)$
            $- \{(N_L - N_F) \ln(N_I - N_F) \ln(N_L - N_F)\}]$

Minimizing the free energy as before, we can write for the Frenkel pairs:

2.5.28.- $N_F = (N_L N_I)^{1/2} \exp(-\Delta F_{iv} / 2RT)$

Note that ΔF applies to both interstitial and vacancy sites. Using 2.5.23. and 2.5.24., we can get the fraction of Frenkel defects as:

2.5.29.- $X_F = \exp(\Delta F_{iv} / 2RT$

since the free energies are approximately the same, i.e.- $\Delta F_{iv} \cong \Delta F_i + \Delta F_V$

c. Schottky Defects

The calculation of Schottky defects follows the same method already given. We use cv for cation-vacancy and av for the associated anion vacancy. The free energy of this defect is then:

2.5.30.- $\Delta F_{Sh} = N_{cv} \Delta E_{cv} + N_{AV} \Delta E_{AV} - T (N_{cv} \Delta S^v{}_{cv} + N_{AV} \Delta S^v{}_{AV})$
$$- RT \ln \frac{N!}{N_{cv}!(N - N_{cv})!} \times \frac{N!}{N_{AV}!(N - N_{AV})!}$$

Again, we have two combinatorial factors, one each for the associated cation vacancies. Here, the total number of lattices sites, $N \gg N_{Sh} = N_{cv} = N_{AV}$. Minimizing the free energy gives:

2.5.31.- $N_{Sh} / (N - N_{Sh}) = \exp(-\Delta F_{Sh} / 2RT)$

and: $\qquad X_{Sh} = \exp(-\Delta F_{Sh} / 2RT)$

You will note that this approach uses Statistical Mechanics to approximate the thermodynamic constants for the number of defects present.

V. Defect Equilibria

Just as chemical reactions can be described and calculated in terms of thermodynamic constants and chemical equilibria, so can we also describe defect formation in terms of equilibria. This is given as follows:

2.5.32- Law of Mass Action:

$$bB + cC \leftrightharpoons dD + eE \ ; \qquad K = å^d_D \ å^e_E \ / \ å^b_B \ å^c_C$$

Using this equation, we can calculate the numbers of defects for various defects in the MX crystal as:

2.5.33.- Frenkel Defects (for the MX crystal)
$$M_x \leftrightharpoons M_i + V_M \ ; \qquad K_F = å_{M_i} \ å_{V_M} \ / \ å_M$$

2.5.34.- Schottky Defects (for the MX crystal)
$$MX \leftrightharpoons V_M + V_X \ ; \qquad K_{Sh} = å_{V_M} \ å_{V_M} \ / \ å_{M_X}$$

Note that we have specified the equilibrium constants in terms of the activity, å, of the **associated defects.** We can also write thermodynamic equations for these defects:

2.5.36.- Chemical Thermodynamics

$$\Delta G = \Delta H - T \ \Delta S = - RT \ln K \ ; \quad K = \exp \Delta S/R \cdot \exp - \Delta H/RT$$

2.5.37.- Defect Thermodynamics

$$K_d = \exp \Delta S_d /R \cdot \exp - \Delta H_d /RT$$

where d refers to the specific defect.

We may summarize the knowledge we have already developed for the Schottky and Frenkel defects:

1. We have shown by Statistical Mechanics that we can calculate numbers of defects present at a given temperature.

2. There is an Activation Energy for defect formation. In many cases, this energy is low enough that defect formation occurs at, or slightly above, room temperature.

3. Defects may be described in terms of thermodynamic constants and equilibria. The presence of defects changes both the local vibrational frequencies in the vicinity of the defect and the local lattice configuration around the defect.

One question we may logically ask is how are we to know what types of defects will appear in a given solid? The answer to this question is given as follows:

IT HAS BEEN FOUND:

"There are two associated effects on a given solid which have opposite effects on stoichiometry. Usually, one involves the cation site and the other the anion site. Because of the differences in defect-formation-energies, the concentration of other defects is usually negligible".

Thus, if Frenkel Defects predominate in a given solid, other defects are usually not present. Likewise, for the Schottky Defect. Note that this applies for **associated defects**. If these are not present, there will still be **2 types** of defects present, each having an opposite effect upon stoichiometry. **Thus, we conclude that intrinsic defects usually occur in pairs.** This conclusion cannot be overemphasized. The following discussion shows how this occurs in the real world of defects in solids.

2.6.- DEFECT EQUILIBRIA IN VARIOUS TYPES OF COMPOUNDS

Up to now, we have been concerned with the MX compound as a hypothetical example of the solid state. We will now undertake more concrete examples as found in the real world, using the concepts developed for the simple MX compound. For the sake of simplicity, we restrict ourselves to binary compounds, that is- one cation and one anion. An example of a ternary compound is ABX_S , where A and B are different cations, and S is a small whole number.

Our example of a binary compound will be:

MX_S

We will distinguish between four states for this hypothetical compound, to wit:

stoichiometric vs: non-stoichiometric

non-ionized vs: ionized

I. Stoichiometric Binary Compounds of MX_S

In the real world of defect chemistry, we find that in addition to the simple defects, other types of defects appear, depending upon the type of crystal we are dealing with. These may be summarized as shown in the following. According to our nomenclature, V_M is a vacancy at an M cation site, etc. The first five pairs of defects given above have been observed experimentally in solids, whereas the last four have not.

2.6.1- Defects in the MX_S Compound

	PAIRS OF DEFECTS
Schottky	$V_M + V_X$
Frenkel	$V_M + M_i$
Anti-Frenkel	$X_i + V_X$
Anti-Structure	$X_M + M_X$
Vacancy-Structure	$V_M + M_X$
Structure-Vacancy	$V_X + X_M$
Interstitial	$M_i + X_i$
Interstitial-Structure	$M_X + X_i$
Structure-Interstitial	$M_i + X_M$

This answers the hypothesis posed above, namely that **defects in solids occur in pairs.** Study these defect-pairs carefully so that you become familiar with them. They represent the type of structure defects found in most solids. We have now introduced into our nomenclature a distinction between structure and anti-structure defects. What this means is that stacking faults can sometimes result in X_M and M_X defects, which are high-energy defects since they exist as cations (anions) surrounded by a complete positive (negative) charge. In fact, this would appear to be contrary to established modes of charge compensation in a solid. However, it has been noted that these defects generally exist as a vacancy plus the anti-structure defect, not as two anti-structure defects together.

Table 2-1 summarizes the various pairs of defects possible for binary compounds.

TABLE 2-1
POSSIBLE DEFECTS IN THE MX_S COMPOUND

1. Constant Number of Sites

Defect Type	Defect Pair	Example	Defect Equation	Equilib. Constant
Schottky	$V_M + V_X$	TiO	$0 \leftrightarrows V_M + S\, V_X + \alpha V_i$	$K_{Sh} = V_M\, V_X{}^S\, V_i{}^\alpha$
Frenkel	$V_M + M_i$	ZnO	$M_M + \alpha V_i \leftrightarrows V_M + M_i$	$K_F = V_M\, M_i / M_M\, V_i{}^\alpha$
Anti-Frenkel	$X_i + V_X$	LaH$_2$	$X_X + \beta\, V_i \leftrightarrows X_i + V_X$	$K_{AF} = X_i\, V_X / X_X\, V_i{}^\beta$
Anti-Struct.	$X_M + M_X$	AuZn	$M_M + X_X \leftrightarrows M_X + X_M$	$K_{AS} = M_X\, X_M / M_M\, X_X$

2. Excess Number of Sites

Defect Type	Defect Pair	Example	Defect Equation	Equilib. Constant
Vac.-Struc.	$V_M + M_X$	NiAl	$\delta M_M \leftrightarrows \delta M_X + (1-\delta) V_M$	$K_{VS} = \dfrac{M^\delta{}_X V_M{}^{1+\delta}\, V_i{}^\alpha}{[M_M{}^\delta + \alpha]\, V_i}$
Structure-Vacancy	$V_X + X_M$	-----	$\delta X_X \leftrightarrows$ $X_M (1+\delta) V_X + \alpha V_i$	$K_{SV} = X_M\, V_X{}^{1+\delta}\, V_i{}^\alpha / X_x{}^\delta$
Interstitial	$M_i + X_i$	-----	$\delta(M_M + S\, X_X) \leftrightarrows$ $M_i + S\, X_i - (1+\delta+\alpha) V_i$	$K_i = M_i X_i\, S\, V_i{}^{1+\delta+\alpha} /$ $M_M{}^\delta\, X_X S\delta$
Interst.-Struct.	$M_X + X_i$	----	$[M_M + (1+\delta) X_X \leftrightarrows M_X$ $+ (1+\delta)\, X_i - (1+\delta+\alpha) V_i$	$K_{IS} = M_X X_i{}^{1+\delta} V_i{}^{(1+\delta+\alpha)} /$ $(1+\delta) X_i - (1+\delta+\alpha) V_i]$
Struct-Interst.	$M_i + X_M$	----	$[(1+\delta) M_M + S\, X_X \leftrightarrows$ $(1+\delta) M_i + \delta\, XM]$	$K_{SI} = M_i\, X_m{}^d /$ $M_M{}^{1+\delta}\, X_M{}^\delta$

Equilibria are given along with the appropriate equilibrium constant. You will observe that these pairs of defects span the possible **combinations** of defects, taken two at a time. The first four are based upon a constant number of sites, whereas the last five are based upon an excess in the number of sites available. This excess we call "δ". Note that we are not speaking of the ratio of cations to anions, i.e.- stoichiometry, but of an excess of cations or anions to the normal concentration of cations or anions.

II. Defect Concentrations in MX$_S$ Compounds

It is of interest to be able to determine the number of intrinsic defects in a given solid. As we have shown, pairs of defects predominate in any given solid. Thus, the number of each type of intrinsic defects, N_i (M) or N_i (X), will equal each other, For Schottky defects in the MX$_S$ crystal, we have:

2.6.2.-
$$N_i (V_M) \;=\; N_I \, (\, S \, V_X)$$

This makes our mathematics simpler since we can rewrite the Schottky equation of Table 2-1 as:

2.6.3.-
$$0 \leftrightharpoons N_{i\,(V_M)} + S \, N_{i\,(V_M)} + \alpha \, V_i$$

Here, we have expressed the concentration as the ratio of defects to the number of M- atom sites (this has certain advantages, as we will see). We can than rewrite the defect equilibria equations of Table 2-1 in terms of numbers of intrinsic defect concentrations. These are given as follows:

2.6.4.- Equilibrium Constants = Function of Numbers of Intrinsic Defects

SCHOTTKY:	$K_{Sh} = N_i \, (S \, N_i)^2 = S^S \, N_i^{S+1}$
FRENKEL:	$K_F \;= N_i^2 \, / \, (1 - N_i) \, (\alpha - N_i)$
ANTI-FRENKEL:	$K_{AF} = N_i^2 \, / \, (S - N_i) \, (\alpha - N_i)$
ANTI-STRUCT:	$K_{AS} \;= N_i^2 \; / \, (1 - N_i)(S - N_i)^S$
VAC.-STRUCT:	$K_{VS} \;= S^S \, (S + 1)^{S+1} \cdot N_i^{2S+1} \, / \, (S - N_i - S \, N_i)^S$
STRUCT-VAC:	$K_{SV} \;= (S+1)^{S+1} \, N_i^{S+2} \, / \, (S - N_i - S \, N_i)$
INTERSTITIAL:	$K_I \;\;= S^S \, N_i^{S+1} \, / \, (\alpha - N_i - S N_i)^{\,\alpha + S + 1}$

Some of these equations are complicated and we need to examine them in more detail so as to determine how they are to be used. Equation 2.5.16. given above shows that intrinsic defect concentrations will increase with increasing temperature and that they will be low for high enthalpies of defect formation. This arises because the entropy effect is a positive exponential while the enthalpy effect is a negative exponential.

Consider the following practical example:

TiO is cubic with the NaCl structure. A sample was annealed at 1300 °C. Density and X-ray measurements revealed that the intrinsic defects were Schottky in nature ($V_{Ti} + V_O$) and that their concentration was 0.140. In this case, S = 1 so that:

2.6.5.- $\qquad K_{Sh} = 0.0196 = 2 \times 10^{-2}$

This crystal is quite defective since 1 out of 7 Ti-atom-sites (0.14^{-1}) is a vacancy, and likewise for the oxygen-atom-sites.

Another example is:

CeH$_2$. From thermodynamic measurements, it was found that the intrinsic defects were Anti-Frenkel in nature, i.e.- ($H_i + V_H$). An equilibrium constant was calculated as:

2.6.6.- $\qquad K_{AF} = 3.0 \times 10^{-4}$

at a temperature of 600 °C. This compound has the cubic fluorite structure with **one** octahedral interstice per Ce atom. Therefore, a = 1, and S = 2 for CeH$_2$. We can therefore write:

2.6.7.- $\qquad k_{AF} = N_i^2 / (2 - N_i)(1 - N_i) = 3.0 \times 10^{-4}$

$\qquad$ **or** $\qquad N_i = 2.4 \times 10^{-2}$ (600 °C.)

This means that 1 out of 42 hydride atoms is interstitial, and 1 out of 84 hydride-atom-sites is vacant.

Let us review what we have covered concerning stoichiometric binary compounds:

1. We have shown that defects occur in pairs. The reason for this lies in the charge-compensation principle which occurs in all solids.
2. Of the nine defect-pairs possible, only 5 have actually been experimentally observed in solids. These are: Schottky, Frenkel, Anti- Frenkel, Anti-Structure, Vacancy-Structure.
3. We have given defect-equations for all nine types of defects, and the Equilibrium Constant (EC) thereby associated. However, calculation of these equilibria would require values in terms of **energy** at each site, values which are sometimes difficult to determine.

A better method is to convert these EC equations to those involving **numbers of each type of intrinsic defect,** as a ratio to an intrinsic cation or anion. This allows us to calculate the actual number of intrinsic defects present in the crystal, at a specified temperature.

III.   Non-Stoichiometric Binary Compounds

We will now extend our treatment of intrinsic defects to the non-stoichiometric non-ionized compounds, as represented by:

2.6.9.- $MX_{S \pm \delta}$

where δ is a small increment. The question is: "How do we obtain non-stoichiometry in the solid?". Consider a compound governed by either or both the following equilibria:

2.6.10.- $X_X \rightleftharpoons 1/2\ X_2\ (gas)\Uparrow + V_X$

2.6.11.- $M(external\ phase) \rightleftharpoons M_M + S\ V_X + V_{Mi}$

One example might be a halide crystal which has become non-stoichiometric due to its being heated to a temperature sufficient to cause a small amount of the halide to become volatile.

Another case might be an oxide, heated in the presence of excess metal, e.g.- $ZnO + Zn$.

For a non-stoichiometric crystal, the concentration of each point defect, in each conjugate pair, **is no longer equal.** If there is an excess of V_M , X_i , or X_X , then the compound will have a surplus of X (or deficiency of M, which is the same thing) over the ideal stoichiometric composition. This is called a positive deviation from stoichiometry. Conversely, for a negative deviation, there will be an excess of V_X , M_i , or M_M . This explains the plus and minus in equation 2.6.9. In terms of the above given defects, δ may be expressed as shown in the following Table:

TABLE 2-2.

Non-Stoichiometry, δ , as a Function of Specific Types of Defects

in $MX_{S \pm \delta}$ Binary Compounds

Vacancy Formation	Defect Equation	Equilibrium Constant
V_X from external M	$M \leftrightarrows M_M + S[V_X] + \alpha V_i$	$K_{V_X} = M_M[V_X]^S V_i^\alpha / M$
V_M from gaseous X	$1/2 X_2 \leftrightarrows S X_X + [V_M] + \alpha V_i$	$K_{V_M} = X_X^S [V_M] V_i^\alpha / p_{X_2}^{1/2S}$
	{if $X_X \approx 1$ and $V_i = \alpha$,	then $\delta = S[V_M]/ 1 - V_M$}
X-Interstitials	$1/2\, X_2 + \alpha V_I \leftrightarrows [X_i]$	$K_{Xi} = [X_i] /(\alpha - X_i) \, p_{X_2}^{1/2}$
		or: $\{p_{X_2}^{1/2} = 1/K_{Xi} \cdot \delta/ \alpha - \delta\}$
M-Interstitials	$M_M + S X_X + (1+\alpha) V_i \leftrightarrows$ $[M_i] + S/2\, X_2$	$K_{M_i} = [M_i] \, p_{X_2}^{S/2} / M_M X_X^S$ $[\alpha - M_i]^{(1+\alpha)}$
		or: $p_{X_2}^S = K_{Mi} (\alpha(S + \delta) +$ $\delta)^{1+\delta} /(- \delta (S+\delta)^\alpha)^{1/S}$ (from gaseous X_2)
X-Substitutionals (from gaseous X_2 on an M-site)	$1/2(S+1) X_2 \leftrightarrows S X_X + \alpha V_i$ $\cdot V_i^{(1+a)}$	$K_{X_M} = [X_M] \, \alpha^\alpha / p_{X_2}^{(S+1/2)}$ {or: $p_{X_2}^{(S+1/2)} = 1/ K_{X_M} \cdot$ $\delta/(1+S+\delta)^{1/S+1}$}
M-Substitutionals (gaseous X_2 formed)	$(S+1) X_X + M_M + a V_i \leftrightarrows$ $((S+1)/2)\, X_2 + M_x$	$K_{M_X} = p_{X_2}^{((S+1)/2)} [M_X]/$ $(1 - M_X)^{S+1} \alpha^\alpha$ {or: $p_{X_2}^{((S+1)/2)} =$ $(K_{M_X})^{1/S+1} (S+1)(S+d)/$ $[(-\delta)(S+1+\delta)^S]^{1/S+1})$

Note the various mechanisms which give rise to the specific combinations of defects. These mechanisms have been thoroughly studied as a function of specific compounds. It is sufficient for us, at this point, to observe which defect equations govern both the equilibria and the non-stoichiometry of the general compound, $MX_{S \pm \delta}$.

The equation used to calculate the non-stoichiometry factor, δ, in the general case is:

2.6.12.- $\delta = (X_i - V_X) + S(V_M - M_i) + (S+1)(X_M - M_X)/1 + M_i + M_X - V_M - X_M$

We can express relationships between defect formation, the influence of various **external factors,** and the equilibrium constant thereby related. We do this in terms of δ, the degree of non-stoichiometry, as given in Table 2-2. Even though these equations are rather formidable-looking, we shall be able to use them to good advantage.

Note that each case corresponds to the influence of a **reacting** external factor on a stoichiometric solid, which contains intrinsic defects. These factors produce **additional defects** because of nonstoichiometry and charge-compensation. The **defect produced** is enclosed in **brackets** in Table 2-2.

Consider this factor carefully by again examining Table 2-2. Also given is the reaction producing the defect, with its corresponding equilibrium constant. In most cases, the deviation, δ, is presented in terms of the equilibrium constant and the partial pressure of the external gaseous reactant.

Thus, **if the number of defects produced can be measured** and an equilibrium constant calculated, then δ can be determined both as a function of partial pressure, p_{X_2}, and temperature (see 2.5.22.).

IV. Defect Concentrations in $MX_{S \pm \delta}$

We now proceed as we did for the stoichiometric-case, namely to develop

defect-concentration equations for the non-stoichiometric case, i.e.-
$MXS_{\pm\delta}$. Consider the effect of Anti-Frenkel defect production. From Table
2-1, we get K_{AF}, with its associated equation, k_{AF}. In Table 2-2, we use
K_{Xi} for X-interstitials. Combining these, we get:

2.6.13.- $\qquad K_{AF} = K_{V_X} \cdot K_{Xi} = N_i^2 / (S-N_i)(\alpha - N_i)$

When both V_X and X_i coexist in the lattice, the deviation from
stoichiometry (from 2.6.12.) becomes:

2.6.14.- $\quad \delta = [X_i] - [V_X]$

Using the equilibrium constant of 2.6.13., i.e.-

2.6.15.- $\quad K_{V_X} = px_2^{1/2} [V_X] /X_X = px_2^{1/2} [V_X] / S - [V_X]$

and the appropriate one from Table 2-2 (i.e.- K_{Xi}), we get (assume for
simplicity that $S = \alpha = 1$) :

2.6.16.- $\quad \delta = \alpha\, px_2^{1/2} K_{Xi} /px_2 K_{Xi} + 1 - S K_{V_X} / K_{V_X} + px_2^{1/2}$

We can rearrange terms in 2.6.16. to obtain: $K_{Xi} (1- \delta) px_2 - \delta(K_{V_X} K_{Xi} + 1)$
$px_2^{1/2}$ and if : $K_{V_X}^2 (1- \delta) = 0$, we can, by using 2.6.13. and $N_i \ll 1$, obtain:

2.6.17.- $\quad N_i^2 (1- \delta) px_2 - \delta K_{V_X} px_2^{1/2} - K_{V_X}^2 (1- \delta) = 0$.

Solving for px_2 yields :

2.6.18.- $px_2 = K_{V_X}^2 (\delta^2 + 2 N_i (1- \delta^2) \pm \delta [\delta^2 +4 N_i (1-\delta^2)]^{1/2} /2 N_i^4 (1-\delta)^2$

Since at stoichiometric composition, δ must equal zero, this rather
formidable equation reduces to:

2.6.19.- $\quad p_{X2}^o = K_{V_X}^2 / N_i^2$

where $p_{X_2}^o$ is the pressure of X_2 gas in equilibrium with the MX$_S$ crystal at the stoichiometric composition. This gives us the opportunity to divide 2.6.18. by 2.6.19. to obtain:

2.6.20.- $p_{X_2}/p_{X_2}^o = \delta^2 + 2 N_i (1- \delta^2) \pm \delta [\delta^2 + 4 N_i (1-\delta^2)/2N_i^2 (1-\delta)^2$

We can therefore calculate δ in terms of the ratio of p_{X_2} to $p_{X_2}^o$ and N_i, shown as follows:

2.6.21.-

Effect of External Pressure of X_2 Gas on Non-Stoichiometry of the Hypothetical Compound, MX $_{S\pm\delta}$

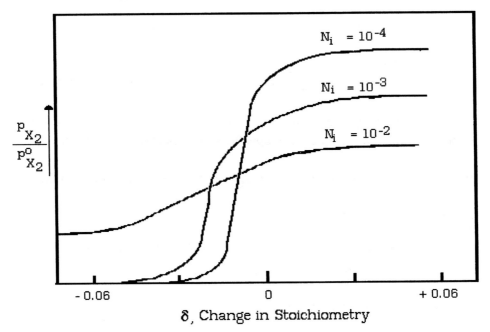

δ, Change in Stoichiometry

Although we will not treat the other types of pairs of defects, it is well to note that similar equations can also be derived for the other intrinsic defects. What we have shown is that **external reactants** can cause further changes in the non-stoichiometry of the solid.

V. Ionization of Defects

We have already covered, albeit briefly, non-ionized stoichiometric and non-ionized non-stoichiometric intrinsic-defect compounds. Let us now consider the ionization of defects in these compounds. In the MX_S compound, if we remove some of the X-atoms to form V_X, the electrons from the removed X-atom (or from the bond holding the X-atom in the crystal) are left behind for charge compensation reasons. At low temperatures, these electrons are localized near the vacancy but become dissociated from the point defect at higher temperatures. They become free to move through the crystal, and we say that the intrinsic defect has become ionized. We can write the following equations for this mechanism:

2.6.22.-   VACANCIES

$$V_X \leftrightarrows V_X^+ + e^- \qquad\qquad K_{V_X}^* = |V_X|^+ |e^-|/V_X$$

$$V_M \leftrightarrows V_M^- + p^+ \qquad\qquad K_{V_M}^* = |V_M^-| |p^+|/V_M$$

In the equilibrium constant equations, each symbol is actually a concentration, i.e.- numbers of specified defects and electrons, etc.

In a like manner, we write for interstitials and anti-structure defects:

2.6.23.-   INTERSTITIALS

$$M_i \leftrightarrows M_i^+ + e^- \qquad\qquad K_{Mi}^* = |M_i^+| |e^-| / M_i$$

$$X_i \leftrightarrows X_i^- + p^+ \qquad\qquad K_{Xi}^* = |X_i| |p^+|/ X_i$$

2.6.24.-   ANTI-STRUCTURE

$$X_M \leftrightarrows X_M^+ + e^- \qquad\qquad K_{X_M} = |X_M^+| |e^-|/X_M$$

$$M_X \leftrightarrows M_X^- + p^+ \qquad\qquad K_{M_X} = |M_X^-| |p^+|/ M_X$$

A useful example for understanding the above equations, and the ionization of defects, is that of cobaltous oxide, CoO. External oxygen pressure will affect stoichiometry and produce cobalt vacancies, but the vacancies are ionized at room temperature:

2.6.25.- $1/2\ O_2 \leftrightarrows O_o + V_{Co}^{\cdot} + \pi^+ + \alpha\ V_i$

The π^+ is free to migrate throughout the lattice. However, at pressures below 10^{-6} (obtained by application of a vacuum), the Co vacancies become **doubly-ionized:**

2.6.26.- $V_{Co}^{\cdot} \leftrightarrows V_{Co}^{2-} + \pi^+$

This illustrates how the ionized defect arises.

We have already illustrated how vacancies arise through the use of a **PLANE NET.** (SEE 2.2.1.). Therein, we used MX as the molecule to build the NET. Let us now return, using the ions, M^{2+} and X^{2-} as the ionic forms with which to build the NET. This is shown as follows:

2.6.27.-

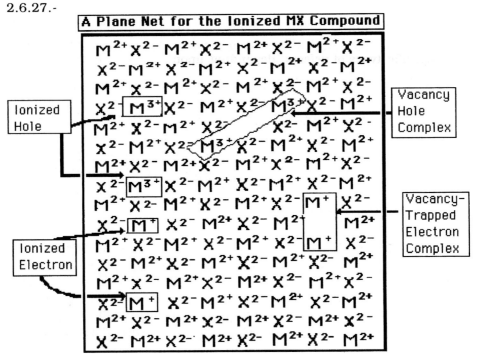

2.7.3.-

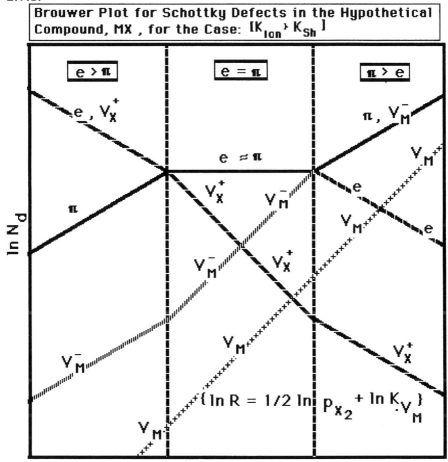

We can readily see that concentrations are now expressed in powers of R, and that by taking logarithms , we get slopes of $\pm 1/2$ or ± 1. As shown in **2.7.3.,** we can now differentiate between 3 regions of defect concentrations, namely:

$$\text{REGION I} \quad - \quad V_X^+ > V_M$$
$$\text{REGION II} \quad - \quad V_X^+ = V_M^-$$
$$\text{REGION III} \quad - \quad V_M^- > V_X^+$$

At large positive deviations from the stoichiometric composition (REGION III), $V_M^- \gg e$ and $p \gg V_X^+$. In this REGION, 2.7.1.-g becomes:

2.7.7. - $\ln V_M^- = \ln p$

In the vicinity of the stoichiometric composition (REGION II), the concentration of defects depends upon whether $K_S > K_{ion}$, or vice-versa. The former usually holds for large band-gap ionic compounds. Then, 2.7.1.- g becomes:

2.7.8.- $\ln V_M^- = \ln V_X^+$

We can also show that the change from near stoichiometry to a negative deviation of stoichiometry (REGION I) occurs when:

2.7.9.- $(R \, K_S \, K_{V_X}^* \, / \, K_{ion})^{1/2} = (K_S \, K_{ion} \, / \, K_{V_X}^* \, R)^{1/2}$

This gives us:

2.7.10.- $R_{II \sim I} = K_{ion} \, / \, K_{V_X}^*$

and in a like manner:

2.7.11.- $R_{II \sim III} = K_S \, / \, K_{V_X}^*$

Because the stoichiometry is rigidly defined in REGION II by the condition:

2.7.12.- $V_M + V_M^- = V_X + V_X^+$

changes in R or p_{X_2} do not greatly affect deviation from stoichiometry. But, there are large changes in both $\Sigma \, e^-$ and $\Sigma \, p^+$ in this region. Thus, the stoichiometric composition is probably best defined when $e = p$.

This occurs, as shown in 2.7.3., at the value:

2.7.13.- $R_o = (K_S K_{ion})^{1/2} / K_{V_X}{}^*$

An example of a compound where $K_S \gg K_{ion}$ is **KBr.** Krœger (1964) obtained values (at 600 °C.) of:

$$K_S = 8 \times 10^{-14}$$

$$K_{ion} = 3 \times 10^{-35}$$

If we subtract 2.7.9. from 2.7.10. , we find that the Br_2 gas - pressure changes by 10^{43} (since $R \approx p_{Br_2}{}^{1/2}$) over Region II. Moreover, the deviation from stoichiometry, δ , changes only by 10^{-3}, or remains essentially constant. However, Σ e⁻ (or Σ p⁺) changes by a factor of approximately 10^{21} over Region II. In electronic semi-conductors, the condition: $K_{ion} \gg K_S$, usually prevails. We usually get a Brouwer analysis like that of 2.7.3.

Now, consider the case where ionization is the norm. This case, shown in 2.7.14. (next page), is for the hypothetical semi-conductor alloy MX, where M acts as a cation , and X acts more like an anion. That is, when we get ionization of defects, M loses an electron and X is positively ionized. An example could be GaAs. In this case, electronic charge is relatively insensitive (Region II) to composition, whereas the deviation from stoichiometry, d , varies considerably. In the positive-deviation direction (Region III), the major defects are $V_M{}^-$, p , and V_M .

This gives the relation:

2.7.15.- $V_M \leftrightharpoons V_M{}^- + p^+$

In the negative-deviation direction (Region I), e and $V_M{}^-$ predominate.

For the major defects of this system, we have the conditions shown in 2.7.16. on the next page.

2.7.14.-

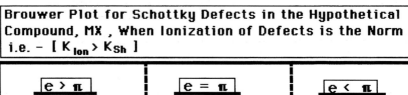

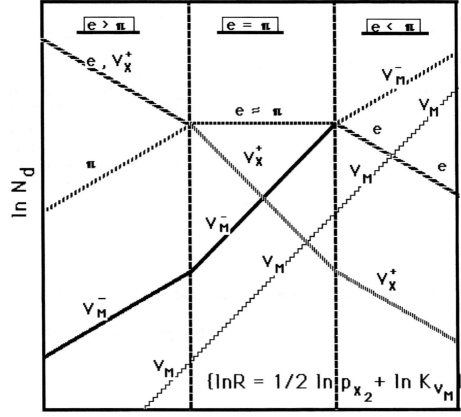

2.7.16.   K$_{ion}$ >> K$_{sh}$

           Region I -   e > p
           Region II -   e ≈ p
           Region III -   p > e

Note that we have rather well defined the defects present for this type of semi-conductor, using relationships defined by Brouwer's method. It should be apparent that this method adds considerable power to our ability to analyze intrinsic defects where ionization is the norm.

2.8.- ANALYSES OF REAL CRYSTALS USING BROUWER'S METHOD- COMPARISON TO THE THERMODYNAMIC METHOD

The silver halide series of compounds have been extensively studied because of their usage in photographic film. In particular, it is known that if silver bromide is incorporated into a photographic emulsion, any incident photon will create a Frenkel defect. When the film is developed, the Ag_i^+ is reduced to Ag metal. These localized atoms act as nuclei to cause metal crystal formation at the points "sensitized" by the photon action. (Note that this description is an oversimplification of the actual mechanism. Nevertheless, it should be apparent that a knowledge of defect chemistry of the compound, AgBr, should prove to be very important in understanding the chemistry of photographic films).

I. The AgBr Crystal with a Divalent Impurity, Cd^{2+}

Consider the crystal, AgBr. Both cation and anion are monovalent, i.e.- Ag^+ and Br^- . The addition of a divalent cation such as Cd^{2+} should introduce vacancies, V_{Ag} , into the crystal, because of the charge-compensation mechanism. To maintain electro-neutrality, we prefer to define the system as:

2.8.1.- $(1-\delta)\ Ag^+\ Br^-\ :\ \delta\ Cd^{2+}\ S^=$

Fortunately, AgBr is easy to grow as a single crystal, using Stockbarger Techniques. Possession and measurement of a single crystal greatly facilitates our measurement of defects. The imperfections we expect to find are:

2.8.2.- V_{Ag} , Ag_i , e^- , p^+ , Cd_{Ag} , and $[Cd_{Ag} , V_{Ag}]$

The last defect is one involving two nearest neighbor cation sites in the lattice.

The following table gives the defect reactions governing this case:

TABLE 2-3

DEFECT REACTIONS IN THE AgBr CRYSTAL CONTAINING Cd^{2+}

a.	0	$\leftrightarrows$ $e^- + p^+$
b.	$Ag_{Ag} + \alpha V_i$	$\leftrightarrows$ $V_{Ag} + Ag_i$
c.	$Ag_{Ag} + \alpha V_i$	$\leftrightarrows$ $V_{Ag}^- + Ag_i^+$
d.	Ag_i	$\leftrightarrows$ $Ag_i^+ + e^-$
e.	$1/2\ Br_2$	$\leftrightarrows$ $Br_{Br} + V_{Br}$
f.	Cd_{Ag}	$\leftrightarrows$ $Cd_{Ag}^+ + e^-$
g.	$Cd_{Ag} + V_{Ag}^-$	$\leftrightarrows$ $[Cd_{Ag}, V_{Ag}]$
h.	$Cd_{Ag}^+ + V_{Ag}^-$	$\leftrightarrows$ $[Cd_{Ag}, V_{Ag}]$

i. <u>For Constant Cd</u>

$$Cd_T = Cd_{Ag} + Cd_{Ag}^+ + [Cd_{Ag}, V_{Ag}] = K$$

j. <u>For Electro-neutrality</u>

$$e^- + V_{Ag}^- = p^+ + Ag_i^+ + Cd_{Ag}^+$$

We would normally plot $\ln N_d$ vs: $\ln K_{Ag_i} + \ln K_{V_{Ag}}$. However, we find it more convenient to plot $\ln N_d$ vs: $1/T$. The reason for this is as follows. Experimentally, we find that if we fix the Cd^{2+} content at some convenient level, it is necessary to anneal the AgBr crystals at a fixed temperature for times long enough to achieve complete equilibrium. If the temperature is changed, then both type and relative numbers of defects may also change.

Thus, we plot $\ln N_d$ vs: $1/T$, as in the Brouwer diagram of 2.8.3., given on the next page. At low temperatures (Region III), singly-charged defects predominate, i.e.- $2 V_{Ag}^- = Ag_i^+ + Cd_{Ag}^+$. At the junction of III - II, the charged moieties begin to cluster to form Cd_{Ag}^+ and V_{ag}^- . These in turn may form the complex:

2.8.4.- $\qquad Cd_{Ag}^+ + V_{Ag}^- \leftrightarrows \qquad [Cd_{Ag}, V_{Ag}]$

At the same time, the concentration of Ag_i^+ drops dramatically.

2.8.3.-

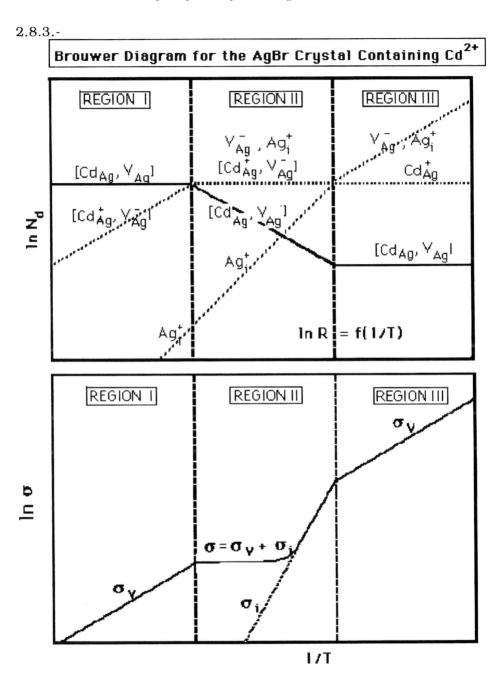

Brouwer Diagram for the AgBr Crystal Containing Cd^{2+}

At the juncture of II - I in the Brouwer diagram, the complex clearly dominates.

While the above results show how intrinsic defects are affected by temperature, we still do not know how the electrons and holes vary as a function of temperature. (Note that temperature, as specified, is a **preparation** temperature, not **measurement** temperature. Measurement of intrinsic conductivity, σ , is shown at the bottom of 2.8.3. At low temperatures, conductivity due to vacancies, σ_V , appears to be the major contributor:

2.8.5.- $V_{Ag} \leftrightarrows V_{Ag}^- + p^+$

Conductivity decreases at higher temperatures because Ag_i^+ concentrations decrease. Finally, at the very high temperature region (Region I), conductivity is relatively low, and approaches zero, because the complex [Cd_{Ag} , V_{Ag}] predominates as the number of charged moieties, ($Cd_{Ag}^+ + V_{Ag}^-$) decreases.

Varying the Cd^{2+} content in the AgBr crystal affects the relative defect ratios, as shown in the following diagram, shown as 2.8.6. on the next page. Again, we can identify 3 Regions as a function of Cd- concentration:

2.8.7.-   REGION I - $Ag_i^+ > V_{Ag}^-$

                 REGION II - $Ag_i^+ @ V_{Ag}^-$

                 REGION III - $V_{Ag}^- > Ag_i^+$

Region I of 2.8.7. corresponds closely to that of Region III of 2.8.5. Teltow (1949) also studied this crystal. He measured conductivity of AgBr crystal containing various amounts of Cd^{2+} , as a function of measurement temperature. Up to 175 °C., he obtained conductivity curves similar to those of the middle of 2.8.6. But, as the measurement temperature increased, the pronounced dip seen in 2.8.6. tended to flatten out. At the highest measurement temperature of 410 °C., the conductivity was flat. He concluded that elevated temperatures preclude the formation of clusters and/or complexes, so that conductivity due to Ag_i^+ remains the

2.8.6.-

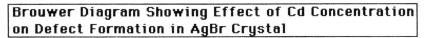

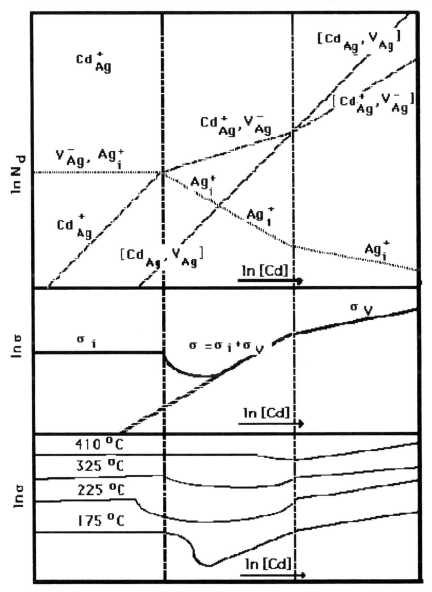

major contributor to the conductivity, as according to the defect reaction "δ" of Table 2-3.

LET US NOW SUMMARIZE WHAT WE HAVE COVERED TO DATE:

1. By rewriting the equilibrium constants of Table 2-1 and 2.7.20. (ionization of vacancies) as logarithms, we obtained linear relations among the set of defect equations.

2. By defining sets of defects as a ratio, R, we can then plot the ratios so as to show how the relative numbers vary as a function of the type of defect present in the chosen crystal lattice. This is the Brouwer Method.

3. We also illustrated the method for a AgBr crystal containing Cd^{2+} . The set of defect reactions were given, so as to illustrate the possible defects present. Then, a Brouwer diagram illustrated the numbers and types of defects actually present as a function of Cd^{2+} content in the crystal.

II. Defect Disorder in AgBr- A Thermodynamic Approach

To illustrate yet another approach to analysis of defect formation, consider the influence of Br_2 - gas upon defect formation in AgBr. The free energy of formation, ΔG, is related to the reaction:

2.8.8.- $Ag^o + 1/2\, Br_{2\,(g)} \leftrightharpoons AgBr_{(g)}$ $\{\Delta G_{AgBr}\}$

This can be rewritten as:

2.8.9.- $å_{Ag}\; p^{1/2}_{Br_2} = \exp \Delta G_{AgBr} / RT$

It makes no difference as to which of the activities we use. If we now fix p_{Br_2} at some low value, we find that the **possible** defects in our AgBr crystal, as influenced by the **external factor,** p_{Br_2}, will be:

2.8.10.- Ag_i^+ , V_{Ag}^- , Br_i^- , V_{Br}^+ , e^- and π^+

where we use π for the positive charge to differentiate between pressure, p, of the external gas. Because of the high electrostatic energy required to maintain them in an ionic crystal such as AgBr, we can safely ignore the following **possible** defects:

2.8.11.- Ag_{Br}^+ , Ag_{Br}^{++} , Br_{Ag}^- , $Br_{Ag}^=$.

If we have thermal disorder at room temperature (I do not know of any crystal for which this is not the case), then we can expect the following defect reaction relations:

2.8.12.- Ag_i^+ = V_{Ag}^-

                  Br_i^- = Ag_i^+

                  V_{Ag}^- = V_{Br}^+

                  Br_i^- = V_{Br}^+

At equilibrium, the following equations arise:

2.8.13.- Ag_{Ag} + $1/2\, Br_{2\,(g)}$ ⇋ $AgBr_{(s)}$ + V_{Ag}^- + π^+

                  K_d = $V_{Ag}^- \cdot \pi^+ / p^{\,1/2\,Br}{}_2$

This gives us a total of eight (8) concentrations to calculate. They involve the following crystal defects:

2.8.14.- Ag_{Ag} , Br_{Br} , Ag_i , Br_i , V_{Ag}^- , V_{Br}^+ , e^- , π^+

Our procedure is to set up a site balance in terms of lattice molecules, i.e.- for AgBr:

2.8.15.- Ag_{Ag} , + V_{Ag}^- + e^- = 1 $(e^- \equiv Ag_{Ag}^-)$
                      Br_{Br} + V_{Br}^+ + π^+ = 1 $(\pi^+ \equiv Br_{Br}^+)$

Since Br_2 (gas) is the driving force for defect formation, we need also to consider deviation from stoichiometry, δ . Thus, we also set a $Ag_{1-\delta}$ Br balance:

2.8.16.- Ag_{Ag} + Ag_i^+ + e^- = $1 + \delta$

                      Br_{Br} + Br_i^- + π^+ = 1

To maintain electroneutrality:

2.8.17.- Ag_i^+ + V_{Br}^+ + π^+ = Br_i^- + V_{Ag}^- + e^-

We also set up the following equations:

2.8.18.- Ag_{Ag} + αV_i $\leftrightharpoons$ Ag_i^+ + V_{Ag}^- K_e = Ag_i^+ V_{Ag}^- / V_i^α

2.8.19.- Br_{Br} + αV_i $\leftrightharpoons$ Br_i^- + V_{Br}^+ K_g = Br_i^- V_{Br}^+ / V_i^α
and
2.8.20.- e^- + p^+ $\leftrightharpoons$ 0 K_b = (e^-) (π^+)

Note that we have distinguished between three (3) situations, to wit:

        a. Electroneutrality

        b. Thermal Disorder

        c. Non-stoichiometry (excess cation)

These are the eight equations (2.8.12. to 2.8.20.) required to calculate the defect concentrations arising from the effects of the external factor, p_{Br_2} . From measurements of conductivities, transfer numbers (electromigration of charged species), lattice constants and experimental densities, it has been shown that Frenkel defects predominate (Lidiard - 1957). This means that:

2.8.21.- $\quad\quad\quad Ag_i^+ , V_{Ag}^- \;\gg\; Br_i^- , V_{Br}^+$

Furthermore, $V_{Ag}^- \equiv Ag_i^+$ so that in terms of our equilibrium constants we get:

2.8.22.- $\quad$ FOR FRENKEL DEFECTS: $K_e \gg K_g$ and $K_e \gg K_d$

Thus, we need only to consider the above two (2) defects, namely - V_{Ag}^- and Ag_i^+ , since they are the major contributors to non-stoichiometry. By calculating $p^o_{Br_2}$ as before (when $\delta = 0$, see 2.7.22 & 2.7.23.), we can express our overall defect equation as:

2.8.23.- $p^{1/2}_{Br_2} / (p^o_{Br_2})^{1/2} = \{\delta/2\varepsilon + [(1+\delta/2\varepsilon)^2]^{1/2}\}\{\delta/2\beta + [1+(\delta/2\beta)^2]^{1/2}\}$

Because of the conditions given in 2.8.16., the first half of the equation can be set equal to one. **Note that we are using** ε , β , **and** γ **as the equilibrium constants . i.e. -**

2.8.24.- $\quad\quad\quad \varepsilon \equiv K_\varepsilon^{1/2} \quad\quad \beta \equiv K_\beta^{1/2} \quad\quad \gamma \equiv K_\gamma^{1/2}$

In the remaining part of the equation, $\delta \gg \beta$. By taking logarithms, we can then obtain:

2.8.25.- $\quad\quad\quad 1/2 \ln p_{Br_2} / p^o_{Br_2} = \ln \delta - \ln \beta$

This result then leads us to a plot of the effect of partial pressure of Br_2 on the deviation from stoichiometry , d , for the AgBr crystal, as shown in 2.8.26. on the next page (this work is due to Greenwood - 1968).

For $\beta = 0$, there is a point of inflection where the slope of the line is defined by the equilibrium constant, i.e.- $\beta = K_\beta^{1/2}$.

2.8.26.-

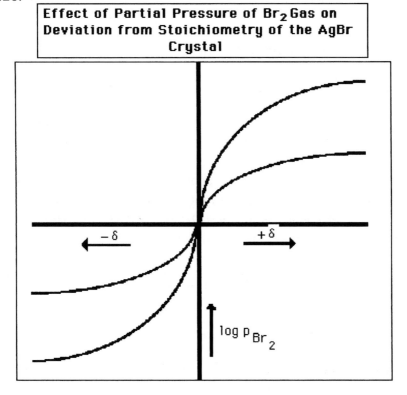

Effect of Partial Pressure of Br₂ Gas on Deviation from Stoichiometry of the AgBr Crystal

The larger this value, the flatter is the curve. All relations regarding the defects can now be derived.

The major defects turn out to be:

2.8.27.- $V_{Ag}^{-} \leftrightarrows Ag_i^{+}$

$$\pi+ \leftrightarrows K_\delta (1/K_e)^{1/2} \, p^{1/2}_{Br_2}$$

$$e^- \leftrightarrows (K_e)^{1/2} (1/K_e) \, K_b (1/ p_{Br_2})^{1/2}$$

This shows that both $P+$ and e^- are minority defects dependent on p_{Br_2}.

The following gives the standard enthalpies and entropies of these defect reactions, according to Krœger (1965):

2.8.28.-	DEFECT REACTION	ΔS	ΔH
		Cal./mol/°K	Kcal./mol.

$Ag_{Ag} \leftrightarrows Ag_i^+ + V_{Ag}^-$ (Frenkel)   25.6   29.3

$Ag_{Ag} + Br_{Br} \leftrightarrows V_{Ag}^- + V_{Br}^+$ - 13.3   36

$0 \leftrightarrows e^- + p^+$ 25   78

$1/2\ Br_2 + Ag_{Ag} \leftrightarrows AgBr + V_{Ag}^- + p^+$ 4.9   25.4

It is apparent that the Frenkel process coupled with the electronic process are the predominating mechanisms in forming defects in AgBr through the agency of external reaction with Br_2 gas.

A final comment: we can use these thermodynamic values to calculate the equilibrium constants according to:

2.8.29.- $K_i = exp - \Delta G_i^o / RT$

and can also obtain the activity of the silver atom in AgBr from 2.8.29. By using equation 2.8.9., we can show :

$$For\ \alpha_{Ag} = 1\ ,\ @\ T = 277\ °C.\ ;\ \delta = + 10^{12}$$
$$For\ p_{Br_2} = 1\ atm.\ @\ 277\ °C.\ ;\ \delta = - 10^{-7}$$

where the plus or minus indicate an excess or deficit of the silver atom in AgBr. This result is due to Wagner (1959).

2.9.- SUMMARY AND CONCLUSIONS

Let us now summarize the major conclusions reached regarding the defect solid. You will note that we have investigated the following hypothetical compounds: MX, MX_S and $MX_{S\pm\ \delta}$. But, when we investigated crystals in the real world, we found that actual defects in such solids did not conform entirely to those of our hypothetical compounds.

Nonetheless, in order to comprehend and form a foundation to understand how defects affect the properties of actual solids, it was necessary to study those hypothetical compounds. The following is a summary of the conclusions we reached regarding the defect solid state:

1. The charge compensation mechanism represents the single most important mechanism which operates within the defect solid.
2. We have shown that defect equations and equilibria can be written for the MX_S compound, both for the stoichiometric and non-stoichiometric cases.
3. The conclusion that we reach is that defect formation is favored in the solid because of the entropy factor. It is much more difficult to obtain a "perfect" solid, so that the defect-solid results. We have also shown that the intrinsic defects can become ionized.
4. Although more than one defect reaction may be applicable to a given situation, only one is usually favored by the prevailing thermodynamic and electrical conditions. Thus, we conclude that intrinsic defects usually occur in pairs. This conclusion cannot be overemphasized.
5. We have shown that defects occur in pairs. The reason for this lies in the charge-compensation principle which occurs in all solids.
6. Of the nine defect-pairs possible, only 5 have actually been experimentally observed in solids. These are: Schottky, Frenkel, Anti- Frenkel, Anti-Structure, Vacancy-Structure.
7. There can be no doubt that both Schottky and Frenkel defects are thermal in origin.
8. We have shown by Statistical Mechanics that we can calculate numbers of defects present at a given temperature.
9. There is an Activation Energy for defect formation. In many cases, this energy is low enough that defect formation occurs at, or slightly above, room temperature.

10. Defects may be described in terms of thermodynamic constants and equilibria. The presence of defects changes both the local vibrational frequencies in the vicinity of the defect and the local lattice configuration around the defect.

11. There are two associated effects on a given solid which have opposite effects on stoichiometry. Usually, one involves the cation site and the other the anion site. Because of the differences in defect-formation-energies, the concentration of other defects is usually negligible.

12. In examining the defect state of real crystals such as AgBr, we find that we can write, using equilibrium constants and defect thermodynamics derived from Statistical Mechanics and classical Thermodynamics, valid equations for the numbers and types of various associated defects present. However, we also find that we cannot solve for the value of the unknown quantities in a set of simultaneous equations since the equations are not linearly solvable. The equations can be written, but the set of equations cannot be easily solved. It is for this reason that we have resorted to graphical method like that of Brouwer, even though it is not entirely satisfactory in its solutions to the numbers and types of defects present in real crystals.

Thus, it should be clear that lattice defects in the solid state is the normal state of affairs and that it is the defects which affect the physical and chemical properties of the solid.

2.10. - THE EFFECTS OF PURITY (AND IMPURITIES)

Our study has led us to the point where we can realize that the primary effect of impurities in a solid is the formation of defects, particularly the Frenkel and Schottky types of associated defects. Thus, the primary effect obtained in purifying a solid is the **minimization** of defects. Impurities, particularly those of differing valences than those of the lattice, cause charged vacancies and/or interstitials. We can also increase the reactivity of a solid to a certain extent by making it more of a defect crystal by the addition of selected impurities.

It is not so apparent as to what happens to a solid as we continue to purify it. To understand this, we need to examine the various grades of purity as we normally encounter them. Although we have emphasized inorganic compounds thus far (and will continue to do so), the same principles apply to organic crystals as well. COMMERCIAL GRADE is usually about 95% purity (to orient ourselves, what we mean is that 95% of the material is that specified, with 5% being different (unwanted-?) material. Laboratory or "ACS-REAGENT GRADE" averages about 99.8% in purity.

2.11.1.- GRADES OF PURITY FOR COMMON CHEMICALS

GRADE	%	ppm IMPURITIES	IMPURITY ATOMS PER MOLE OF COMPOUND.
Commercial	95	50,000	3.0×10^{22}
Laboratory	99.8	2000	1.2×10^{21}
Luminescent	99.99	100	6×10^{19}
Semi-conductor	99.999	10	6×10^{18}
Crystal Growth	99.9999	1	6×10^{17}
Fiber-Optics	99.999999	0.01	6×10^{15}

The GRADES listed above are named for the usage to which they are intended, and are usually minimum purities required for the particular application. Fiber-optic materials are currently prepared by chemical vapor deposition techniques because any handling of materials introduces impurities. Furthermore, this is the only way found to date to prepare the required materials at this level of purity.

The frontiers of purity achievement of solids presently lie at the fraction of parts per billion level. However, because of Environmental Demands, analytical methodology presently available far exceeds this. We can now analyze metals and anions at the femto level (parts per quadrillion= 10^{-15}) if we wish to do so.

Nevertheless, it is becoming apparent that as high purity inorganic solids are being obtained, we observe that their physical properties may be

different than those usually accepted for the same compound of lower purity .

The higher-purity compound may undergo solid state reactions somewhat differently than those considered "normal" for the compound. If we reflect but a moment, we realize that this is what we might expect to occur as we obtain compounds (crystals) containing far fewer intrinsic defects.

It is undoubtedly true that many of the descriptions of physical and solid state reaction mechanisms now existing in the literature are only partially correct. It seems that part of the frontier of knowledge for Chemistry of The Solid State lies in measurement of physical and chemical properties of inorganic compounds as a function of purity. A case in point is that of the so-called "Nano-Technology", the vanguard of research into chemical and physical properties of materials in the research community today.

2.11.- Nanotechnology and The Solid State

In the next chapter, we shall examine the methods of characterizing solids including: the properties of individual particles (including single crystals); the solid state reactions that are used to form various solids; and methods used to describe an assembly of particles (particle size). We will find that most solid materials are composed of particles in the 1- 300 μm. range. This is 1-300 x 10^{-6} meters. Most inorganic materials are produced having particles in this size range. These are the familiar powders such as coal dust, inorganic chemicals, silt and fine sand, and even bacteria.

Current research defines nano-technology as the use of materials and systems whose structures and components exhibit novel and significantly changed properties when control is achieved at the atomic and/or molecular level. What this means is that when a given material is produced having particle sizes at fractions of a μm (micron), it displays novel properties not found in the same material whose particles are larger than 1.0 μm (micron). Nanotechnology involves dimensions where atoms and molecules, and interactions between them, influence their chemical and physical behavior. Authentic nano-particles are so small that there are many more atoms on the surface of each particle than the normal particle of 1.0 μm. Particles of 1.0 μm, i.e.- 1000 nm or 1000×10^{-9} m, may seem small but those atoms on the surface of each particle are only about 0.0015% or 15 in a million of the atoms composing the lattice. A nano-particle with dimensions of 10 nm. brings the surface atoms to about 15% of the total atoms composing the particle. At this size range, quantum physics and quantum effects determines the primary behavior of such particles.

Consider that atoms have a size range of about 1-2 Å. Most inorganic solids, with the exception of halides, sulfides (and other pnictides), arc based upon the oxygen atom, i.e.- oxide = $O^=$, whose atomic radius does not change even when sulfates, phosphates and silicates are formed. Oxide has an atomic diameter of 1.5 Å or 0.15 nm. = 0.00015 μm. Nano-particles are clumps of 1000 to 10,000 atoms. The latter would be a particle of 0.15 μm. in diameter. They can be metal oxides, semi-conductors, or metals with novel properties useful for electronic, optical, magnetic and/or catalytic uses.

When light meets particles this small, it behaves differently. One example is TiO_2 (titanium dioxide), which has been used as an ultra-violet absorber for sun-screen products. The usual product is applied to the skin as a white-reflecting cream. The process for making titanium dioxide varies but usually employs $TiCl_4$ and its hydrolysis under controlled conditions. When particles of 50 nm. are formed, the sun-screen cream now is transparent since the particles absorb and scatter visible light much less than the larger particles previously used. However, the ultraviolet light absorption is not changed, only the reflection of white light.

As we shall see in the next chapter, particles are formed first as "embryos" which are minute particles of the nano-particle class. These then grow into "nuclei" which then grow into particles. The science of particle growth has been a major source of our understanding of particles. As we have already shown, lattice defects, due to thermal effects, are the norm when a crystal grows to sizable proportions. However, when nanocrystals are formed, the numbers of embryos allowed to form, with corresponding nuclei, are controlled. The nuclei growth is then confined to atomic dimensions. Much of this growth forms by "Spontaneous Assembly". That is, when nano-particles are formed, atoms are added one at a time to form the embryo and then the nucleus. It is the size of the nucleus that is restricted.

I submit that the prediction given in the previous section, i.e.- see p. 107, has already been realized. That is, nanoparticles form by self-assembly of atoms (ions) into **defect-free** crystals. It is this lack of intrinsic defects that give such particles their unique chemical and physical properties. Note that if normal growth were allowed to proceed further, then we would have the normal defect-crystal.

Suggested Reading

1. A.C. Damask and G.J. Dienes, *Point Defects in Metals*, Gordon & Breach, New York (1972).

2. G.G. Libowitz, "Defect Equilibria in Solids", *Treatice on Solid State Chem.-* (N.B. Hannay- Ed.), **1**, 335-385, (1973).

3. F.A. Krœger, *The Chemistry of Imperfect Crystals*, North-Holland, Amsterdam (1964).

4. F.A. Krœger & H.J. Vink in *Solid State Physics, Advances in Research and Applications* (F. Seitz & D. Turnbull-Eds.), pp. 307-435 (1956).

5. J.S. Anderson in *Problems of Non-Stoichiometry* (A. Rabenau-Ed.), pp.1-76, N. Holland, Amsterdam (1970).

6. W. Van Gool, *Principles of Defect chemistry of Crystalline Solids*, Academic Press, New York (1964).

7. G. Brouwer, "A General Asymmetric Solution of Reaction Equations Common in Solid State Chemistry", Philips Res. Rept., **9** , 366-376 (1954)

8. A. B. Lidiard, "Vacancy Pairs in Ionic Crystals", *Phys. Rev.*, **112**, 54-55 (1958).

9. J.S. Anderson, "The Conditions of Equilibrium of Nonstoichiometric Chemical Compounds, *Proc. Roy. Soc. (London)* , **A185**, 69-89 (1946).

10. N.N. Greenwood, *Ionic Crystals, Lattice Defects & Non-Stoichiometry*, Butterworths, London (1968).

11. Hayes and Stoneham, "Defects and Defect Processes in Non-Metallic Solids"- J. Wiley & Sons, New York (1985).

CHAPTER 3

The Solid State: Mechanisms of Nucleation, Solid State Diffusion, Growth
of Particles and Measurement of Solid State Reactions

In the previous Chapter, we examined intrinsic defects in the solid. In
this chapter, we will examine solid state reactions as related to nucleation
and diffusion mechanisms. We will also investigate how particles grow
during reactions to produce solids. Solids are generally created as
crystallites (particles) unless special methods to produce large single
crystals are employed. When we have particles, they will undergo solid
state reactions and change form, habit and structure if heated. The solid
state reaction occurs at the boundary of **direct contact** between adjacent
crystallites.

We will also examine methods of characterizing such solid state reactions,
including DTA (Differential Thermal Analysis) and Thermogravimetric
Analysis, i.e.-TGA). We will reserve the discussion of how size of particles
and a particle size distribution is determined until the next chapter.

If we wish to create a phosphor, there are two general methods to do so.
One is by **precipitation** from an aqueous (rarely non-aqueous) solution. The
other is by **solid state reaction**. Sometimes, both methods are employed.
The phenomena concerning particle growth in solution differ quite widely
from those governing growth of particles caused by solid state reactions.
Yet they retain many similarities. The formation and growth of nuclei
during particle formation and growth in either precipitation or solid state
reactions will be of major concern in this chapter.

Both single crystal and crystallite formation employ very similar methods
but differ in technique. They include formation from:

Single Crystal	Crystallites (Powders)
Liquid solvent	Vapor
Vapor	Melt
Melt	Flux
Molten salt	Precipitation from solution

Obviously, the major difference in the single-crystal and polycrystalline (crystallite) state is a matter of size. For the single-crystal, the size is large ($\geq$ 10 cm), whereas in the polycrystalline state, the size of the crystals is small (10 μm = 0.001 cm.) The methods for obtaining one type of crystal or the other differ considerably.

Most solids that we encounter in the real world are in the form of powders. That is, they are in the form of discrete small particles of varying size, i.e.- crystallites. Each particle has its own unique diameter and size. Additionally, their physical proportions can vary in shape from spheres to needles. For a given powder, all particles will be the same shape. However, their sizes can be altered by the method used to create them in the first place. Methods of particle formation include:

> Solid State Reaction
> Precipitation
> Condensation or Evaporation

These methods involve the formation of nuclei and the rate of their growth. We will examine such mechanisms as they relate to both **solid state** reactions and **precipitation** as methods for obtaining discrete particles (crystallites). Still other processes exist **after** particles have formed, including sequences in particle growth like sintering.

Nucleation can be homogeneous (no outside influences) or heterogeneous (by specific outside coercion). In precipitation processes to form a distribution of particle sizes, it is probably a combination of both mechanisms. Since nucleation differs in either precipitation or solid state reactions, we will address each separately. But, before we can investigate nucleation in solid state reactions and compare it to nucleation during precipitation processes to form a solid, we need to determine how solid state reactions occur.

3.1.- SOLID STATE REACTIONS AND NUCLEATION MECHANISMS

Solid state reaction mechanisms include phase changes, formation of

while the latter will be very slow. The **Kirchendall Effec**t deals with this manifestation.

Suppose we have a situation where A reacts with B to form a solid solution, AB. Let us further suppose that the diffusion reaction is:

3.1.71.- $A_i + V_A = B_i + V_B$

In our case, we will ignore charged species. Let us also suppose that:

3.1.72.- $D_{V(A)} \gg D_{V(B)}$ so that: $dc_{V(A)}/dt \gg dc_{V(B)}/dt$

Given these restraints, we obtain a rather queer result. The mechanism gives rise to the **creation of new sites across the diffusion zone** and actually causes a **deformation** in the solid because the V_A defects pile up and finally become annihilated because of clustering. **Actual holes appear** in the solid due to vacancy diffusion, vis-

3.1.73.-

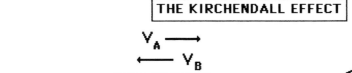

Note that in this case, we are not forming a new compound through solid state reaction, but are forming a solid solution of A and B. The Kirchendall Effect has been observed many times, but occurs most often in reactions between metals to form alloys.

There are three (3) types of diffusion-controlled reactions possible for heterogeneous solid state reactions. They are given in 3.1.74. on the next page:

3.1.74.- TYPES OF DIFFUSION REACTIONS

Simple Diffusion:	$\beta = k_1 \, (t)^{1/2}$
Phase-Boundary Controlled	$\beta = k_2 \, t$
Material Transport	--------

Note that these equations relating diffusion reactions resemble the mechanisms already given for nuclei growth in 3.1.41.

For simple diffusion-controlled reactions, we can show the following holds:

3.1.75.- $x^2 = 2 \, D \, c_0 \, t \, V_M + (a)^{1/2}$

where c_0 = concentration of constituents at interface; V = volume of product AB per mol of reactants; a = surface layer thickness at interface when t = o. It is also well to note that the final volume of the product, AB, may not be the same as that of the reactants, vis.-

3.1.76.-

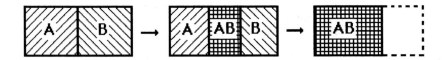

For phase-boundary controlled reactions, the situation is somewhat different. Diffusion of species is fast but the reaction is slow so that the diffusing species pile up. That is, the reaction to rearrange the structure is slow in relation to the arrival of the diffusing ions or atoms. Thus, a phase-boundary (difference in structure) focus exists which controls the overall rate of solid state reaction. This rate may be described by:

3.1.77.- $dx/dt = k_1 \, (s_t / V_0) \, - 1 - (1-x)^n = k_1 t / r_0$

where s_t = instantaneous surface area; V_0 = original volume of particles reacting; r_0 = original radius of particles; and n = 1, 1/2 , 1/3 for a 1-, 2-, or 3- dimensional reaction. Material transport frequently involves an external gaseous phase, but **a general formula has not evolved.**

The above equations summarize the three (3) diffusion mechanisms, one of which usually predominates in any given case (see 3.1.74.). But these equations contain quantities that are hard to measure, or even estimate. A much better way is to follow the method of **Hancock and Sharp** (~ 1938). If one can measure x, the amount of reactant formed in time, t, then the following equation applies:

3.1.78.- $-\log (\ln(1-x)) = m \ln T + \ln k$

	range of m
simple diffusion	0.57 - 0.62
nuclei growth	1.00 - 1.15
phase boundary	1.25 - 3.00

If [- log (ln(1-x)] is plotted against ln t, one obtains a value for the slope, m, of the line which allows classification of the most likely diffusion process. Of course, one must be sure that the solid state reaction is primarily diffusion-limited. Otherwise, the analysis does not hold.

V.- Analysis of Diffusion Reactions in the Solid State

Let us now turn to diffusion in the general case, without worrying about the exact mechanism or the rates of diffusion of the various species. As an example to illustrate how we would analyze a diffusion-limited solid state reaction, we use the general equation describing formation of a compound with spinel (cubic) structure and stoichiometry (Spinel is the name applied to both one particular compound, i.e.- $MgAl_2O_4$, and a group of compounds all having the same cubic structure and composition).

3.1.79.- $AO + B_2O_3 \Rightarrow AB_2O_4$

Here, A and B are two different metallic ions and O is the oxygen atom (as the oxide). Spinel is a cubic **mineral** composed of magnesium aluminum oxide or any member of a group of rock-forming minerals. All of these are metal oxides with the general composition AB_2O_4 , in which A may be magnesium, calcium, iron, zinc, manganese, or nickel. B may be aluminum, tin, chromium, or iron, and O is oxygen (Note that A is divalent and B is trivalent and sometimes quadrivalent).

Blasse (1964) listed close to 200 spinels having either a "normal" or "inverted' spinel structure. What this means is that the cations normally occupying the "A" site would occupy the "B" site would be exchanged, depending upon the ionic radius of the two cations. Thus, if we could make Mg^{2+} smaller in radius, and Al^{3+} were made larger, we would have: Al_2MgO_4 as an inverted spinel. In normal spinels, the divalent cations occupy tetrahedral sites while the trivalent cations occupy the octahedral sites. The inverted state depends upon which cations are involved and their relative ionic size. Thus, we have two cation "sub-lattices" in the spinel lattice, the tetrahedral or A-sublattice and the octahedral B-sublattice.

a. ANALYSIS OF DIFFUSION REACTIONS IN SPINEL

There are two (2) different cases that we can distinguish, both of which represent possible diffusion mechanisms in spinel. These are: Ion diffusion or Gaseous. Let us consider the ion-diffusion mechanism first. This is shown in the following diagram.

3.1.80.-

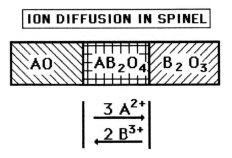

Here, we have shown A^{2+} and B^{3+} diffusing through AB_2O_4. The actual reactions involved are:

3.1.81.-
$$3\ A^{2+}\ +\ 4\ B_2O_3\ \Rightarrow\ 3\ AB_2O_4\ +\ 2\ B^{3+}$$
$$2\ B^{3+}\ +\ 4\ AO\ \ \ \ \Rightarrow\ AB_2O_4\ \ +\ 3\ A^{2+}$$

$$4\ AO\ +\ 4\ B_2O_3\ \ \Rightarrow\ 4\ AB_2O_4$$

Notice that the partial reactions in 3.1.81. are balanced both as to material and charge. There are at least two other possible mechanisms, as shown in the following diagram:

3.1.82.-

ION DIFFUSION IN SPINEL

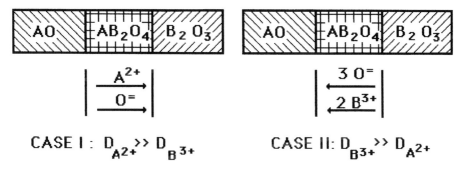

CASE I : $D_{A^{2+}} \gg D_{B^{3+}}$ CASE II: $D_{B^{3+}} \gg D_{A^{2+}}$

In one case, the diffusion of A^{2+} is much faster than B^{3+} and in the other the opposite is true. Note that charge-compensation of migrating species is maintained in all cases by the diffusion of oxide species.

We can also illustrate gaseous transport reaction in the same manner, as shown in 3.1.83., given on the next page. The partial reactions are:

3.1.84.-
$$O^= \ \Rightarrow\ 1/2\ O_2\ +\ 2\ e^-$$
$$A^{2+}\ +\ 2\ e^-\ \ +\ 1/2\ O_2\ +\ B_2O_3\ \ \Rightarrow AB_2O_4$$

In this mechanism where $D_A{}^{2+} \gg D_B{}^{3+}$, transport of external oxygen gas is involved in the solid state, accompanied by electronic charge diffusion.

There is still one more mechanism to consider. That is the one where the diffusion rate of B^{3+} is much greater than that of A, i.e.- $D_B{}^{3+} \gg D_A{}^{2+}$.

3.1.83.-

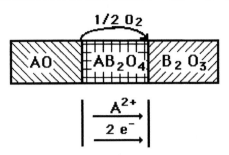

External Gaseous Transport

Then transport of O_2 is just opposite that shown in 3.1.84., and the reactions are:

3.1.85.- $6\ O^= \Rightarrow 3/2\ O_2 + 6\ e^-$

$AO + 3/2\ O_2 + 6\ e^- + 2\ B^{3+} \Rightarrow AB_2O_4$

We have now presented **all of the possible** diffusion reactions in spinel synthesis. The next step will be to determine which of them is the valid one.

To start, we measure formation rate in air, i.e.- in pure oxygen gas, and then in an inert gas. If the rates do not differ significantly, then we can rule out gaseous transport mechanisms. There are other tests we can apply, including electrical conductivity, transference numbers and thermal expansion. Although these methods have been investigated in detail, we shall not present them here. It suffices to say that:

> **"For the most part, in oxygen-dominated hosts, diffusion**
> **by cations (small) prevails".**

Thus, we write the solid state diffusion reaction for spinel as follows in 3.1.86. on the next page. In this case, Frenkel pair diffusion predominates and **is faster** than any other possible mechanism in the spinel lattice.

3.1.86.-
$$3 \, Mg^{2+} + 4 \, Al_2O_3 \Rightarrow 3 \, MgAl_2O_4 + 2 \, Al^{3+}$$
$$\underline{2 \, Al^{3+} \quad + 4 \, MgO \Rightarrow \quad MgAl_2O_4 \quad + 3 \, Mg^{2+}}$$
$$4 \, MgO \; + 4 \, Al_2O_3 \Rightarrow 4 \, MgAl_2O_4$$

Frenkel Pair: $\quad Al_{Al} \; + \infty V_i \leftrightharpoons \quad V_{Al} \quad + \; Al_i^{3+}$

b. PHOSPHORS BASED ON SPINELS

The spinel group is divided into three immiscible series, the spinel in which B is aluminum (aluminum-spinel), the chromite (chromium-spinel) series in which B is chromium; and the magnetite (iron-spinel) series, in which B is iron.

For reasons which will become clear later, we cannot use chromium-spinels or iron-spinels as a phosphor base. Thus, we are limited to aluminum, tin and the like. We will limit our discussion to tin and gallium-based compounds, i.e.- Mg_2SnO_4 and Mg_2GaO_4.

Both of these materials can be "activated" by Mn^{2+} which also forms a spinel, i.e.- Mn_2SnO_4. The latter as a phosphor, i.e.- $Mg_2GaO_4:Mn^{2+}$, is used as the source of green light in most copying machines today. It is interesting to note that the cubic spinel structure is formed by a majority of dibasic cations like Mg^{2+} and that many trivalent or quadrivalent cations can form oxide-based anions like stannate, aluminate, titanates, vanadates and the like.

To prepare a phosphor, we need to follow a certain series of steps. We will go into more detail in another chapter, but here we only present a bare outline of the preparation of the phosphor- $Mg_2SnO_4 : Mn^{2+}$, where the actual formula is a solid solution of: $Mg_2SnO_4 \bullet Mn_2SnO_4$

As shown in the last chapter, "luminescent-grade" materials must have no more than a total of 100 ppm of impurities. The reaction that we will use to prepare our phosphor is:

3.1.87.-
$$2 \, (MgO + MnO) + SnO_2 \Rightarrow \quad (Mg,Mn)_2SnO_4$$

We will also use a magnesium compound that will decompose upon heating to form the oxide, thus ensuring that the solid state reaction will proceed to 100% completion, as we have already illustrated. However, magnesium does not form a carbonate directly so we have to use a hydroxy-carbonate, i.e.- $3MgCO_3 \bullet Mg(OH)_2 \bullet 3 H_2O$. A suitable phosphor formulation is given in the following:

3.1.88.-   2.500 mol of MgO (as $3MgCO_3 \cdot Mg(OH)_2 \cdot 3 H_2O$)
                  1.000 mol SnO_2
                  0.025 mol $MnCO_3$

Since we intend to use this phosphor in a fluorescent lamp, we need to control its ultimate ultraviolet absorption properties. It is axiomatic that lamp phosphors must not contain an excess of a UV-absorbing constituent. In our case, this is SnO_2. We thus add an excess of MgO. However, the magnesium compound only produces about 45 gm of MgO per 100 gm. of the hydroxy-carbonate. That is, it has an "assay" of 44.6%. The above formulation thus needs to be adjusted to compensate for this factor. Additionally, we would normally assay all of the other components as well. This is done by firing them to obtain the requisite oxides which are actually the reacting components in the above reaction of 1.3.86. Note that in some cases, the assay reflects only the amount of adsorbed water, i.e.- SnO_2 would have an assay of about 99.4%.

The procedure is as follows:

1. Weigh out 100.8 gm of MgO (corrected, this is 226.0 gm of hydroxy-carbonate), 134.7 gm of SnO_2 and 2.125 gm of $MnCO_3$.
2. Mix these powders thoroughly by use of a hammermill and/or other solid state blender.
3. Prefire the mix in a covered silica crucible at 2000 °F (1,093 °C) in air for 2.0 hours.
4. Allow to cool in air
5. Crush the "cake", blend and remix it.
6. Refire in a reducing atmosphere of 80% nitrogen and 20% hydrogen for 1.5 hours.

7. Allow to cool in a nitrogen atmosphere (Note that a "tube" furnace is normally used to fire phosphors in controlled atmospheres).

8. Remove the phosphor cake, crush it and blend thoroughly.

9. Since we have an excess of MgO in the fired product, we need to remove it. A 3 liter solution of 1.0 molar acetic acid is prepared.

10. Heat the acetic acid solution to 80 °C and add the phosphor powder slowly with stirring.

11. Stir for about 10 minutes and then allow the particles to settle.

12. Pour off the supernatent liquid and resuspend the powder in deionized water.

13. Allow to settle and resuspend 2 more times.

14. Filter the particles in a Büchner funnel and dry the wet cake overnight at 105°C.

15. Sieve the dried phosphor cake through a 325 mesh screen.

The resulting phosphor has a green emission band, as shown in the following diagram:

3.1.89.-

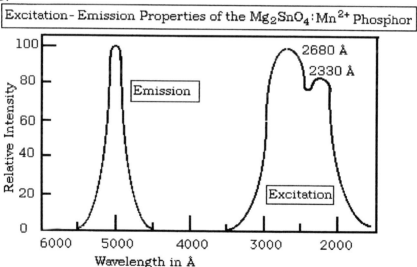

In this case, we see the emission band which peaks at 5000 Å and two excitation bands at 2330 Å and 2680 Å.

What this means is that these UV-radiation wavelengths will excite the phosphor which then emits visible green light. When the phosphor is prefired in air, the resulting phosphor does not respond to UV excitation. However, cathode-ray excitation (an electron-beam like a television tube) produces the same green emission. Emission occurs from Mn^{2+} centers in the spinel structure. These centers are not intrinsic defects as such since the divalent manganese is able to substitute directly at the Mg^{2+} sites in the spinel structure. The only difference is the radius of the two cations at the tetrahedral site. It is because of this difference that increasing the Mn^{2+} concentration leads to less efficient, i.e.- "duller", phosphors.

That is, there is an "optimum" activator concentration (as is true for all phosphors). However, UV excitation is not obtained until the prefired phosphor is subjected to a reducing atmosphere. It has been determined that if the reacting oxides are prefired directly in a reducing atmosphere, a black mass is obtained containing tin in the metallic state.

The two UV excitation peaks are due to Sn^{2+} centers in the phosphor. If one looks at the excitation peaks of the $Sr_2P_2O_7:Sn^{2+}$ phosphor, the peaks are identical. No other cations having optical absorption in the UV are present other than Sn^{2+} (Mn^{2+} does not). Thus, the UV absorption results in excited Sn^{2+} centers which transfer this excitation energy to Mn^{2+} centers which emit the characteristic green emission.

It is clear that some of the Sn^{4+} cations in the B-site SnO_4 groups are reduced but no one has determined the extent of such reduction. It is apparent that only a small fraction of such groups are affected. Otherwise the lattice structure would not be maintained. Note that a significant difference in ionic radius occurs when stannic tin is reduced to stannous tin. X-ray analysis shows no difference between the air-fired and the reduced phosphors. K. Th. Wilke studied this phenomenon in 1957 and proposed the following mechanism, given as 3.1.90. on the next page.

The fact that oxygen vacancies are formed is an important factor in our analysis of the defect chemistry of this phosphor.

illustrate these reactions by a graph. This is shown in the following diagram:

3.1.108.-

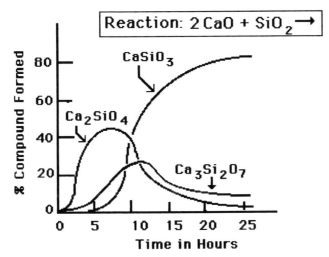

In the sequence of diffusion reactions, we note that Ca_2SiO_4 is formed immediately, followed by $Ca_3Si_2O_7$. Both **begin to disappear** when $CaSiO_3$ begins to form. Near the end, $CaSiO_3$ becomes the major phase present, but we never reach the point where just ONE COMPOUND remains. We **always** obtain a mixture! This point cannot be overstressed.

The overall reaction mechanisms are diffusion-controlled, and the total reaction is given as follows in 3.1.109. Although these partial solid state reactions are written to show the formation of the metasilicate, we already know that none of these reactions come to completion. There are competing **side-reactions.**

3.1.109.-

$$8\ CaO + 4\ SiO_2 \quad\Rightarrow\ 4\ Ca_2SiO_4$$
$$Ca_2SiO_4 + SiO_2 \quad\Rightarrow\ 2\ CaSiO_3$$
$$3\ Ca_2SiO_4 + SiO_2 \quad\Rightarrow\ 2\ Ca_3Si_2O_7$$
$$2\ Ca_3Si_2O_7 + 2\ SiO_2 \quad\Rightarrow\ 6\ CaSiO_3$$

$$\overline{8\ CaO\ +\ 8\ SiO_2 \qquad =\ 8\ CaSiO_3}$$

Thus, we start with a 2:1 stoichiometry of CaO and SiO_2, but end up with a stoichiometry which is mostly 1:1, that of the metasilicate. Another way to look at this phenomenon is that the diffusion conditions favor the formation of the **metasilicate**, and that the "excess Ca" is taken up in the formation of compounds having a "calcium-rich" stoichiometry (in comparison to that of the metasilicate).

However, if we start with a stoichiometry of 1:1 CaO to SiO_2, we discover that the same compounds form as before, and a mixture of compounds is **still obtained.** The only difference is that there are smaller amounts of Ca_2SiO_4 and $Ca_3Si_2O_7$ present! The only conclusion that we can draw is that diffusion-controlled solid state reactions tend to produce mixtures of compounds, the relative ratio of which is related to their thermodynamic stability **at the reaction temperature.** Obviously then, if we change the temperature of reaction, we would expect to see somewhat different mixtures of compounds produced.

Let us now look briefly at another similar system where we will start with a stoichiometry of : 1.00 BaO to 1.00 SiO_2 (in mols). The mixture would be expected to react as:

3.1.110.- $BaO + SiO_2 \Rightarrow BaSiO_3$

with the diffusion condition: Ba^{2+} , $O^=$ >> Si^{4+} . We show the solid state reaction behavior again by a chart, given as follows in 3.1.111. on the next page. One might think that since Ba^{2+} is a much larger ion than Ca^{2+} , it would diffuse slower. Such is not the case as can be seen by comparing the x- axis of 3.1.108. with that of 3.1.111. in terms of reaction time. It is remarkable that in this case , we started with a 1:1 stoichiometry and ended up with a compound that has a 2:1 stoichiometry!

The series of diffusion-controlled reactions are, for the case of:
$$1.0\ BaO + 1.0\ SiO_2 \Rightarrow :$$

3.1.112.-   (a) $2\ Ba^{2+} + 2\ O^= + SiO_2 \Rightarrow Ba_2SiO_4$

          (b) $Ba_2SiO_4 + Si^{4+} + 2\ O^= \Rightarrow BaSiO_3$

This gives us the following diffusion diagram:

3.1.123.-

This situation gives rise to a complicated set of diffusion conditions:

> a. The charged vacancy is one of the migrating species.
> b. The charged vacancy combines with Ni^{3+} to form a defect compound.
> c. Since $D_{Ni^{3+}} >> D_{V(Ni)}$, Ni^{3+} diffuses in an opposite direction to Al^{3+} , as shown in the following set of actual diffusion reactions, shown as 3.1.124. on the next page.

Note that the charged **vacancy** diffuses as one of the **reacting** species to form the defect compound. This situation is quite common in the solid state chemistry of compounds containing multivalent cations. The trivalent Ni^{3+} also gives rise to a new compound, $NiAlO_3$. Yet the same reactions given in 3.1.120.. are also OPERATIVE.

Thus, the overall reaction actually taking place is a combination of 3.1.120. and the following:

3.1.124.-

$$\begin{array}{c} \xrightarrow{\quad Ni^{3+} \quad} \\ \xrightarrow{\quad O^{=} \quad} \\ \xrightarrow{\quad V^{-}_{Ni} \quad} \end{array} \qquad Ni^{3+} + O^{=} + Al_2O_3 + V^{-}_{Ni} \rightarrow NiV_{Ni}Al_2O_4$$

$$\begin{array}{c} \xrightarrow{\quad Ni^{3+} \quad} \\ \xleftarrow{\quad Al^{3+} \quad} \end{array} \qquad \begin{array}{l} Ni^{3+} + Al_2O_3 \rightarrow NiAlO_3 + Al^{3+} \\ Al^{3+} + 4\,NiO \rightarrow NiAl_2O_4 + 3\,Ni^{2+} \end{array}$$

It is well to note that these types of defect solid state reactions are prevalent in many solid state reactions.

3.2- HOMOGENEOUS NUCLEATION PROCESSES - PARTICLE GROWTH

We will now examine homogeneous nucleation that may occur during a solid state reaction. One example is a decomposition reaction where a compound is heated to decompose it to form another material. In one example, a carbonate decomposes to form an oxide: $BaCO_3 \Rightarrow BaO$. Nuclei formation and growth is homogeneous since only heat is involved in the decomposition reaction, not another compound or external reactant. We have already stated that homogeneous nucleation can be contrasted to heterogeneous nucleation in that the former is random within **a single** compound while the latter involves more than one phase or compound.

Homogeneous nucleation also applies to precipitation processes where homogeneous nucleation must occur **spontaneously** before precipitation can occur.

To begin, we know that a change in free energy must occur as a change in phase (nucleation) occurs. This will be related to the total volume, V, and the total surface area, Σ, vis-

3.2.1- $\Delta G = V_{\Delta g} + \Sigma \sigma$

where ΔG is the free energy per unit volume and σ is the surface energy of the individual nuclei. If we have spherical nuclei, then:

3.2.2- $\Delta G = 4/3 \, \pi \, r^3 \, \Delta g + 4\pi \, r^2 \, \sigma$

If the total free energy, ΔG, does not change with nuclei radius, i.e.- $d(\Delta G)/dr$ is defined as zero, then:

3.2.3.- $r^*_{crit} = -2\sigma / \Delta g$

where r^*_{crit} is the critical radius for nucleus formation.

We can combine equations to obtain:

3.2.4.- $\Delta G^* = 16\pi \sigma^3 / 3(\Delta g)^2$

where ΔG^* is the change in free energy at the critical radius for nucleus formation. If the following holds:

3.2.5.- $\Delta g = \Delta h(T-T_0) / T_0 = \Delta h (\Delta T/ T_0)$

,where Δg and Δh refer to individual free energy and heat of transformation of nuclei, then:

3.2.6.- $\Delta G^* = 16 \pi \sigma^3 T_0^2 / 3 (\Delta h)^2 \Delta T^2$

Since we know that the number of nuclei can be described by a Boltzmann distribution, then:

3.2.7.- $N^* = N \exp \Delta G^* / kT$

We define dN/dt as a frequency (t is time):

3.2.8.- $dN/dt = S^* f_e \exp \Delta G_D / kT$

where G_D is the free energy of the defect (nucleus), S^* is the critical surface interface, and f_e is the **interfacial energy.** Combining the two above equations gives us:

3.2.9.- $I \equiv dN^*/dt = NS^* f_e \exp (- \Delta G^* / kT + \Delta G_D / kT)$

The quantity, $NS^* f_e$, is approximately equal to 10^{36} per cc. per second. We can solve this graphically, as shown in the following, given as 3.2.10. on the next page.

It can be seen in **A** that the value of r^* is remarkably constant over a wide range of temperatures, but that it starts to approach infinity at some

3.2.10.-

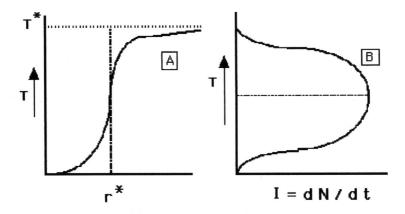

critical temperature, T^*. In contrast, the production of nuclei in **B** is maximum at some particular temperature.

Let us now examine some growth kinetics for the general case, that is, the growth of nuclei after they have formed. The general equation for growth after formation is given as:

3.2.11.- $dx/dt = k_1 (1-x)^n$

and:

$x = 1 - \exp (- k_1 t)^t$

where all terms in 3.1.11. are the same as already given. We solve these equations by a Taylor series, i.e.-

3.2.12.- $-\ln(1-x) = 1 + kt/2! + kt^2/3! + kt^3/4!$

If we now define τ as an instantaneous **moment of time**, i.e.-

3.2.13.- $(dx/dt)_{t=\tau} = K n e^{-k_1 t}$
This can be arranged to:

3.2.14.- $(dx/dt)_{t=\tau} = nk_1 t^{n-1} (1-x)$

If x is small, then:

3.2.15.- $dx/dt = nk_1 t^{n-1}$

This can be integrated to an equation similar to that of 3.1.11.

Let us now examine transformation kinetics for the homogeneous case of **nucleation and growth.**

We assume:

    a. The nucleation is random.
    b. The nucleation and growth rate are independent of each other.
    c. The spheres grow until impingement.

Then, if spheres are nucleated at time, τ , the time of growth becomes a function of $(t - \tau)$.

If we now define x_{ext} as the **extrinsic amount** per nucleus before growth (note that it is not a radius; in fact, it is the same x given in 3.2.11.), then we have:

3.2.16.- $x_{ext} = \int 4/3\ \pi\ U^3\ (t - \tau)\ I\ d\tau$

where $U \equiv$ volume of the nuclei. Note the similarity of this equation to 3.1.18. If we then further define x_f as the **final amount of volume per sphere** at impingement where growth stops, i.e.- $x_f = 1 - x_{ext}$,then we get the Johnson-Mehl equation for **slow nucleation**:

3.2.17.- $x_f = 1 - \exp\{(- 1/3\ \pi\ I\ U^3\ )\ t^4\}$

If the nucleation is **fast,** the Avrami equation is operative:

3.2.18.- $x_f = 1 - \exp\{(- 1/3\ \pi\ N_v\ U^3\ )\ t^3\}$

where N_V is defined as the number of nuclei per unit volume.

These two equations have been used to describe homogeneous nucleation, particularly in precipitation processes.

3.3.- NUCLEATION IN PRECIPITATION REACTIONS

Before we begin a discussion of nucleation during a precipitation, we need to explain some of the intricacies of precipitation. We first start with two solutions, one containing the cation of interest and the other the anion. Both involve soluble substances like nitrates or chlorides for the cations and acids or sodium compounds (or the like) for the anions. When the two solutions are combined, a precipitate will form if its aqueous solubility, as defined by its solubility product, is less than either of the cation and/or anion solubilities. What this means is that if we use two soluble solutions and combine them, a precipitate will form if it is insoluble in the aqueous solution. There are many examples of precipitate formation. We need to define the effect of exactly how the precipitation should be carried out in terms of phosphor preparation.

Consider two solutions. One contains, for example, $CaCl_2$ and the other $(NH_4)_2HPO_4$. if we combine these, we will obtain a precipitate of $CaHPO_4$. The solubility product of $CaHPO_4$ is $K_{SP} = 1.2 \times 10^{-7}$ mol/L. However, the exact nature of this precipitate depends upon how the process is performed. The following diagram, given as 3.3.1. on the next page, illustrates this factor. You will note that we have shown two processes, both using a tank containing either $CaCl_2$ or NH_4HPO_4. If we follow the method shown on the left, the first Ca^{2+} cations added will see a large excess of HPO_4^{2-} anions. The precipitate will thus contain a very slight excess of phosphate, adsorbed on the surface of the so-formed particles. As the ratio of Ca^{2+}/HPO_4^{2-} in the tank approaches 1.00, the precipitate composition then approaches a 1.000:1.000 ratio in the solid. The same mechanism applies to the case shown on the right except that a slight excess of Ca^{2+} is obtained in the final precipitate.

This effect has a decisive effect upon the precipitate obtained.

3.3.1.-

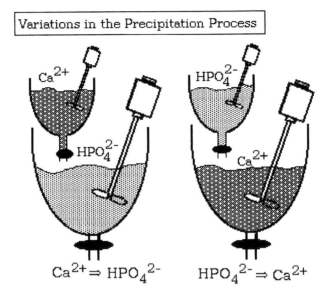

Even though the amount of excess Ca^{2+} is so small that it is difficult to measure, the phosphor prepared from such a precipitate is inferior in its properties to that prepared from the precipitate containing a small excess of phosphate. As we have indicated in the previous chapter, nucleation processes can be homogeneous (no outside influences) or heterogeneous (by specific outside constraints). In precipitation processes to form a particle size distribution, it is probably a combination of both mechanisms. It should be clear that as Ca^{2+} is added to HPO_4^{2-}, the first to form is an embryo of $CaHPO_4$. This rapidly changes into a nucleus which then grows until all of the nutrient (in this case, addition of calcium ion) is used up to form the precipitate. Finally, the particles are so large that they impinge upon each other. We will discuss this in more detail in the following section where particle growth can be related to both precipitation and solid state reaction.

3.4. - SEQUENCES IN PARTICLE GROWTH

We have already shown that embryo formation leads to nucleation, and that this nucleation **precedes** any solid state reaction or change of state.

In a like manner, nuclei must form in order for any precipitation process to proceed. Once formed, these nuclei then grow until impingement of the growing particles occurs. Impingement implies that all of the nutrient supplying the particle growth has been used up. This mechanism applies to both solid state reaction and precipitation processes to form product particles. Then a process known as **Ostwald Ripening** takes over. This process occurs even at room temperature.

The sequences in particle growth are:

3.4.1.-   Embryo formation
                     Nucleation
                     Nuclei Growth
                     Impingement
                     Ostwald Ripening (coarsening)
                     Sintering
                     Formation of Grain Boundaries

We have already covered the first three in some detail. Impingement involves the point where the growing particles actually touch each other and have used up all of the nutrient which had originally caused them to start growing in the first place. This mechanism occurs in both precipitation and solid state reaction mechanisms. It is for this reason that we did not discuss precipitation until we examined solid state reaction mechanisms.

Ostwald ripening usually occurs between particles, following impingement, wherein larger particles grow at the expense of the smaller ones. One example of this would be if one had a precipitate already formed in solution. In many cases, the smaller particles redissolve and reprecipitate on the larger ones, causing them to grow larger. Interfacial tension between the particles is the driving force, and it is the surface area that becomes minimized. Thus, larger particles having lower surface area **increase** at the expense of numerous smaller particles which have a relatively high surface area.

Ostwald ripening differs from nuclei growth in that the relative size and numbers of particles change, whereas in nuclei growth, the numbers of particles growing from nuclei do not change. Sintering, on the other hand, is an entirely different process, and usually occurs when external heat is applied to the particles.

I. - <u>Ostwald Ripening of a Calcium Phosphate Product</u>

It should be clear that the precipitation process needs to be controlled carefully in order to produce a material composed of particles of a desired size and configuration. Both Ostwald ripening and sintering can be utilized to obtain a particle of desired size, dimensions and particle habit. Industrial technologists have taken advantage of these particle forming and altering mechanisms. One example of this type of particle growth is described as follows.

In fluorescent lamps, a layer of phosphors is applied to the inside of a glass tube by means of a suspension of particles, i.e.- the halophosphate phosphor having a composition of: $Ca_5F,Cl(PO_4)_3$: Sb^{3+} : Mn^{2+} (plus minor amounts of other phosphors to achieve certain lamp "colors"). It has been determined that the lamp brightness and duration of light output (maintenance) is highly dependent upon how well the internal surface of the glass is covered by the particles. By maintaining precipitation conditions so that small thin squares of $CaHPO_4$ result from the precipitating solution, a maximum coverage of the glass is achieved. Precipitation occurs by adding a solution of $(NH_4)_2HPO_4$ to a solution of $CaCl_2$. The resulting precipitate at 20° C. is $CaHPO_4 \cdot 2H_2O$ which consists of very fine particles and is not very crystalline. As a matter of fact, the particles were usually ill-defined and bordered upon amorphous.

The solution temperature is then raised so that Ostwald ripening can occur. During this process, $CaHPO_4$ becomes the stable form at the solution temperature of 80° C. As the larger crystallites begin to grow, the smaller ill-defined crystallites, having a much larger surface area, dissolve and reprecipitate upon the larger ones. The resulting single-crystal squares, i.e.- □, have an average size of about 25μm. It has been determined that diamond-shaped crystals can also be obtained in more concentrated solutions. What this means is that Ostwald ripening causes the crystals to grow in the x and y- direction, but that a higher concentration in the "mother-liquor" containing the precipitate, $CaHPO_4$ $2H_2O$, also grows in the z-direction of the lattice comprising the crystals. The solid state reaction to form the halophosphate phosphor is:

3.4.2.- $6\ CaHPO_4 + 3\ CaCO_3 + CaF_2\ \Rightarrow\ 2\ Ca_5\ F\ (PO_4)_3$

where we have not shown the Sb_2O_3 and $MnCO_3$ added as "activators". The halophosphate thus produced follows the crystal habit of the major ingredient, in this case that of $CaHPO_4$ habit. When applied, the thin squares lie flat and overlap on the glass surface. During sintering at about 1200 °C, the $CaHPO_4$ does not disintegrate but undergoes an internal rearrangement to form $Ca_2P_2O_7$ while maintaining the same crystal habit. The $CaCO_3$ disintegrates into small particles (like $BaCO_3$) while CaF_2 exhibits a sublimation pressure at the firing temperature. The resulting solid state reaction to form the halophosphate product thus depends upon the crystal habit of the major ingredient, $CaHPO_4$.

II - Sintering and Sintering Processes

Sintering of particles occurs when one heats a system of particles to an elevated temperature. It is caused by an interaction of particle surfaces whereby the surfaces fuse together and form a solid mass. It is related to a solid state reaction in that sintering is governed by diffusion processes, but no **solid state reaction,** or change of composition or state, takes place.

The best way to illustrate this is to use pore growth as an example. An example is shown in 3.4.3. as follows on the next page. When a system of particles, i.e.- a powder, is heated to high temperature, the particles do not undergo solid state reaction, unless there is more than one composition present. Instead, the particles that are touching each other will fuse together and form **one** larger particle, as shown above. As can be seen, voids arise when the particles fuse together. The void space will depend upon the original shape of the particles.

3.4.3.-

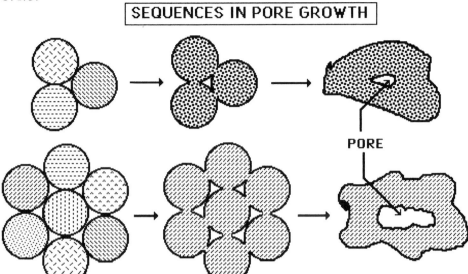

SEQUENCES IN PORE GROWTH

PORE

Here, we have shown spherical particles which produce only a few voids. Additionally, an overall change in the total volume of the particles, and that of the fusion product, occurs. Mostly, the change is negative, but in a few cases, it is positive. Experimentally, this change is a very difficult problem to measure. What has been done is to form a long thin bar or rod by putting the powder into a long thin mold and pressing it in a hydraulic press at many tons per square inch. One can then measure the change of volume induced by sintering, by measuring a change in length as related to the overall length of the rod. It has been found that the sintering of many materials can be related to a power law, and that the **shrinkage**, ΔL, can be described by:

3.4.4.- $$\Delta L \, / \, L_0 \;\; = \; k \, t^m$$

where k is a constant, L_0 is the original length, t is the time of sintering, and the exponent, m , is dependent upon the material being investigated. it has also been further demonstrated that a considerable difference exists between the sintering of **fine** and **coarse** particles, shown on the next page as 3.4.5.

3.4.5.-

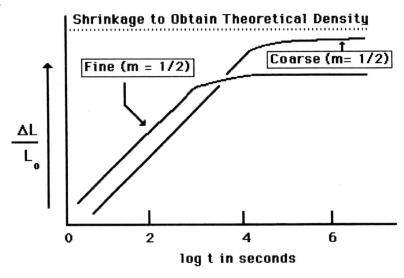

In this case, the "fine" particles and the "coarse" particles were separated so that the difference in size between individual particles was minimized. That is, most of the individual particles in each fraction were almost the same size. Both the fine and coarse particles have a sintering slope of 1/2 but it is the coarse particles which sinter to form a solid having a density closest to theoretical density. This is an excellent example of the effect of pore volume, or void formation, and its effect upon the final density of a solid formed by powder compaction and sintering techniques. Quite obviously, the fine particles give rise to many more voids than the coarser particles so that the attained density of the final sintered solid is much less than for the solid prepared using coarser particles. It is also clear that if one wishes to obtain a sintered product with a density close to the theoretical density, one needs to start with a particle size distribution having particles of varied diameters so that void volume is minimized.

III.- Grain Growth

The next subject we will discuss is that of grain growth. The simplest way to illustrate this factor is through the sintering behavior of aggregates, as shown in the following diagram:

3.4.6.-

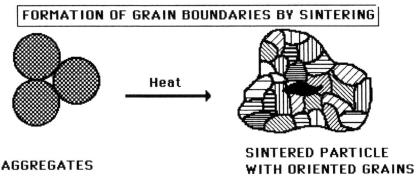

In this diagram, two steps are implicit. An aggregate is defined a large particle, composed of many small particles, It is the aggregates, made up of very fine specks, which sinter to form larger grains (particles). But, since many particles are growing simultaneously, growth occurs until impingement, with the formation of boundaries between the particles (grains). It is these grains which form the final sintered whole. Note that the crystallographic orientation of each grain differs from that of its neighbors.

When a system of **very fine** particles is formed, the interfacial tension is high due to the very high surface area present. Agglomerates (weak surface energy interchange) will form, or aggregates (strong surface energy interchange) can result, especially if the Ostwald ripening mechanism is slow, or is inhibited. Immediate removal of a precipitate from its "mother liquor" is one example where the likelihood of aggregate formation is enhanced. Sintering then produces both pore growth and grain growth. This mechanism also applies to powder compaction processes where aggregates may be present in the powder. Sintering then leads to grain growth as well.

Even where a metal is melted and then cast, nucleation leads to formation of many fine particles in the sub-solidus (partially solidified) state. This leads to grain growth in the solid metal, thereby lowering its strength. Sometimes, special additives are added to the melt to slow nucleation during cooling, thereby increasing the strength of the metal product. As we stated, once a particle forms, it may undergo certain reactions involving the solution from which it was formed. This includes formation of "embryos", "nuclei" in solution and growth of crystals from a precipitated nucleus. Precipitated particles usually grow from a nucleus to which ions are added in a regular manner to form a three-dimensional structure. Such crystals cease growing when the "nutrient" (the material which serves to form the particle) becomes depleted in the solution. If a multitude of nuclei are formed initially, then their overall size of growth will be limited. Such particles are known as "crystallites". Each will consist of several grains, having a differing orientation of the crystal lattice, within each individual particle, as shown in the following diagram:

3.4.7.-

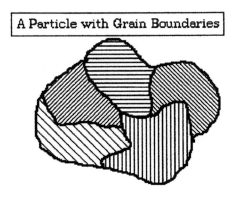

A Particle with Grain Boundaries

Note that this involves a precipitation mechanism where aggregates form. Grain boundaries form junctions between grains within the particle, due to vacancy and line-defect formation. This situation arises because of the 2nd Law of Thermodynamics (Entropy). Thus, if crystallites are formed by precipitation from solution, the product will be a powder consisting of many small particles. Their actual size will depend upon the methods used to form them. Note that each crystallite can be a single-crystal but, of necessity, will be limited in size.

Another example is our old friend, $BaCO_3$. If we fire this solid compound in air at a very high temperature for a long time, we get several changes. First, it decomposes to very fine particles of BaO. These fine particles have a large surface area, and with continued firing sinters to form larger particles. Eventually, the particles get big enough, and the porosity decreases to the point where grain boundaries begin to form between particles. The grains sinter together to form a large particle with many grain boundaries. It should be again be emphasized that each grain in the large particle is essentially a small single crystal with its lattice oriented in a slightly different direction from that of it neighbors.

Let us now consider the thermodynamics of sintering. There are two types of sintering which are distinguished by the change in volume which occurs. These are:

3.4.8.-   NO SHRINKAGE : $dV/dt = 0$

          WITH SHRINKAGE : $dV/dt = f(V)$

As we have already said, the change in volume, from initial state to final state, can be positive or negative, but is usually negative. The driving force is a **decrease** in Gibbs free energy, ΔG. It is related to both the interfacial tension (surface energy), g , and the surface energy of the particles (which is related to their size), vis-

3.4.9.- $\Delta G = g \ dA$

Consider a more familiar example, that of a droplet sitting upon the surface of a **liquid.** The droplet has a radius, r, and there are n-moles of liquid within it with a molal volume, V. To form the droplet requires an amount, nV, of the liquid, where V is the fractional **molar** volume of the droplet.

This gives us:

3.4.10.- $nV = 4/3 \ \pi \ r^3$

and the change in free energy to form the droplet is:

3.4.11.- $dG = \Delta G \ dn = g \ dA$

If we now differentiate these equations, we obtain the equations given in the following:

3.4.12.- Volume of Droplet: $n = 4\pi r^3 / 3 V$ so that: $dn = 4\pi / V r^2 dr$
and:
Surface Area of Droplet: $A = 4\pi r^2$ so that: $dA = 8\pi r dr$

We can put all of the equations together so as to yield the Kelvin equation for change in free energy as a function of the radius of the **spherical** particle, vis-

3.4.13.- $\Delta G = 2 g V / r$

One can do the same for a cubic particle, in fact for any shape factor. Since the chemical potential, μ , is related to ΔG, we use the following:

3.4.14.- $\mu - \mu_0 \quad = \Delta G = RT \ln p/p_0$
                    $2 g V / r = \quad RT \ln p/p_0$
or: $p = p_0 \exp (2 g V / r RT)$

If we now define $\Delta p = p - p_0$, then:

3.4.15.- $p/p_0 \quad = \quad 1 + \Delta p/p_0$

Mathematically, $\ln (1 + \Delta p/p_0) \cong \Delta p/p_0$ (within 5% if $\Delta p/p_0 \geq 0.1$). Thus we get the approximate equation:

3.4.16.- $\Delta p/p_0 \quad = \quad 2 g V / r \cdot 1/RT$

It is this equation which has been used more than any other to evaluate sintering. To evaluate it use, consider the following example:

Alumina: Al_2O_3 is a very refractory compound. It melts above 1950 °C. and is not very reactive when heated. If we attempt to sinter it at 1730°C., we find the following values shown in the following to apply:

8) Precipitation is usually homogeneous. Formation of embryos which grow into nuclei which grow into particles suspended in the mother-liquor will vary according to the precipitation: temperature, concentration of reacting components and method of addition.

9. Particles can grow into larger particles either due to Ostwald Ripening (in solution) or by sintering (solid state).

10. Particles that are too small cannot sinter to grow into larger particles.

11. Particles that are formed by either precipitation or sintering form grain boundaries within each individual crystallite. This is due to consolidation of the very small crystallites, having a large surface area, into larger crystallites, having a decreased surface area.

We have now examined the factors involved in solid state reactions and particle growth. Nucleation, particle growth and diffusion mechanisms were the main theme of our discussion of solid state reactions, but we have not addressed how we can obtain a measure of solid state reactions as they occur. That is, how do we measure how the reaction occurs? The next section shows how DTA and TGA are applied to the decomposition and reaction of solids.

3.5. - METHODS OF MEASUREMENT OF SOLID STATE REACTIONS

If we wish to follow a solid state reaction from initial compound(s) to final product(s), there are only a few methods we can use. Of primary importance is x-ray identification since we must know what we started with, and what we end up with. We have already discussed the x-ray method in some detail and how one goes about using the method.

If a solid is stable at room temperature, it will remain in that state until it is heated. We find that two effects occur simultaneously, a thermal change and a weight change. As an example, consider $CaCO_3$. When it is heated to about 800 °C., it forms calcium oxide, CaO, by solid state reaction, vis:

3.5.1.- $$CaCO_3 + heat = CaO + CO_2 \Uparrow$$

The arrow indicates that the gas formed, CO_2, is volatile and that a **weight loss** occurs. The **orthorhombic** structure of $CaCO_3$ changes to the **cubic** form of CaO. Thermal energy is required to rearrange the atoms. What has actually happened is that we have exceeded the bonding energy of one compound ($CaCO_3$) by increasing the vibrational energy of the atoms to the point where chemical bonds are broken. This occurs at about 840 °C. and CO_2 gas is formed which is stable (and volatile). When we cool the product, we find that we have CaO. Because this change requires heat which is absorbed, the overall process is **endothermic.** If it had released heat during the change, it would be called **exothermic.**

Measurement of the weight change is called thermogravimetric analysis (TGA) whereas measurement of the thermal change accompanying the structural metamorphosis is called thermal analysis (TA). ALL CHANGES IN PHASE involve a release or absorption of calories. One reason for this is **that each solid has its own heat capacity.** That is, there is a characteristic heat content for each material which depends upon the atoms composing the solid, the nature of the vibrations within it, and its structure. The total heat content, or enthalpy, of each solid is defined by:

3.5.2.- $\Delta H = \int C_p \, dT$, where : $C_p \equiv (\partial q / \partial T)_p$

Thus, as we go from one solid to another, we see a change in caloric content.

I. Differential Thermal Analysis

In 1821, Seebeck discovered that by joining two wires of different chemical composition together to form a loop (two junctions), a direct current (DC) would flow in the circuit, namely-

3.5.3.-

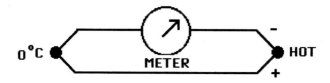

Seebeck used antimony and copper wires and found the current to be affected by the measuring instrument (ammeter). But, he also found that the **voltage** (EMF) was directly proportional to the **difference** in temperature of the two junctions. Peltier, in 1834, then demonstrated that if a current was induced in the circuit of 3.53., it generated **heat** at the junctions. In other words, the SEEBECK EFFECT was found to be reversible. Further work led to the development of the thermocouple, which today remains the primary method for measurement of temperature. Nowadays, we know that the SEEBECK EFFECT arises because of a difference in the electronic band structure of the two metals at the junction. This is illustrated as follows:

3.5.4.- The Fermi Level at a Junction of Two Dissimilar Metals

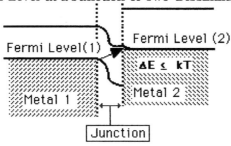

In this diagram, we show the band model structure at the juncture of two metals, each of which has its own Fermi Level. Flow of electrons is indicated by the arrow. Since the height of the Fermi Level is proportional to temperature, then the EMF generated is a function of temperature also. It is thus apparent that a thermocouple (TC) will consist of a negative and a positive "leg".

The common thermocouples in use today are listed in Table 3-3 along with the temperature range over which they are useful. Also listed is the approximate EMF generated over this range, as well as the nature of each "leg", i.e.- positive or negative. The compositions of the alloys used to make the thermocouples listed in Table 3-3 are given on the next page as 3.5.5. on the next page. In differential thermal analysis, i.e.- DTA, we use one thermocouple "bucked" against another of the same composition to

TABLE 3-3

USEFUL TEMPERATURE RANGES FOR COMMON THERMOCOUPLES

Composition	Code	Output Range (millivolts)	Useful Temp. Range, °F.	Useful Atm.*
(+)Copper-Constantan(-)	T	-5.28 to 20.81	- 300 to 750	A, N, R
(+)Iron-Constantan(-)	J	-7.52 to 50.05	- 300 to 1600	R
(+)Chromel-Alumel(-)	K	-5.51 to 56.05	-300 to 2300	A, N
(+)Chromel-Constantan(-)	E	0 to 75.12	32 to 1800	A, N, R
(-)Platinum-Pt(10%Rh)(+)	S	0 to 15.979	32 to 2900	A , N
(-) Pt- Pt (13% Rh)(+)	R	0 to 18.636	32 to 3100	A , N
(-)Tungsten(5%Re) - Tungsten(26%Re) (+)	C	0 to 38.45	32 to 5000	N, R

* A = air or oxidizing; N = neutral ; R = reducing

3.5.5.- Compositions Used to Make Thermocouples

CHROMEL:	90% Ni - 10% Cr
ALUMEL:	95% Ni - 5% Al , Si , Mn
CONSTANTAN:	57% Cu - 43% Ni

produce a "net" EMF. What this means is that either the positive (or negative) legs of both thermocouples are electrically connected so that the net EMF at any given temperature is zero. Only if one thermocouple temperature differs from that of the other does one obtain an EMF response.

3.5.6.- A Thermocouple Used for Differential Thermal Analysis

TC(1) = TC(2)

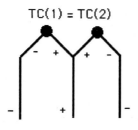

If we put a sample next to one thermocouple and a "standard" next to the other, we can follow any thermal changes that may take place as the sample is heated since each TC generates its own EMF as the temperature changes.

Thus, we put a reference material, R, directly in contact with the "(1)" thermocouple junction (hereinafter, we will refer to this thermocouple junction as "TC-1") and a sample, S, at the other, i.e.- TC-2, and can detect thermal changes occurring in the sample as compared to the reference. Note that if both TC-1 and TC-2 of 3.5.6. are at the same temperature, **no EMF is generated.** Actually what we are measuring are changes in heat flow as related to C_p (see 3.5.2.). For inorganic materials, the best reference material to use is $\alpha- Al_2O_3$. Its heat capacity remains **constant** even up to its melting point (1930 °C.). In DTA, we want to measure ΔC_p , but find that this is actually:

3.5.7.- $\qquad [C_{p(S_f)} - C_{p(S_i)}]dT_S \pm [C_p]dT_R$

where S_i is the initial state and S_f is final state for a given solid state reaction of the sample, S (No reaction occurs for R).

It should be apparent that we must maintain an equal **heat flow into both R & S simultaneously**, at a uniform rate. If we raised the temperature by steps, we would find that the actual heat flow in both R & S lags behind the furnace temperature considerably, as shown in the following diagram, shown as 3.5.8. on the next page.

But if we program the temperature in a linear manner, the sample temperature also increases linearly. The rate of heating generally used for most inorganic materials ranges between about 2° to 20°/min. while that for organic compounds lies between about 15° to 100°/min.

Thus, any signals from the DTA thermocouple will mirror differences in the heat capacity between the sample and the standard material. Since we use α-Al_2O_3 as the reference material (which has a constant heat capacity from room temperature to its melting point of 1830 °C), the DTA

3.5.8.-

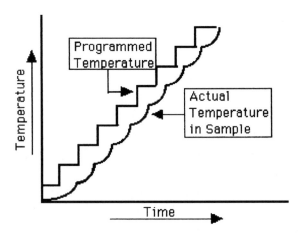

thermocouple output will reflect changes occurring in the sample itself Exothermic and/or endothermic peaks result due to changes in heat capacity as the sample reacts upon heating.

Thus, $CaCO_3$ reacts to form CaO and an endothermic peak can be measured (since heat is required to cause $CaCO_3$ to decompose). The parameters associated with the DTA method include:

3.5.9.-   PARAMETERS ASSOCIATED WITH THE DTA METHOD

1. $dT/dt = k$
2. Sample Size
3. Rate of Heating
4. Degree of Crystallinity of Sample
5. Effects of External Atmosphere

In the following diagram, shown as 3.5.10. on the next page, the arrangement of the sample, S, and the reference, R, across the differential TC is shown, and a typical DTA analysis is also given. Note that at low temperatures, the DTA peaks are endothermic, that is, heat is absorbed. Such peaks are similar to those obtained when water-of-hydration is lost, or when the solid state reaction undergoes a loss of water such as the following reaction, shown as 3.5.11. on the next page:

3.5.10.-

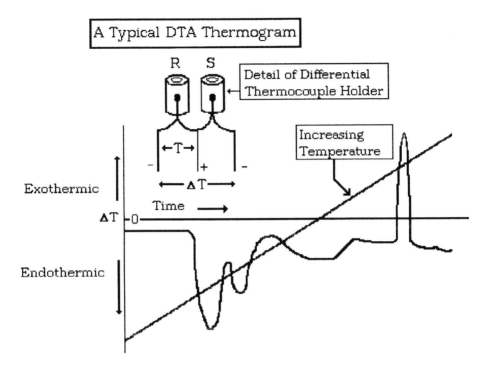

3.5.11.- $CaHPO_4 \cdot 2 H_2O \Rightarrow CaHPO_4 + 2 H_2O \Uparrow$

The **broad** endothermic peak shown in the above diagram is similar to that we might see for a change in composition such as:

3.5.12.- $2 CaHPO_4 \Rightarrow Ca_2 P_2 O_7 + H_2O \Uparrow$

The **exothermic** peak may be a change in structure such as:

3.5.13.- $\beta - Ca_2 P_2 O_7 \rightarrow \alpha - Ca_2 P_2 O_7$

In general, solid state decomposition reactions occur as endothermic peaks (ΔH is negative and heat is absorbed) while phase changes, i.e.- changes in structure, occur as exothermic peaks (ΔH is positive and heat is evolved). Note that temperature change is programmed to be linear with time, in the above diagram.

The components of a simple DTA Apparatus are shown in the following diagram:

3.5.15.-

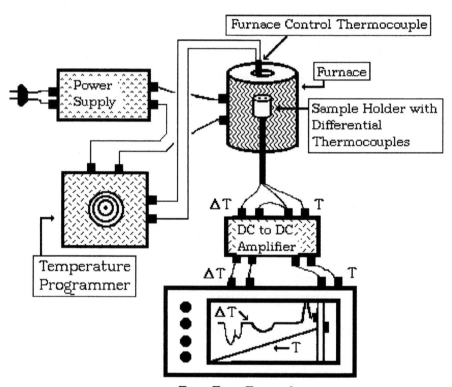

Two Pen Recorder

Note that you could build your own DTA apparatus if you so wished. The procedure to obtain a DTA run is given as 3.5.16. on the next page. The sample size (a function of its crystallinity) is generally determined by trial and error. About 500 milligrams is usually sufficient. The reason for this is practical. One **must** maintain a constant heat flow across the DTA head.

TABLE 3-4

THERMOMETRIC FIXED POINTS

FIXED POINTS*	TEMPERATURE	
	°C.	° F.
B. P. of O_2	- 183.0	- 297.3
Sublimation Point of CO_2	- 87.4	- 109.2
F.P. - Hg	- 38.9	- 38.0
Triple Point of Water	0.01	32.0
Ice Point	0.00	32.0
B.P. - Water	100.0	212.0
Triple Point of Benzene	122.4	252.4
B. P. of Naphthalene	218	424.3
F.P. of Sn	231.9	449.4
B.P. of Benzophenone	305.9	582.6
F.P. of Cd	321.1	610
F.P. of Pb	327.5	621.5
F.P. of Zn	419.6	787.2
B.P. of S	444.7	832.4
F.P. of Sb	630.7	1167.3
F.P. of Al	660.4	1220.7
F.P. of Ag	961.9	1763.5
F.P. of Au	1064.4	1948
F.P. of Cu	1084.5	1984.1
F.P. of Pd	1554	2829
F.P. of Pt	1772	3222
M.P. of Ir	2493	4520
M.P. of Ta	2988	5410
M.P. of W	3438	6220
* F.P. = Freezing Point M.P. = Melting Point B.P. = Boiling Point		

We can immediately write (see 3.5.2., 3.5.18. & 3.5.19.) :

3.5.26.- $C_{p(S)}$ $dT_S / dt = K_S (T_B - T_S) + k ((T_R - T_S) + d(\Delta H) / dt$

and:

$$C_{p(R)} \, dT_R / dt = K_R (T_{SB} - T_R) + k (T_S - T_R)$$

The equation for R is simplified because R is thermally inert and there is no change in enthalpy involved. To simplify matters further, we define:

3.5.27.- $\Phi_S \equiv C_{p(S)} / K_S$

$\Phi_R \equiv C_{p(R)} / K_R$

$H_S \equiv k / K_S$

$H_R \equiv k / K_R$

Making these substitutions, we get:

3.5.28. - $\Phi_S \, dT_S / dt + (1 + H_S) T_S - H_S T_R = T_B + (1 / K_S) d(\Delta H) / dt$

and:

$$\Phi_R \, dT_R / dt + (1 + H_R) T_R - H_R T_S = T_B$$

Now **if** $k = 0$, i.e.- there is no heat exchange between R & S (as in a properly designed apparatus), and **if** :

3.5.29.- $T_S = T_0 + t (dT / dt)$

as it will be if we are programming the temperature. Then we can define T_0 as being equal to zero, so as to obtain the following equations:

3.5.30.- $\Phi_S \, dT_S / dt + T_S = t (dT / dt) + (1 / K_S) d(\Delta H) / dt$

and:

$$\Phi_R \, dT_R / dt + T_R = t (dT / dt)$$

If we are not in a region where a solid state reaction is taking place, then $d(\Delta H) / dt = 0$.

The **change** in baseline temperature (see above) is now:

3.5.31.- $\Delta T_B = dT/dt\ (\Phi_S - \Phi_R) = T_S - T_R$

In other words, the change in baseline temperature is caused by a difference in the relative temperatures of sample and reference, which is related to their relative heat capacities. Thus, one needs to choose the reference material very carefully. If we subtract the equations in 3.5.30., we can get:

3.5.32.- $\Phi_S\ d(T_S - T_R)/dt + (T_S - T_R) \cdot (dT_R/dt) = (1/K_S)\ d(\Delta H)/dt$

This can be rearranged to:

3.5.33.- $\Phi_S\ d\,\Delta T_S/dt + \Delta T \cdot (dT/dt) = (1/K_S)\ d(\Delta H)/dt$

ΔT is the difference in temperature between the reference and the sample. In other words, it is ΔT that creates the DTA peak. Since we are not measuring **absolute** values of the temperatures, we can define a **relative** temperature:

3.5.34.- $\Delta T_{Rel} = \Delta T - \Delta T_B = \Delta T - (T_S - T_R)$

Using this relation, we get:

3.5.35.- $\Phi_S d\Delta T_{Rel}/dt + \Delta T_R + (dT/dt - dT_S/dt)(\Phi_S - \Phi_R) = (1/K_S)\ d(\Delta H)/dt$

If we choose a suitable reference material (such as α-Al_2O_3 for inorganics), then dT_R/dt will be **equal** to dT/dt. Our equation is thus simplified to:

3.5.36.- $\Phi_S \int (d\,\Delta T_{Rel}/dt) + \Delta T_{Rel} = (1/K_S) \int d\,\Delta H/dt$

However, the first term is equal to zero, so:

3.5.37.- $\int d\,\Delta H = K_S \int \Delta T_{Rel}\ dt$

This is what we started out to prove, i.e.- ΔH equals the area of the peak times the total heat flow to the sample.

II. Differential Scanning Calorimetry

Since ΔH is proportional to the area of the DTA peak, one ought to be able to measure heats of reaction directly, using the equation: 3.5.22. Indeed we can and such is the basis of a related method called Differential Scanning Calorimetry (DSC), but only if the apparatus is modified suitably. We find that it is difficult to measure the area of the peak obtained by DTA accurately. Although one could use an integrating recorder to convert the peak to an electrical signal, there is no way to use this signal in a control-loop feed-back to produce the desired result.

A more practical way to do this is to control the rate of heating, i.e.- dT/dt, and provide a **separate** signal to obtain a heating differential. One such way that became the basis of DSC is shown in the following diagram:

3.5.38.-

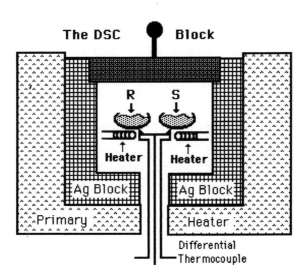

The apparatus consists of a DSC-head within a furnace, like the DTA apparatus. However, there is also a silver block which encloses the DSC head as well. This ensures complete and even heat dispersion. There are individual heaters for both the reference (R) and sample (S) pan-holders. What is measured is the **current required** to keep the differential thermocouple balanced, i.e. $\Delta T = 0$. This signal can be amplified and recorded.

We use the same approach for DSC as we did for DTA. We start with the thermal heat flow equation which is similar to Ohm's Law, vis-

3.5.39.- $dQ/dt = T_B - T_S / r$

We can define the heat change involved with the sample as dh/dt so as to get the equation:

3.5.40.- $dh/dt = C_{p(S)} [dT_S/dt] - dQ/dt = C_{p(S)} [dT_S/dt] - [T_S - T_B] / r$

We are using the same terminology for DSC as we did for DTA. Following the methods given above, we arrive at:

3.5.41.- $dq/dt = (C_{p(S)} - C_{p(R)}) \, dT_B/dt + 1/r\{dT_B/dt \cdot t\}$

We can thus "interpret" a DSC peak in terms of this equation, as we did for the DTA peak, as shown in the following diagram:

3.5.42.-

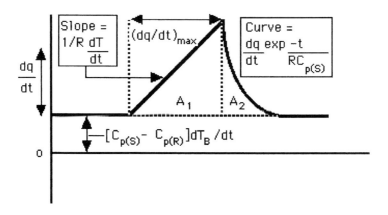

As can be seen, the initial part of the reaction up to $(dq/dt)_{max}$ is linear whereas the curve becomes exponential past the peak. This is due to the difference in C_p between the reactant and product. Thus, $A_1 \neq A_2$.

Actually, the reaction peak in 3.5.42. is an idealized curve since the baseline is a function of the **difference** in heat capacities between reference to sample and reference and product. Usually, we have a different baseline, vis:

3.5.43.-

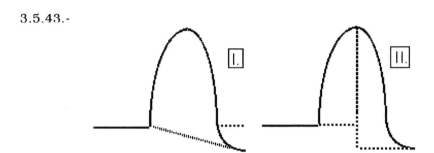

This presents a problem since it is difficult to estimate the area of the peak. One **cannot** simply extend the baseline as in Case I. A much better solution is that shown in Case II, wherein the two very asymmetrical peaks, i.e.- the initial and final parts of the overall thermal reaction taking place, are delineated.

III.- Utilization of DTA and DSC

We will now compare these two methods of thermal analysis of solids.

a. Applications of DTA

One of the major uses of DTA has been to follow solid-state reactions as they occur. All decomposition reactions (loss of hydrates, water of constitution, decomposition of inorganic anions, e.g.- carbonate to carbon dioxide gas, etc.) are *endothermic and irreversible*. Likewise are the *synthesis* reactions such as CaO reacting with Al_2O_3 to form calcium aluminate, $CaAl_2O_4$. Phase changes, on the other hand, are **reversible,** but may be endothermic or exothermic.

Thus, if we follow a solid state reaction by DTA and obtain a series of reaction peaks, it is easy to determine which are phase changes by recording the peaks obtained during the **cooling** cycle. Whereas DTA data are qualitative, those from DSC are quantitative and give information concerning the heat change (change in enthalpy) accompanying the exothermic or endothermic reaction. For example, one can obtain a value for melting of a solid state reaction product in terms of calories/gram or Kcal./mole.

DTA is especially suited in the construction of unknown phase diagrams of binary compounds. A hypothetical phase diagram and the DTA curves which would be used to construct it are shown in the following diagram, shown as 3.5.44. on the next page. In this diagram, the endothermic (Endo) peaks point to the right while the exothermic (Exo) peaks point to the left.

Consider a system with two components, A and B (see 3.5.44.). They form an incongruently melting compound, AB. The compound, AB, forms only a limited solid solution with A. Most of the composition range is a two-phase region, with a eutectic. The DTA runs are superimposed on the specific composition points of the diagram. Thus at (1) on the diagram (about 10% A and 90% AB), we see one Exo and two Endo peaks. The Exo peak is the point where the two-phase mixture, A + AB, changes to a single phase, i.e. - a solid state solution of A in AB. Further on, an Endo peak indicates the melting point of AB, and finally that of A. At (2), we see only the two melting points, first that of AB and then A. But at (3), only the melting point of the eutectic is seen, that is, both A and AB melt at the same temperature. In our Phase Diagram, the compound, AB, melts **incongruently,** that is - it decomposes at its melting point. Therefore, at (6), a double peak is seen representing the decomposition of AB and the melting of A. However, B melts at a later time. Note that one can pinpoint changes in the phase diagram quite accurately by running a DTA thermogram at specific composition points.

3.5.44.-

Construction of a Phase Diagram by Use of Several DTA
Thermograms (DTA Runs 1 through 7)

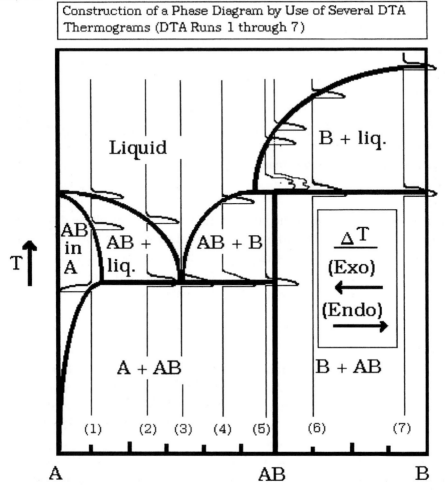

For the most part, the thermal changes observed are specific, but it is
wise to cool reversibly, while observing the DTA peaks in cooling so as to
be sure exactly what the original peak represents.

Still another use to which DTA has been employed is the characterization
of amorphous materials. The following, given as 3.5.45. on the next page,
shows a typical DTA thermogram obtained when a powdered sample of
glass is run.

3.5.45.-

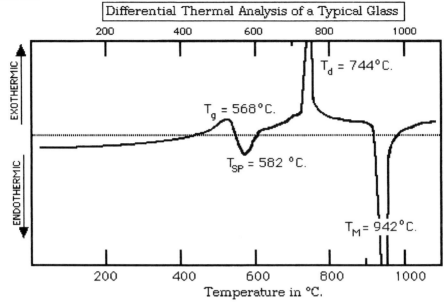

Note that nearly all of the characteristic "glass points" can be determined. These are:

T_g = Glass transition temperature
T_{SP} = Glass softening point temperature
T_D = Glass devitrification temperature
T_M = Melting temperature of crystallized product

The only one that is not readily accessible by DTA is the expansion coefficient. It is determined by use of a thermal expansion apparatus, i.e.- a dilatometer.

b. Uses of DSC

The greatest use for DSC has turned out to be for characterization of organic polymers. It has been found that most polymers are amorphous and have a characteristic T_g, i.e.- a "glass" transition temperature which leads to a "crystalline" phase. Other common uses include determination of melting points, boiling points, T_g, % crystallinity and oxidative stability. In obtaining boiling points by DSC, it is necessary to use a closed pan having an extremely small hole to allow the vapor to escape. The hole is made by use of a laser and should not be more than 50-80 μ in diameter. When a semi-volatile material is heated, an equilibrium will be established between material in the gas phase and in the condensed phase. As the material is being heated, the pressure exerted by the volatile phase, i.e.- the vapor pressure, increases. The rate of heating is important and should be kept between about 6-10 °C. per minute. It is important to have the two phases in equilibrium as the sample increases in temperature, hence the use of an escarpment in the sample pan to retard escape of the volatile material. This arrangement allows the vapor produced to sweep out the air and replace it. At the temperature where the vapor pressure of the sample exceeds the total pressure of its surroundings, the material boils. If the outside pressure is kept constant, there will be an endothermic heat flow associated with condensed phase material entering the vapor phase. As the temperature increases, the rate of boiling also increases. When all of the material is in the vapor stage, it remains in that state until all of the material has boiled off. If the hole is not small enough, then all of the material will be evaporate and be lost before the equilibrium condition is attained. If an equilibrium between vapor and material is not achieved, then the boiling point measured will not be the true boiling point.

The pan has a small hole on the top to limit the amount of water escaping at the boiling point. Keep in mind that the heat flow (which is related to the degree of vapor change achieved) is low in the beginning, but rises rather fast as the boiling point is reached. The flat **leading edge** of the endotherm represents the point where the sample temperature is constant at the boiling point.

If one measures the boiling points at several pressures, including that of atmospheric pressure, one can then extrapolate to obtain the vapor pressure of a material at ambient temperature. This is done using the Clausius-Clapeyron equation:

3.5.46.- $E_o \equiv RT^2 \, (d \ln k_1 \, / dt)$

and: $E_o = \Delta H^* + RT$

As an example of how the data are obtained, the following diagram shows the behavior of water at 1.0 atmosphere when it is subjected to the above conditions:

3.5.47.-

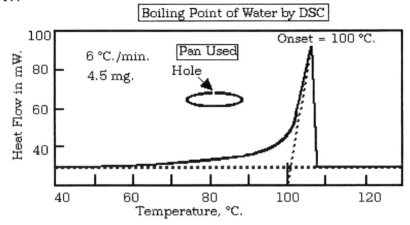

The flat **leading edge** of the endotherm represents the point where the sample temperature is constant at the boiling point. If one measures the boiling points at several pressures, including that of atmospheric pressure, one can then extrapolate to obtain the vapor pressure of a material at ambient temperature.

IV. Thermogravimetry

Thermogravimetric analysis (TGA) measures changes in weight of a sample being heated. A typical apparatus Thermogravimetric analysis (TGA) is shown on the following page as 3.5.48. The weight is monitored in real time and changes, either gains or losses, are evident immediately. The apparatus itself consists of the sample situated within a crucible, which is enclosed within a temperature-controlled furnace. The sample, plus crucible, is counterbalanced on a sensitive ANALYTICAL BALANCE. Weight changes are directly plotted on a two-pen recorder. Weight readout is usually accomplished by one of two methods, a linear transducer or a capacitance change between two flat plates, one of which is free to move with the balance swing. A resistance-capacitance tank

3.5.48.-

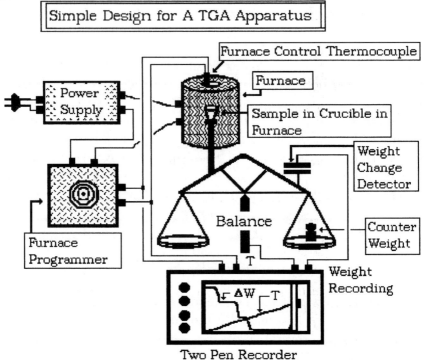

Simple Design for A TGA Apparatus

Two Pen Recorder

circuit completes the electronics, producing a readable voltage. We usually employ a crucible to hold the powder sample, although flat pans are also suitable. Furnace temperature is controlled in a linear manner and recorded. In some cases, the TC is mounted directly at the sample position. It is important that sample and furnace temperatures be nearly equal, so as to record accurate weight losses and gains. Operational parameters for TGA are:

3.5.49.-   1. Sample Size (buoyancy)
                 2. Sample Closure
                 3. Heating Rate
                 4. Heating Mode
                 5. External Atmosphere

The SAMPLE SIZE is important because most balances have a limited range of weighing, as well as a limited sensitivity, i.e.- milligrams per gram of weight detectable. Many balances feature automatic counter-weight loading. If a sample is **fluffy** and a large crucible is used, then the **buoyancy factor** must be accounted for. At high temperatures, i.e.- > 800 °C., the density of air within the furnace is sufficiently **lower** than that of the outside, so that the apparent weight of the sample plus crucible appears lower than it actually is. And, the larger the crucible, the more air is displaced within the furnace. For 10.000 grams of total weight, the buoyancy factor will be about 0.002, enough that a correction needs to be made for precision work. **Sample closure** is important since it affects the *rate* of solid state reaction. Consider the following solid state reaction:

3.5.50.- $CaCO_3$ ⇆ $CaO + CO_2 \Uparrow$

If the gas is restricted from escaping, then the equilibrium is shifted **to the left** and the $CaCO_3$ does not decompose at its usual temperature. A **higher** temperature is required to effect decomposition.

Likewise, an **external atmosphere**, such as CO_2 in the above reaction, restricts the apparent temperature of decomposition, the rate of reaction, and sometimes, the mode of decomposition (depending upon the nature of the compound under investigation). $CaCO_3$ normally decomposes at 860 °C. In a 1.0 atmosphere pressure of CO_2, the decomposition temperature is raised to about 1060 °C.

The HEATING RATE is important from a practical aspect. Usually, we are measuring furnace temperature and cannot program the temperature too fast, for fear that the sample temperature will lag the furnace temperature by too great a degree. It is also for this reason that we use as small a sample as is practical to obtain a weight loss or gain which the balance can discriminate. This again depends upon the sensitivity of the balance and the nature of the reaction we are examining. In general, weight gains (due to oxidation) require more sensitivity and larger sample sizes than those of decomposition. Usually, we restrict heating rates to $\geq$ 15 °C and use a heating rate of about 6-8 °C/ min. at most. The HEATING MODE to be used depends upon the results we wish to achieve. A typical TGA run is given in the following diagram:

3.5.51.-

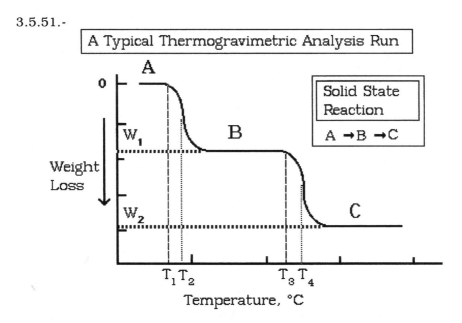

In the solid state reaction depicted, A begins to decompose to B at T_1 and the reaction temperature for decomposition is T_2, with a weight loss of W_1. Likewise, the reaction of B to form C begins at T_3 and the reaction temperature (where the rate of reaction is maximum) is T_4. Note that the weight loss becomes constant as each reaction product is formed and the individual reactions are completed.

If we program the temperature at 6 °C/min., we would obtain the results in 3.5.50. This is called **dynamic thermogravimetry.** However, if we set the furnace temperature just slightly greater than T_2, we would obtain a reaction limited to that of A decomposing to B, and thus could identify the intermediate reaction product, B. This technique is called **isothermal thermogravimetry.**

Thus, we can follow a solid state reaction by first surveying via dynamic TGA. If there are any intermediate products, we can isolate each in turn, and after cooling (assuming each is stable at room temperature) can identify it by x-ray analysis. Note that we can obtain an **assay** easily:

3.5.51.-   Assay $\equiv$ final weight / original weight

If there is a gaseous product, we can also identify it by converting weight loss to mols/mol of original reactant. In 3.5.50. above, 1.00 mol of CO_2 is expected to be lost per mol of reacting $CaCO_3$. The actual number of mols lost depends upon the original sample size.

The most recent TGA apparatus includes what is called EGA, i.e.- effluent gas analysis. Most often, this consists of a small mass spectrograph capable of identifying the various gasses most often encountered in TGA. Gaseous weight losses can be classified according to the nature of the effluent gases detected. These include the following:

3.5.52.-   Gaseous Products = CO_2 , N_2O_4 , H_2O , SO_2 , SO_3 , CO.
$$\text{Water - of hydration}$$
$$\text{- of constitution}$$
$$\text{- adsorbed}$$

Regardless of how well a sample has been dried, it will always have an adsorbed monolayer of water on the surface of the particles. If we run the DTA carefully, we will see a small endothermic peak around 100 °C. Additionally, if we run the TGA properly, we will see a small loss plateau before the major losses begin. As a matter of fact, **if we do not see the loss of adsorbed water**, either the apparatus is not operating properly, or we do not have sufficient sensitivity to observe the reactions taking place.

The different types of water which can be present during any inorganic solid state reaction is easily illustrated by the following example:

Dibasic calcium orthophosphate can be formed as a dihydrate or as an anhydrous product, depending upon the conditions used in its precipitation:

3.5.53.- $CaHPO_4 \cdot 2 H_2O$ - brushite

$CaHPO_4$ - monetite

Brushite reacts to form monetite which then reacts to form pyrophosphate:

3.5.54.- $2 CaHPO_4 \cdot 2 H_2O = 2 CaHPO_4 + 2 H_2O \Uparrow$

$2 CaHPO_4 = Ca_2 P_2 O_7 + H_2 O \Uparrow$

We can illustrate the type of calculations needed in order to determine the parameters involved in TGA runs. In the following Table, presented on the next page, a typical problem that one encounters in TGA is shown and the calculations needed to produce the desired results.

We start with the reactions given in 3.5.54. for the reactions of calcium phosphate, since this also illustrates how assays are calculated. The steps include:

1. Determine assay

2. By subtracting actual assay from theoretical assay, obtain amount of water actually adsorbed on particle surfaces

3. Determine losses incurred by stages, if one wishes to determine the actual reactions occurring during the solid state reaction

In the following, we present a typical problem one encounters in TGA, and the calculations needed to produce the desired results. In this case, we start with a known material for which we have already determined the fired product by x-ray analysis.

Table 3-5
A TYPICAL PROBLEM IN TGA

METHOD	ORIGINAL	FIRED PRODUCT
X-ray Analysis:	brushite	$Ca_2P_2O_7$
Weight:	17.311 gram	12.705 gram
Molecular Weight:	172.09	254.11

and Carroll method has been shown by Fong and Chen to be the only one which gives satisfactory answers to **known** reactions, whether zero order, 1st order, 2nd order, or even higher. Even fractional orders of reaction may be determined. This method can be used with either DTA or TGA data. The following is a description of the Freeman-Carroll method applied to DTA data:

b. The Freeman-Carroll Method Applied to DTA

Consider the following DTA peak in which ΔT is plotted vs: time, t, :

3.5.64.-

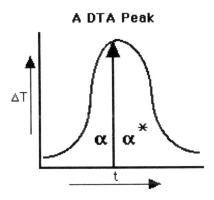

A DTA Peak

At any time, α is the fraction decomposed while $\alpha *$ is the fraction which has yet to react. We can set up equations as before:

3.5.65.- $d\alpha / dt = k_1 (1-\alpha)^n = Z/\phi \exp -E/RT (1-\alpha)^n$

The first part of the equation is the *general* kinetic equation, from which it is easy to obtain the last part (as we have shown). If we perform the mathematical operations of : 1) take the naperian log; 2) differentiation, and then 3) integration, we obtain the following equation:

3.5.66.- $\Delta \ln d\alpha / dt = n (\Delta \ln (1-\alpha) - E/R \Delta (1/T)$

If we now set: $\Delta T/A = d\alpha / dt$ and $\alpha * = A - \alpha$ (where A is the area under the peak) , we can then obtain:

3.5.67.- $\Delta (\ln \Delta T) = n (\Delta \ln \alpha *) - E/R \, \Delta (1/T)$

This allows us to plot: $\Delta (\ln \Delta T) / \Delta (\ln \alpha *)$ vs: $\Delta (1/T \ln \alpha *)$ so as to obtain a straight line. The **slope** is E/R and the intercept is **n** , the order of reaction. Having thus obtained the activation energy and order directly, we can then calculate the reaction **rate.**

c. The Freeman-Carroll Method Applied to TGA

In this case, we define our equation in terms of **weight, w** :

3.5.68.- $-dw/dt = k_1 \, f(w)$

As in the DTA method, we define weight reacted in time, t, as w_t and final weight, w_f . This allows us to define **weight to be reacted** as: $w = w_t - w_f$. Our weight function is then to be defined:

3.5.69.- $f(w) = w^n$

where n is a simple integer, representing the order of reaction. Thus, we obtain in the same manner as for the DTA data:

3.5.70.- $- dw/dt = Z \exp (-E/RT) \, w^n$

We now do the mathematical manipulations in the order given above to obtain:

3.5.71.- $\Delta (\ln (-dw/dt)) - E/R (\Delta (1/T)) + n \, \Delta \ln w$

This allows us to plot, as before:

$$\Delta (\ln - dw/dt) / \Delta \ln w \quad vs: \Delta (1/T) / \Delta \ln w .$$

However, in many cases, it is easier to use **a** , the original weight and calculate: (a - x) and dx/dt , where x is the fraction decomposed at time, t. Then we plot:

3.5.72.- Δ (ln dx/Δ) / Δ (ln (a-x))   vs: Δ (1/T) / Δ ln (a-x)

Thus if we take a well defined curve from either DTA or TGA and find points on the curve, we can calculate **all** of the kinetic parameters. Note that the main concept is to use the **amount left to react** at any given instant. In DTA, this was α * while in TGA, it was (a- x). Both thermal methods give equally satisfactory results.

Recommended Reading

1. W.W. Wendlandt, *Thermal Methods of Analysis*, Interscience-Wiley, New York (1964).

2. P.D. Garn, *Thermoanalytical Methods of Investigation*, Academic Press, New York (1965).

3. W.J. Smothers and Y Chiang, *Handbook of Differential Thermal Analysis*, Chemical Publishing Co., New York (1966).

4. E.M. Barral and J.F. Johnson, in *Techniques and Methods of Polymer Evaluation*, P. Slade & L. Jenkins- Ed., Dekker, New York (1966).

5. W.W. Wendlandt, *Thermochim. Acta* **1** , (1970)

6. D.T.Y. Chem, *J. Thermal Anal.*, **6** 109 (1974) - Part I

7. Chen & Fong, loc. cit. **7** 295 (1975) - Part II

8. Fong & Chen, loc. cit., **8** 305 (1975) - Part III

9. C. Duval, *Inorganic Thermogravimetric Analysis*, 2nd Ed., Elsevier, Amsterdam (1963).

10. C. Keattch, *An Introduction to Thermogravimetry*, Heydon, London (1969).

11. E. P. Manche & B. Carroll "Thermal Methods", Chapter 4 in *Analytical Methods*, , pp. 239-344, Marcel Dekker , New York (1972)

12. Robert E. Reed-Hill, "Physical Metallurgy Principles"-   Van Nostrand, Princeton, New Jersey (1964).

13. A.J. Dekker, "Solid State Physics" -   Prentice-Hall, Englewood Cliffs, New Jersey (1958).

CHAPTER 4
Measuring Particle Size and Growing Single Crystals

In the last chapter, we investigated solid state reactions relating to formation and growth of crystallites (particles). We examined nucleation mechanisms during reaction between solids (heterogeneous) and nucleation mechanisms during precipitation (homogeneous). We will now determine how the size of such crystallites (as powders) are measured. We will also investigate how single crystals are grown from a melt and how stacking faults can arise as a result. Stacking faults are the result of line defects which occur during particle growth, whether it is from a melt, by solid state reaction or during precipitation. Of necessity, we will not be comprehensive in our examination of these topics but will include enough material to familiarize those who do not have the background concerning particle distributions and how single crystals are grown.

However, we reserve description of the formation of solids as thin films to a following chapter. Since the original publication of this volume in 1991, use of thin film technology has grown to be a major part of solid state chemistry. LEDs (light emitting diodes) and diode lasers have become an integral part of our lives as witness the traffic lights now in use. Prior to the development of high-brightness LED lamps, the use of incandescent lamps combined with an appropriate filter in traffic lights was widespread. The LED is in the process, or has, replaced such traffic lamps. The main advantage of the LED traffic light is that it lasts over 25,000 hours before it needs to be replaced. Contrast this to the average replacement time of incandescent lamps of less than 2000 hours. Furthermore, the LED uses milliwatts of electrical power while the incandescent lamp uses watts of power.

4.1.- MEASUREMENT OF PARTICLE SIZES AND SHAPES

Most solids that we actually deal with are powders. The powder is composed of discrete particles, and each particle may be single crystal or contain grain boundaries. The size of particles is of interest to us because many of their physical (and to a certain extent- chemical) properties are dependent upon particle size. For example, the "hiding power" of pigments is dependent upon the size of the pigment particles, and in particular the distributions of sizes of the particles. There is an optimum particle size distribution to obtain maximum hiding power, depending upon the specific application. Pharmaceuticals are another product where particle size and particle size distribution (PSD) are important. The effectiveness of dosage and rate of ingestion by the human body depends upon the proper PSD. Particles are usually defined by their diameter since many are spheroid. We use the micron as a base, i.e.-

4.1.1.-　　　$1\,\mu$ (micron) $= 10^{-6}$ meter $= 10^{-4}$ centimeter $= 10^{-3}$ mm.

The following defines particle ranges that we usually encounter in solid state chemistry:

4.1.2.-

PARTICLE RANGES

Range	Centimeters	Microns, μ	Description
Macro	1.0 - 0.05	10^4 - 500	Gravel
Micro	0.01- 0.0001	100 - 1.0	"Normal"
Sub-Micro	0.0001- 10^{-7}	1.0 - 0.001	Colloidal

It is easier to describe particle size in terms of microns rather than inches, meters or even millimeters. The following diagram, given as 3.1.3. on the next page, summarizes the type of particles that we encounter in the real world. At the top the particle diameters in microns are given. Immediately below are standard screen sizes, including both U.S. and Tyler standard mesh. Screens are made by taking a metal wire of specific diameter and cross-weaving it to form a screen with specific hole sizes in it. Thus, a 400 mesh screen will pass 37μ particles or smaller, but hold up all those which are larger. A 60 mesh screen will pass up to 250μ particles, etc. (U. S. Screen Mesh). This diagram also places and defines the size of most of the particles that we are likely to encounter in the real world. On the left is a comparison of angstroms (Å) and microns up to $1\,\mu$

4.1.3.-

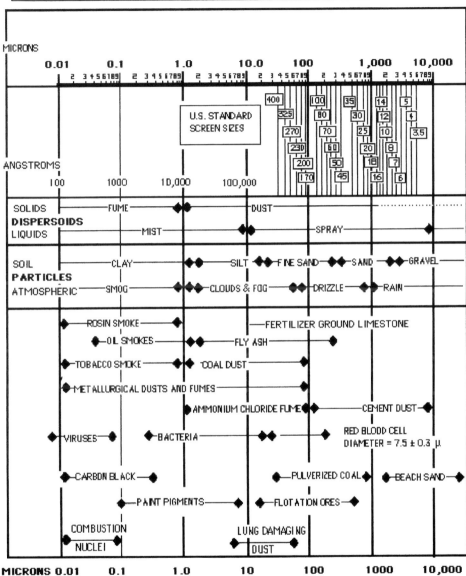

which is $= 10,000$ Å. In the middle of the section, "Equivalent Sizes", is a comparison of sizes and "theoretical mesh". Why the latter term is sometimes used will be discussed later. In "Technical Definitions", size ranges are given for various types of solids in gases, liquids in gases, and those for soil. In addition, we have a separate classification for water in air (fog). Finally, there is a classification for various commercial products and by-products we might encounter, including viruses and bacteria. Note the varied sizes of particles that we normally encounter. This listing is not meant to be all- inclusive but outlines many areas of technical interest where the sizes of particles are important. Now let us begin to more clearly define the properties of particles.

Let us take a 1.00 cm. cube and cut it into **fourths**, as shown:

4.1.4.-

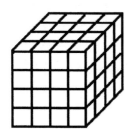

$$\text{NUMBERS} = 4^3 = 64$$

$$\text{SURFACE} = (64)(6)(1/4)^2 = 0.0096 \text{ sq. m.}$$

Now we take the same cube and divide it into 1.0 μ cubes, with the following result:

4.1.5.-   1μ cubes $= 10^4$ cuts in 3-dimensions
          Diameter $= 1.0 \times 10^{-6}$ cubic-meters for each particle
          Number $= (10^4)^3 = 1.0 \times 10^{12}$ particles
          Surface $= (6) (1.0 \times 10^{12}) (1.0 \times 10^{-6})^2 = 6.0$ sq. Meters

We have discovered two facts about particles:

a) there are large numbers present in a relatively small amount of material (we started with a 1-cm. Cube).

b) the total surface area can be quite large and depends upon the size of the particles.

Since the rate of solid state reaction depends upon surface area, we can see that very small particles ought to react much faster than large particles. Another observation is that solid state reactions involving very fine particles ought to be very fast in the beginning, but then will slow down as the product particles become larger. There are other properties of particles which we can think of, as follows:

4.1.6.- PROPERTIES OF PARTICLES

Numbers	Shape	Size
Aggregate	Porosity (pore size)	Size distribution
Agglomerate	Density	Surface area

At this point, we are most interested in the size of particles and how the other factors relate to the problem of size. However, particle shape will determine how we define size. Most of the particles that we will encounter are spheroidal in shape. But, if we discover that we have needle-like (acicular) particles, how do we define their average diameter? Is it an average of the sum of length plus cross-section, or what?

I. Particle Shapes

When we determine the size of a particle, we do so based upon its shape. It is therefore logical that the three-dimensional properties of a particle directly relate to how the distribution of an ensemble of particles is measured. To illustrate exactly what we mean by particle shape, consider the following diagram, given on the next page as 4.1.7.

In this diagram, USP refers to "United States Pharmacopoeia", Vol. USP 24, published in 2000. Note that these shapes are variations in cubic growth in one or more of the x, y, or z directions of the cubic lattice. Although we have shown these shapes as regular in the three crystallographic directions, usually they are not. For example, if one looks at a "flake" under the microscope, the edges are usually not straight but are "ragged". There are many particles which grow according to these examples in which rate of growth is about equal in all crystallographic

4.1.7.-

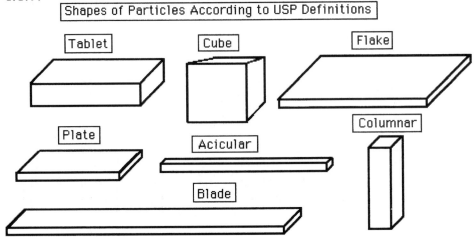

directions. However, these are probably the "exception to the rule". When we have other crystallographic lattices such as tetragonal, hexagonal, monoclinic or triclinic, the rate of growth along the lattice planes becomes complex indeed. The USP organization defines these shapes in terms of length, width and thickness, as follows:

4.1.8.- Definitions of Shape Parameters

- CUBE: particle of similar length, width and thickness in all directions
- FLAKE: thin, flat particle of similar length and width
- ACICULAR: narrow, needle-like particle of similar width and thickness
- COLUMNAR: long thin particles similar to acicular except that the width and thickness are greater
- TABLET: flat particles of similar length and width but thicker than flakes
- PLATE: flat particles similar to tablets but of differing length to width
- BLADE: similar to acicular particles, except wider and longer

In the following diagram, we show how these basic shapes are modified in growth as particles (crystals) grow along certain preferred planes of the crystallographic lattice, i.e.- non-cubic:

4.1.9.-

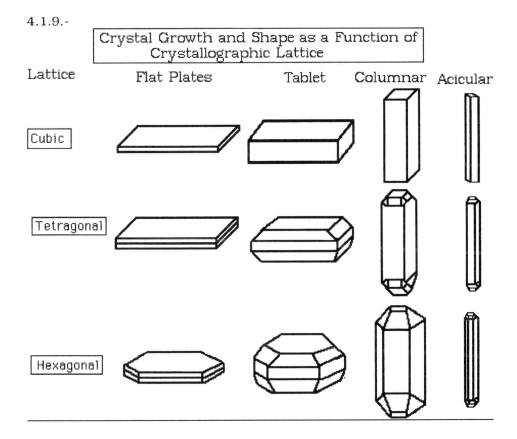

Although these shapes represent regular growth as defined by the crystallographic lattice, it is rare to find such well-defined crystallites. In general, crystal growth will follow the direction of the crystal planes of the lattice and so define the crystal shape. However, rate of crystal growth may well depend upon the {h,k,l} planes involved. For example, cubic YAG, i.e.- $Y_3Al_5O_{11}$, grows very slowly from a melt, especially from a {1,0,0} plane. If growth is induced from a {1,1,1} plane, the growth of the single crystal is accelerated by more than ten times. The reason is that the density of cations along the {1,1,1} plane is much greater than the

{1,0,0} plane. Thus, YAG grows at the rate of about 1.0 cm per hour if growth along the {1,1,1} direction is initiated. However, if neodymium, i.e.- Nd^{3+}, is added to the melt, the rate drops to 0.10 mm per hour. YAG:Nd is the solid state laser workhorse used in industry, comparable to the CO_2 (carbon dioxide) laser used in heavy industry.

Concerning particle shapes, one usually does not observe solitary particles alone but particles that have assembled into more complex arrangements. The USP volume also provides definitions of such particle mixtures, as shown in the following:

4.1.10.- Particle Assemblages and Associations

- Aggregate- mass of adhered particles
- Agglomerate- fused or cemented mass of particles
- Spherulite- radial cluster of particles
- Conglomerate- mixture of two or more types of particles
- Drusy- particle covered with tiny particles
- Lamellar Particles- stacked plates
- Particle Conditions including:
  1. Edges: angular, rounded, smooth, sharp, fractured & irregular
  2. Optical: color, transparency, translucency & opaqueness
  3. Defects: occlusions & inclusions
- Surface Conditions including:
  4. Cracked: partial split, break or fissure
  5. Smooth: free of irregularities, roughness or projections
  6. Rough: bumpy, uneven, not smooth, projections
  7. Porous: having surface openings or passageways
  8. Pitted: having small dimples or depressions on the surface.

To classify particle shapes in terms of their shape, surface condition and condition after they are formed by either solid state reaction or by precipitation is daunting indeed. This is especially true if they are to be used in medicine where the individual particle shape and condition has a major effect upon the efficacy of the medicine. The shape and physical

4.1.17.-

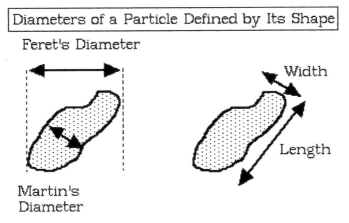

This problem has remained the most serious barrier to obtaining PSD by optical microscopy until recently.

With the advent of computers which can process data at the rate of 700 billion bits per second, this problem has been solved satisfactorily. This instrument (Malvern Instruments, 10 Southville Road, Southborough, Ma. 01772 - Sysmex FPIA-2100) is a fully automated particle **size and shape** analyzer. It uses CCD, i.e.- "charge-coupled device", technology (the optical basis of a digital video camera) to capture a series of images of particles suspended in a liquid medium. Particles are sampled from a dilute suspension which is forced through a "sheath-flow" cell. This insures that the largest area of the particle is oriented toward the video camera and that all of the particles are in focus. The cell is illuminated via a stroboscopic light source and images are captured at 30 Hz. per second. A computer program then processes the images in real time   by the following   steps: digitization,   edge   highlighting,   binarization,   edge extraction, edge tracing and finally storage. The following diagram, given on the next page as 4.1.18., shows how this is accomplished.

All this requires a high speed computer with large memory storage capacity and RAM (something not possible until recently). Image analysis software then calculates the area and perimeter of each of the captured images and then calculates the particle diameter and circularity.

4.1.18.-

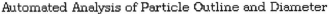
Automated Analysis of Particle Outline and Diameter

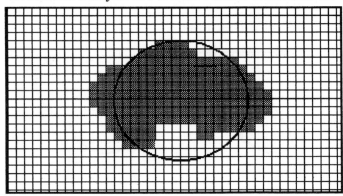

The particle shape is easily recognized in this diagram. The area of the circle is equal to that of the digitized particle so that the particle circularity (shape) can be classified.

Circularity, C, is determined by:

4.1.19.-   C = perimeter of circle / perimeter of particle

Once the measurement is complete, the particle size and circularity can be displayed in both graphical and table form. A typical report includes: particle size distribution, circularity distribution and a scattergram of particle size vs: circularity. All these can be displayed and printed in hard copy. Individual particle images can be displayed and then classified into categories including uni-particle, bi-particle, and agglomerate. This rather new instrument has solved the problem of particle shape and corresponding size through optical digitization methods.

III. Measurement of Particle Size

Fundamental to particles is the *size distribution*. By this, we mean **the number of each size** in a given collection of particles. From a close examination of 4.1.5., it is evident that we must deal with large numbers of particles, even when we have a small starting sample. This problem becomes clearer if we correlate numbers of separated particles with size, starting with a cube, one (1) cm in diameter, as in 4.1.4. , and make the cuts indicated as follows:

4.1.20.-	Size	Number/ cm^3
	1000 μ	1.0×10^3
	500 μ	8.3×10^3
	250 μ	6.4×10^4
	100 μ	1.0×10^6
	50 μ	8.3×10^6
	10 μ	1.0×10^9
	1 μ	1.0×10^{12}

It should be clear that we need to deal with substantial numbers of discrete particles when we have particles of 100μ or smaller. Note that 10μ particles derived from a one centimeter volume number a billion and even those of 100μ equal a **million** in number.

Note that this distribution consists of specific (discrete) sizes of particles. In NATURE, we always have a **continuous** distribution particles. This means that we have **all** sizes, even those of fractional parentage, i.e.- 18.56μ , 18.57μ , 18.58 μ , etc. (supposing that we can measure 0.01 μ differences). The reason for this is that the mechanisms for particle formation, i.e.- precipitation, nucleation and growth, Ostwald ripening, sintering, are *random* processes. Thus, while we may speak of the "statistical variation of diameters" and we use **whole** numbers for the diameters, the actuality is that the real diameters **are** fractional in nature.

Since the processes are random in nature, we find that the use of statistics to describe the properties of a population of particles, and the particle size distribution is well suited for this purpose since it was originally designed to handle large numbers in a population.

A. The Binomial Theorem-Particle Distributions

To describe particle distributions, we will use nomenclature relating to the science of statistical probability. If we are given n-things where we choose x variables, taken r at a time, the individual **probability** of choosing $(1 + x)$ will be:

4.1.21.- $P_i = (1 + x)^n$

This is the BINOMIAL THEOREM. By Taylor Expansion, we can find the total probability, P, for items, n, taken r at a time as:

4.1.22.- $P(r) = \{^n_r\} p^r (1 - p)^{n-r}$

where p^r is the individual probability for a particle taken r at a time and P(r) is the total probability. We define n/r as the combination of particles within a given size range and use the combinatorial theory which applies to statistical mechanics as well to an ensemble of particles:

4.1.22.- $\{^n_r\} \equiv {}_nC_r = n! / r! (n-r)!$

If we now let n approach infinity, we get:

4.1.23. - $P(r) = \Sigma\, p_i = 1 / [\, 2\pi\, np\, (1-p)]^{1/2} \cdot \exp(- x^2 / 2np\, (1-p)$

One will immediately recognize this as a form of the BOLTZMANN equation, or the GAUSSIAN LAW. We can modify this equation and put it into a form more suitable for our use by making the following definitions.

First, we define a mean (average) size of particles in the distribution as $\overline{d}$, given as:

4.1.24.- $\overline{d} = \Sigma\, n\, d_i / \Sigma\, n$

and then define what we call a "standard deviation" as σ, for the distribution of particles. From statistics, we know that this means that 68% of the particles are being counted (34% on either side of the mean, which we denote by $\overline{d}$. This is shown in the following:

4.1.25.- $$\Sigma(\overline{d} \pm d_\sigma) = 0.68$$

where d_σ is the diameter measured to give 34% of the total number of particles. In a like manner:

4.1.26.- $$2\sigma \equiv 0.954 = \Sigma(\overline{d} \pm d_{2\sigma})$$
$$3\sigma \equiv 0.999 = \Sigma(\overline{d} \pm d_{3\sigma})$$

But note that these standard deviations **only** apply of we have a Gaussian distribution. We use these formulas to specify what fraction we have of the total distribution, or to locate points in the distribution.

By further defining:

4.1.27. $$\overline{d} \equiv \Sigma(n \, p_i / n)$$

$$\sigma \equiv (np(1-p))^{1/2} = \text{std. dev.}$$

$$r \equiv np + x = \overline{d} + x$$

we obtain $(d - \overline{d})$ since x is a deviation from the mean. This then brings us to the expression for the GAUSSIAN PARTICLE SIZE DISTRIBUTION:

4.1.28.- $$P_r = \frac{1}{(2\pi)^{1/2} \, \sigma} \exp \frac{(-[d - \overline{d}]^2)}{2\sigma^2}$$

if : $\{ -\infty < d < +\infty \}$

Let us now examine a Gaussian distribution, as shown in 4.1.29. on the next page. What we have is the familiar "Bell-Shaped" curve. This distribution has been variously called:

4.1.30.- Gaussian   Log Normal   Boltzmann   Maxwell - Boltzmann

4.1.29.- Gaussian Distribution

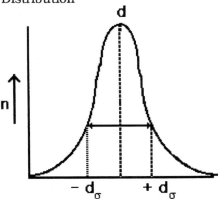

We shall use the term "Log Normal" for reasons which will clear later. It should be now apparent that we use terms borrowed from statistics and a statistical approach to describe a distribution of particles. The two disciplines are well suited to each other since statistics is easily capable of handling large assemblages, and the solid state processes with which we deal are random growth processes which produce large numbers of particles.

Let us now give some examples of the log-normal distribution as shown in the diagram given as 4.1.31. on the next page. In this case, we can compare the parameters given and realize that log-normal distributions can look quite similar or quite different, depending upon the values of the parameters involved.

In Case A, the average particle sizes are all the same. The statistical spread of particle sizes, i.e.- σ, are all equal but the total number of particles differ in the three cases given. In Case B, the average particle size is the same as is σ and the numbers of particle measured.

b. Measuring Particle Distributions

Our next task is to determine how to measure a PSD. Particle sizes are generally measured by the parameters given in 4.1.32. (see next page):

4.1.31.-

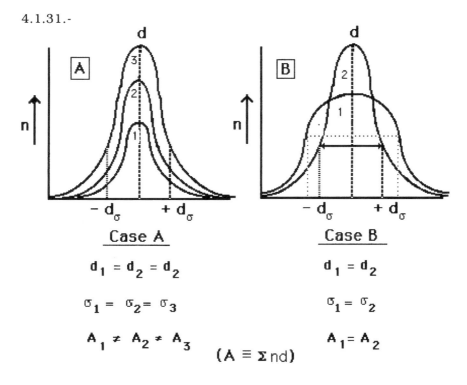

Case A

$$d_1 = d_2 = d_2$$

$$\sigma_1 = \sigma_2 = \sigma_3$$

$$A_1 \neq A_2 \neq A_3$$

Case B

$$d_1 = d_2$$

$$\sigma_1 = \sigma_2$$

$$A_1 = A_2$$

$$(A \equiv \Sigma nd)$$

4.1.32.- Particle Size Parameters

    1. Physical Dimensions
    2. Weight
    3. Volume

We have already discussed measurement of particle size by the use of particle dimensions in Section **II.** where particle shape as relating to particle size was a major concern. Note that even though the shape of the particle may be quite elongated, the stated size will be predicated upon a spheroidal shape. The stated (measured) size of a PSD based upon weight will depend upon how well the particles were separated into fractions. For example, if we use a set of sieves to separate particle fractions, the fractions of particles measured may, or may not, be a function of shape. Acicular particles will not pass through a given screen the same way as particles with regular features even though the average size of spherical particles was measured to be equivalent.

For the most part, the measured size of sieved (screened) particles will be based upon a spheroidal shape since this is easiest to calculate, i.e.- $V = 4/3 \pi r^3$. In the case of measurement by volume, spherical shapes are the only way to specify the apparent size of the particle. Let us now examine how we can obtain a PSD using screens to separate particles into fractions.

Suppose we obtain a sample of beach sand. From 4.1.3., we can see that the particles are liable to range from about 8μ to 2400 μ. In order to generate a PSD, we must separate the particles. But we find that sieves are only available in certain sizes, as also shown in 4.1.3. These are:

4.1.33.-COMPLETE LISTING OF SIEVE SIZES AVAILABLE-(U.S. STD.)

Sieve No.	Microns	Sieve No.	Microns
3.5	5660 ± 3%	35	500 ± 5%
4	4760 ± 3	40	420 ± 5
5	4000 ± 3	45	350 ± 5
6	3360 ± 3	50	297 ± 5
7	2830 ± 3	60	250 ± 5
8	2380 ± 3	70	210 ± 5
10	2000 ± 3	80	177 ± 6
12	1680 ± 3	100	149 ± 6
14	1410 ± 3	170	88 ± 6
16	1190 ± 3	200	74 ± 7
18	1000 ± 5	230	62 ± 7
20	840 ± 5	270	53 ± 7
25	710 ± 5	325	44 ± 7
30	590 ± 5	400	37 ± 7

There are two types of standard screens, the U.S. and the Tyler standard screens. We have given the U.S. screen values. Those of the Tyler are very similar, e.g. - #5 Tyler = 3962μ , # 20 = 833μ, and #400 = 38μ. To screen our sand sample, we choose to use the following US sieves: #10, # 18, #20, #30, #40, #50, #80, #100, #170, #200, #270, #325 & #400. Now if we screen the sand with the #10 screen, all particles larger than 2000 ± 60 μ will be **retained** upon the screen. Now we resieve the part that **passed through** the #10 screen, using the #18 screen, to obtain a **fraction** which is > 1000 μ ± 50μ. We then repeat this procedure to obtain a series of fractions. This is shown as follows:

4.1.34.-   <u>Fractions Obtained</u>   <u>Mean diameter</u> $\overline{d}$

Fractions Obtained	Mean diameter $\overline{d}$
$f_0 > 2000$ μ	?
$1000μ < f_1 < 2000$ μ	1500 μ
$840μ < f_2 < 1000$ μ	920
$590μ < f_3 < 840$ μ	715
$420μ < f_4 < 590$ μ	505
$297μ < f_5 < 420$ μ	358
$177μ < f_6 < 297$ μ	237
$149μ < f_7 < 177$ μ	163
$88 μ < f_8 < 149$ μ	118
$77 μ < f_9 < 88$ μ	83
$53 μ < f_{10} < 77$ μ	68
$44 μ < f_{11} < 53$ μ	49
$37 μ < f_{12} < 44$ μ	41
$f_{13} < 37$ μ	?

We now have 12 fractions that we can use. The mean diameter is apparent from the screens used to separate the fractions. In one case, $\overline{d}_{12}$ = 41 ± 2.8 μ . We can now weigh each fraction and calculate the % weight in each fraction, providing we use the **total weight of all 14 fractions**. Knowing the density of the sand, and assuming the particles to be spheres, we can then obtain the number of particles in each fraction. (Note that we have assumed that all particles small enough to pass through a given screen has done so. In many cases, this is not true, and we have to be cognizant of this error - this factor is the main source of error in the SIEVE METHOD of particle size analysis). There are two general methods for data-reporting, namely:

4.1.35.-   METHOD I = % of total weight
                           METHOD II = Cumulative weight-%

In Method I, we can use the average particle size between the extremes of the sieve-sizes or we can plot a bar-graph showing the spread of the individual fractions. In Method II, we add each fraction to the next and then "normalize" the total to 100%. Each method has its advantages.

In Method I, we calculate the total weight and assign each fraction a %-value. Mean diameters , $\overline{d}$, are easy to obtain in the first method because we know the screen diameters used to separate the fractions. In the second, we add f_2 to f_1 , then f_3 to $f_2 + f_1$, then f_4 to $f_3 + f_2 + f_1$, etc. To obtain $\overline{d}$, we take the average of the added fractions, as follows:

4.1.36.-   _____$\overline{d}$_____

              f_1 1500 μ
              $f_2 + f_1$ 1210
              $f_3 + f_2 + f_1$ 963
              $f_4 + f_3 + f_2 + f_1$ 734

We continue with all the fractions we have. We can now plot the data as:

4.1.37.-

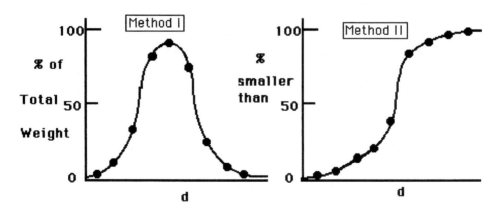

We can see the spread of particles in Method I, but to determine the mean, $\overline{d}$, is a guess. Method II allows us to obtain the mean from the 50% point, that is, the point where 50% of the particles are smaller than, and 50% are larger than. Actually, the data of Method I can also be plotted as shown in the following:

4.1.38.-

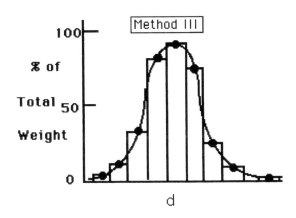

Methods I , II and III are called "Frequency Plot", "Cumulative Frequency Plot" , and "Histogram" respectively.

What this means is that the particle data in Method I is plotted to show the percentage of particles in each particle range. By connecting the dots in the curve, we get an estimate of those particles which we did not directly measure. But, getting the average particle size is difficult.

Method II is more direct since we get $\overline{d}$ and σ directly. The data in Method III, called a "Histogram Plot", have been plotted using a particle spread that is equal from fraction to fraction. If we use the actual spread of particles in each range generated by the use of selected sieves, we get the following diagram shown in 4.1.39. on the next page.

Note that our sand is composed of rather large particles but, more importantly, when we use screens, the particle spread is very large at large particle sizes and very narrow at the smallest size. This illustrates one major factor that we encounter when trying to depict a true particle

4.1.39.-

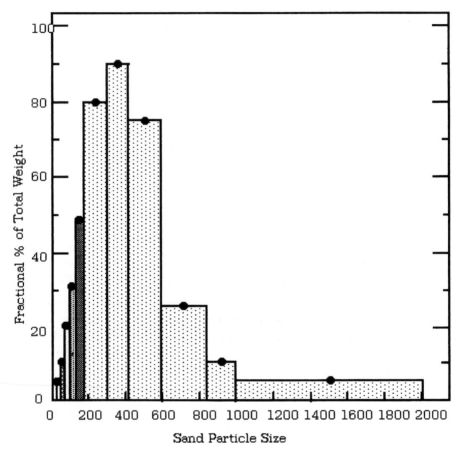

Histogram Plot of Sand Grain Diameters Using Fractions
Derived from Screens

size distribution. Thus, the use of screens for particle size determinations
has seen limited usage.

There is one important point which needs to be emphasized, that is:

"ALL METHODS FOR PRESENTING DATA FROM THE MEASUREMENT
OF PARTICLE SIZE DISTRIBUTIONS, WHETHER INSTRUMENTAL,
SIEVING, SEDIMENTATION, OR PHOTOMETRIC METHODS, MEASURE

deviations from log normality. It is these which supply additional information about the PSD. To plot such a distribution, we use log probability paper. An sample is given in the following diagram:

4.1.45.

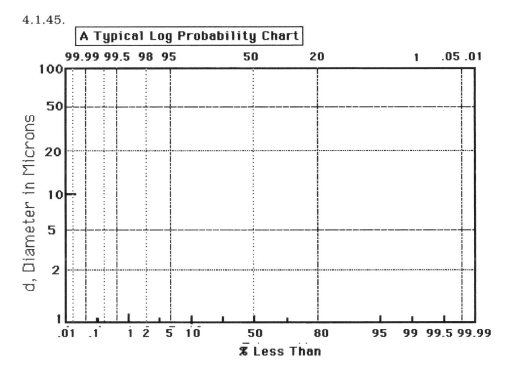

The diameter of the particles is located on the y-axis and is plotted at the proper point on the x-axis of the chart as "% less than" (see Method II of 4.1.16. and 4.1.21, given above).

A truly log normal distribution **will give a straight line** when plotted on this type of paper. This means that the PSD is not **limited**, i.e.- all sizes of particles are present from - ∞ to +∞ .

However, if the PSD is **growth-limited,** it will readily apparent from the graph. Ostwald ripening, WHERE LARGE PARTICLES GROW AT THE

EXPENSE OF SMALL ONES, is one mechanism that gives rise to growth limits. Growth limits are defined in the following:

4.1.47.-

$\overrightarrow{\mathbb{L}}$ – Upper Growth Limit

$\underleftarrow{\mathbb{L}}$ – Lower Growth Limit

$\mathcal{X}_\infty$ – Upper Discontinuous Limit

$\mathcal{X}_0$ – Lower Discontinuous Limit

The discontinuous limit is that in which all of the particles beyond a specific size have been removed or do not exist.

e. Types of Log Normal Particle Distributions

We shall present some examples of particle distributions and how to interpret them.

1. UNLIMITED PARTICLE DISTRIBUTIONS

The following diagram shows a typical PSD where the distribution does not have limits:

4.1.48.-

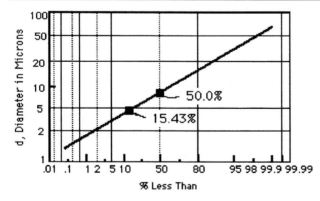

LOG PROBABILITY – UNLIMITED PARTICLE DISTRIBUTION

$\overline{d}$ = 50% prob.

$\sigma = \overline{d} \,/\, d_{15.43}$

$\overrightarrow{\mathbb{L}} = +\,\infty$

$\underleftarrow{\mathbb{L}} = -\,\infty$

This is the type of distribution usually found as a result of most precipitation processes. Following that are distributions that **do** have limits to the "normal" distribution. What this means is that the distributions do conform to the limits defined in 4.1.27.

2. LIMITED PARTICLE DISTRIBUTIONS

If a precipitate is allowed to undergo Ostwald ripening, or is sintered, or is caused to enter into a solid state reaction of some kind, it will often develop into a distribution which has a size limit to its growth. That is, there is a maximum, or minimum limit (and sometimes both) which the particle distribution approaches. The distribution remains continuous as it approaches that limit. The log-probability plot then has the form shown as follows:

4.1.49.-

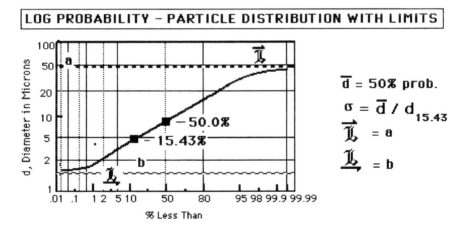

That is, there is a maximum, or minimum limit (and sometimes both) which the particle distribution approaches. The distribution remains continuous as it approaches that limit. In this case, both upper and lower limits are shown. However, one may have **one** or the other, or both. Ostwald ripening tends to use up **all** of the small particles, without limiting the upper size. Then we would have the lower limit but not the upper.

However, there are cases where discontinuous limits apply. This is shown in the following as "Discontinuous Limits".

3. Log Normal Distributions- Discontinuous Limits

Suppose we use a 200-mesh screen to remove particles larger than 74µ and a 400-mesh screen to remove those smaller than 37µ. The PSD would now look as follows:

4.1.50.-

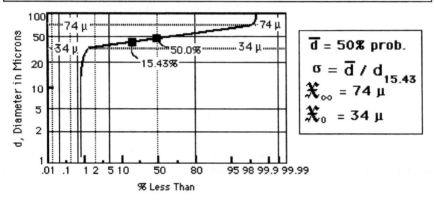

Note that the distribution is linear between 37 and 74 µ but that it abruptly shifts to ± ∞ at these points. This is a particle distribution for which it is impossible to obtain an accurate picture by any other means.

A **frequency plot** of the same data is shown in the following diagram, given as 4.1.51. on the next page.

If we were using this method to present data and got this curve, we would think that the curve was just asymmetrical and that the distribution was not log-normal. Yet, it is obvious from 4.1.50. that it is **log-normal.**

4.1.51.-

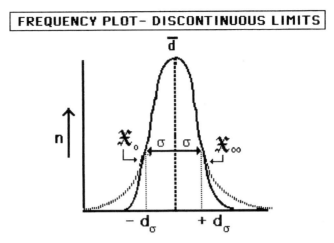

4. <u>Multiple Particle Distributions</u>

Log probability plots are particularly useful when the distribution is **bimodal**, that is, when two separate distributions are present. Suppose we have a distribution of very small particles, say in suspension in its mother liquor. By an Ostwald ripening mechanism, the small particles redissolve and reprecipitate to form a distribution of larger particles. This would give us the case shown as follows:

4,1.52.-

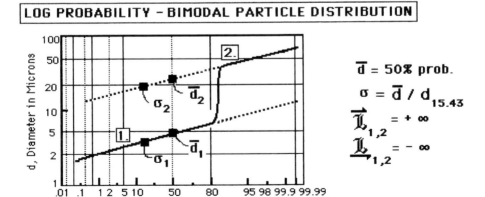

In this illustration, the size distribution parameters of both distributions are readily apparent. A similar determination is almost impossible with other types of data presentation methods. Although a large **discontinuous** gap between the two distributions is shown in 4.1.52., it is rarely the case.

f. Using Particle Distribution Parameters in Industry

As examples of the usefulness of particle distributions, consider the following. Two actual cases encountered in Industry are described in the following examples:

1. FLUORESCENT LAMP PHOSPHOR PARTICLES

Fluorescent lamps are manufactured by squirting a suspension of phosphor particles in an ethyl cellulose lacquer upon the inner surface of a vertical glass tube. Once the lacquer drains off, a film of particles is formed. The lacquer is then burned off, leaving a layer of phosphor particles. Electrodes are sealed on; the tube is evacuated; Hg and inert gas is added; and the lamp ends are added to finish the lamp. Lamp brightness and lifetime are dependent upon the particle size distribution of the phosphor particles. The number of small particles is critical since they are low in brightness output but high in light scattering and absorption. A lamp brightness improvement of 10% is easily achieved by removal of these particles. In 4.1.53. (given on the next page), we show the PSD that resulted when a certain mechanical method was used to remove the "fines".

In this application showing how the log-normal plot can be utilized, we can readily see that the "mechanical" separation of "fines" has created a new particle distribution with $\overline{d}_2 = 2\ \mu$. Even the value of σ_2 differs from that of the major particle distribution. In the "fines" fraction, it appears that the largest particle does not exceed about 5 μ. Needless to say, lamps prepared from this phosphor were inferior in brightness. Armed with this information, one could then recommend that the method of "fines" removal be changed.

4.1.53.-

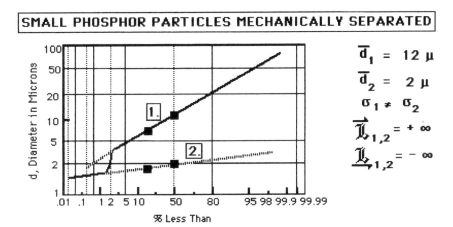

2. TUNGSTEN METAL POWDER

The lifetime of a tungsten filament in an incandescent light bulb depends a great deal on the grain size within the wire used to make the filament. W-metal powder is pressed into a bar as close as possible to theoretical density. The bar is sintered in an inert atmosphere and then "swaged" into a long rod. This rod is then drawn into a fine wire (~ 2- 5 mil = 0.002-0.005") which is wound into the filament form. Metal powder particle size has a major effect upon wire quality since fine particles form **small grains** in the wire while large particles will form **large grains.**

When the filament is operated at 3250 °K to produce light, the wire is hot enough that vacancy migration will occur in the wire. Simultaneously, grain growth and an increase in grain size also occurs. The grain boundaries between large grains will "pin" vacancies, but grain boundaries between small grains do not appear to do so. If they do, the effect is an order of magnitude smaller. As the vacancy concentration builds up at the large grain boundaries (see the Kirchendall effect given in Chapter 3), local electrical resistance increases. Eventually, a "hot spot" develops and the tungsten wire melts at that point, with failure of the filament and the lamp. If only small metal particles are used to make the wire, then the small grains produced must grow into large grains before "hot spots" cause its eventual failure . Therefore, incandescent lamp manufacturers exercise very close control of the tungsten-metal powder PSD and the sintering processes (pore elimination and actual density) used to produce the wire.

A typical PSD for a tungsten metal powder where the large particles have been removed is shown as follows. In this case, the PSD is suitable and particles larger than about 5 μ appear to have successfully removed.

4.1.54.-

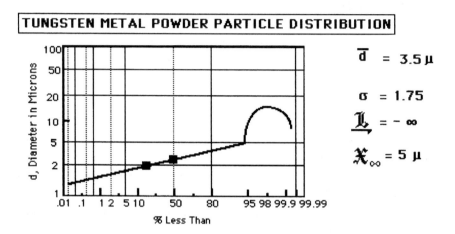

Note that we can easily determine the spread of the particles as well as the average size. It is these two parameters which are most important in the selection (and modification) of tungsten metal powders to be used to manufacture filaments used in all electronic equipment. Let us now examine a method of calculating a PSD.

V. A Typical PSD Calculation

Let us suppose that we have a particle counting instrument which sorts and counts the number of particles at a given particle size. There are several commercial instruments available including the "Coulter Counter" and those based on "Laser-Refractometry". We will use one of these to obtain the following particle data, given as follows:

4.1.55. - EXPERIMENTAL COUNTS AT PARTICLE DIAMETER

d	n Counts	d	n Counts
2.2 μ	2	10 μ	902
3.0	4	15	1402
5.0	28	25	1902
7.0	232	30	1982
9.0	602	40	1998
		50	2002

Having listed these data, our next step is to calculate the **average** size in each particle **interval.**

TABLE 4 - 1
A TYPICAL PARTICLE SIZE DISTRIBUTION CALCULATION

Diameter	n-Counts	Δn	Ave. d	%	Σ %
2.2 μ	0				
		2	2.6 μ	0.1	0.1
4.0	2				
		24	4.5	1.2	1.3
5.0	26				
		206	6.0	10.2	11.5
7.0	232				
		370	8.0	18.5	30.0
9.0	602				
		300	9.5	15.0	45.0
10	902				
		500	12.5	25.0	70.0
15	1402				
		500	20.0	25.0	95.0
25	1902				
		80	27.5	4.0	99.0
30	1982				
		18	35.0	0.9	99.9
40	2000				

For particles less than 2.2 μ, we have 2 particles. At 3.0 μ, we have 4 particles. Therefore, **in the range** 2.2 - 3.0 μ, we have **two (2)** particles whose average size is 2.6 μ. In the range 3.0 - 5.0 μ, we have **four (4)** particles whose average size is 4.0 μ. We have a total of 2000 particles that we have measured and we can calculate the percent (%) of total counts for each size range.

We therefore continue to calculate Δn as shown in the above table. Note that the same methods of arranging size data are used regardless of whether we use screens to separate the particles into size fractions or whether we use PSD instrumentation to do so. This gives the following log-normal PSD plot, as shown in the following:

4.1.56.-

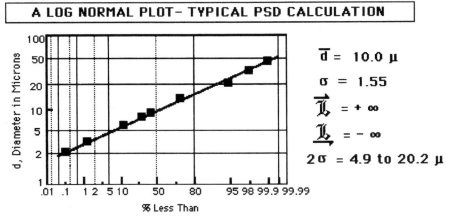

Here, the parameters of the distribution are given. This material appears to have been precipitated, or it may have been obtained by a grinding process, since the parameters indicate a Gaussian unlimited distribution. Particles grown by solid state reaction, including Ostwald ripening, generally have PSD characteristics of the original PSD from which they were formed. That is, there will be limits according to the original distribution from which it has arisen.

If the original PSD had limits, then the progeny PSD will also have such limits. The basis for this behavior lies in the fact that if one starts with a certain size range of particles as the basis of particle reaction and growth, one will end with the same size range of particles in the PSD of the particles produced by the solid state reaction. Such a case is shown in the following:

4.1.57.-

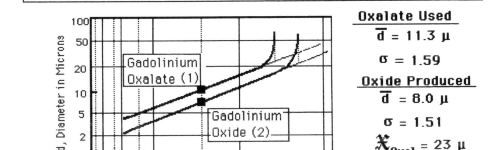

In this case, the oxalate was prepared by addition of oxalic acid to a $Gd(NO_3)_3$ solution to form a precipitate. There is an upper limit of about 23 μ in size. This may be due to the fact that the precipitation was accomplished at 90 °C., or from the fact that rare earth oxalates tend to form very small particles during precipitation which then grow via Ostwald ripening and agglomeration to form larger ones. Nevertheless, it is clearly evident that when the oxalate is heated at elevated temperature (~ 900 °C), the oxide produced **retains** the PSD characteristics of the original precipitate.

VI. Methods of Measuring Particle Distributions

There are four (4) primary methods used to obtain data concerning particle size distributions. These are shown on the next page:

SEDIMENTATION
ELECTRICAL
OPTICAL
ABSORPTION AND PERMEATION

We have already covered the microscopic method of determining particle size. We will now examine other methods in more detail, including the sedimentation method, in order to show that some procedures allow one to obtain information about particles not easily obtained by any other method. For example, the BET method gives an average particle size but, more importantly, allows one to determine the porosity of the particles as well.

A. Sedimentation Methods

There are several particle sizing methods, all based upon sedimentation and Stokes Law. If a particle is suspended in a fluid (which may be gas, or any liquid), the force of resistance to movement by the particle will be proportional to the particle's velocity, v, **and** its radius, r, namely-

4.1.58.- $f = 6\pi r \eta v$

where η is the viscosity of the fluid, providing the particle is spherical. If the particle settles under the influence of gravity, then we can write:

4.1.59.- $f = 4/3 \pi r^3 (\rho_s - \rho_l) g = 6\pi r \eta v$

where ρ_s and ρ_l are the densities of the solid particle and the liquid, respectively. Since the distance settled, D, is: $h = v t$, then:

4.1.60.- $D = \{ 18 h / \alpha t (r_s - r_l) g \}^{1/2}$

This is Stokes Law for sedimentation where we have added α , a shape factor, just in case we do not have spherical particles. You will note that we have already given a formula for Stokes Law in 4.1.42., given above. This equation is more suited for determination of particle size using a sedimentation method.

where r_0 is the actual particle radius, r_e is the **effective** radius (in case the particle is not spheroid), V_p is the volume of the particle, A is the cross-sectional area of the orifice, and a is a constant which depends upon the solvent and solute used to make the conducting solution.

The size of the orifice must be fitted to the **range** of particles present. If the particles are too large, they will not pass through the orifice. Too small and the pulse is not easily detected, Since large numbers of particles can be present, the powder content must be controlled. Too high a particle concentration-density will result in "coincidence counting". That is, two small particles can be counted as one larger one. This phenomena has been addressed and tables are provided to correct for incidence counting. The electronic part of the Coulter Counter has several hundred "channels" to accept pulses. These are then displayed via an oscilloscope as :

4.1.65.-

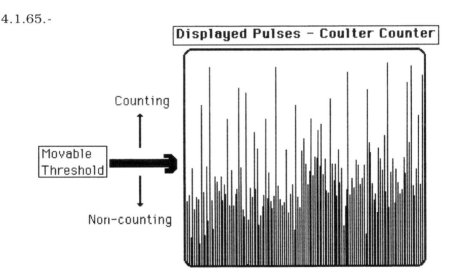

The "threshold" is electrical, **and movable,** so that one counts the large particles first (large pulses) and then moves the threshold lower to count the smaller particles. The threshold is actually a pulse height analyzer which converts heights to numbers. The data obtained are easily converted to a frequency plot or a cumulative plot. Since ΔR response is linear, one can readily calibrate the instrument at one point, using monosized particles such as pollen or specially prepared plastic particles. The only requirement is that the calibrating particles ought to be in the range of the particles to be measured. The Coulter Counter is particularly useful and unique in that steps as small as 1.0 micron may be used if so desired.

The latest Coulter Counter instruments incorporate a microprocessor and the operator can specify the form of the output data to be obtained. The instrument converts count-data to practically any format desired, with the exception of log-probability plots.

c.- Laser Diffractometry

Particle Size by Laser Refractometry is based upon Mie scattering of particles in a liquid medium. There are two primary methods, earlier ones based on sedimentation and later ones upon scattering of light (refractometry). Both employ Mie scattering. The earlier instruments were dependent upon settling time of the particles in the liquid medium. The later instruments utilize an instantaneous detection of refracted light, determined by the refractive index of the particle (and its size) and the liquid medium employed. Up until about 1985, the power of computers supplied with laser diffraction instruments was not sufficient to utilize the rigorous solution for homogeneous spherical particles formulated by Gustave Mie in 1908. Laser particle instrument manufacturers therefore used approximations conceived by Fraunhofer.

The hypotheses made in Mie Theory include:

1. The particle is an optically homogeneous smooth sphere whose real and imaginary refractive indices are both known.
2. The spherical particle is illuminated by a plane wave of infinite extent and of defined wavelength.
3. The real and imaginary refractive indices of the medium surrounding the particle are both known.

The Fraunhofer approximation includes:

1. The incident light-wave is plane, of infinite extent and of known wavelength.
2. The scattering is from a circular aperture in a thin opaque screen.
3. The extinction coefficient for all sizes is 2.0

Fraunhofer rules do not include the influence of refraction, reflection, polarization and other optical effects. Early laser particle analyzers used Fraunhofer approximations because the computers of that time could not handle the storage and memory requirements of the Mie method.

For example, it has been found that the Fraunhofer-based instrumentation cannot be used to measure the particle size of a suspension of lactose (R.I. = 1.533) in iso-octane (R.I. = 1.391) because the relative refractive index is 1.10, i.e.- 1.533/1.391. This is due to the fact that diffraction of light passing through the particles is nearly the same as that passing around the particles, creating a combined interference pattern which is not indicative of the true scattering in the far field where the detector is located. The Mie solution anticipates this.

A laser is required to provide a single wavelength so that photons passing through or around any given particle is diffracted and scattered, free from any other optical interference. The Mie theory is rigorous and is used to predict the scattering from particles whose size range from Angstroms to centimeters.

Mie originally devised his theory to better define the propertics of "fog", i.e.- a dispersion of very small circular drops of water in the atmosphere. It has now been adapted to particle counting and particle distributions. Mie theory requires that the real and imaginary refractive be known or be measured. This may require additional work initially. Experimentation may be required to choose the proper liquid medium and/or dispersing agents initially. This may be especially true if one is trying to determine the PSD of pharmaceutical agents such as new drugs. Laser diffraction particle size instruments have now become one of the major procedures for particle size determination, particularly for the Pharmaceutical Industry, i.e.- organic based particles.

d. Other Methods of Measuring Particle Size

Permeability is one other method for obtaining information about particle diameters. If one packs a tube with a weight of powder exactly equal to its density, and applies a calibrated gas pressure **through** the tube, the pressure-drop can be equated to an average particle size. The instrument based on this principle is called the "Fisher Sub-Sieve Sizer™". Only one value can be obtained but the method is fast and reproducible. The instrument itself is not expensive and the method can be applied to quality control problems of powders. Permeametry is useful in the particle range of 0.5 to 50 μ.

Gas adsorption is one other method sometimes used for determining average particle size. In this case, one is usually interested in the surface area of a powder and calculates the average size of the particles secondarily. The method is called the BET-method after its developers, Brunauer, Emmett and Teller. The procedure is time-consuming and is accomplished as follows. A gas analysis train is used in which gas volumes can be recycled and measured very accurately. A weighed sample is placed in the sample tube and allowed to come to equilibrium. Since all solid materials have a monolayer of water on the surface of the particles (we live in a wet world), the sample is heated to 300 °C. to expel the water, in a nitrogen gas flow. Next, it is cooled to liquid nitrogen temperature. At this point, the sample will adsorb a **monolayer** of N_2-gas molecules on the surface of each particle. By allowing the temperature of the sample to return to room temperature, while measuring the volume of nitrogen gas released, one obtains a value which can be converted to a surface area of the particles being measured. In practice, one recycles the system several times, measuring adsorption and desorption volumes successively so as to obtain an average value. It is found that the **unoccupied** fraction of the surface (1- y) approaches a constant value where it is assumed that y = 0. Nitrogen gas is essential since it is the only gas which forms monolayers easily without the tendency to form more than one layer of gas molecules on the surface. If the gas pressure is kept between strict limits, the BET equation is:

4.1.66.- $pg / (Vg [1- pg] = 1/ (V_{mono} \cdot C) + (C - 1) pg / (V_{mono} \cdot C)$

where: $C = exp \{ (F_{Ads.} - F_{Cond.}) / RT \}$, and pg is the gas pressure. Vg is the gas volume measured; V_{mono} is the gas volume of the monolayer calculated from the known dimensions of the nitrogen-gas molecules; $F_{Ads.}$ is the Gibbs free energy of adsorption, and $F_{Cond.}$ is the free energy of condensation. pg is measured as the **ratio** of the equilibrium pressure at specific temperatures, i.e.- room and liquid nitrogen temperatures.

It has been found that a plot of $pg / \{Vg [1-pg]$ vs: pg is linear for the pressure range of 0.05 to 0.4, with a slope of $(C - 1) / (V_{mono} \cdot C)$ and intercept of $1/ (V_{mono} \cdot C)$. Let us now do a simple calculation using BET data obtained.

Suppose we have a 20.00 gm. sample having a density of 2.0. We measure the surface area as 6.0 m^2. From the area of a sphere, $A = \pi D^2$, and the volume of a sphere, $V = 4/3 \pi D^3$, we find the total volume of **n** spheres to be 10 cc, i.e.- $n\{4/3 \pi D^3\} = 10$. The surface area of $n\{\pi D^2\}$ spheres is 6 m^3. And n, the total number of spheres present is the same in both formulas. Therefore, by substitution, we find D= 10 μ.

If we obtain a particle diameter by some other method and find that it is much smaller than that of the BET method, we infer that the particles are porous. We thus speak of the porosity and need to correct for the pore surface area if we are to make a reasonable estimate of the true diameter by the BET method. Note that the BET method is the only method whereby one can obtain an estimate of the porosity of any given set of particles.

We are now ready to undertake the art and science of growing a single crystal. A great many methods are available and the choice one makes will depend upon the physical properties of the material which we use to grow a single crystal.

4.2.- GROWTH OF SINGLE CRYSTALS

We will now describe the growth of single crystals. That is, how single crystals are obtained. It will become apparent that most single crystals are obtained from a **melt** of the material desired. Although many melt-growth methods were proposed and tried in the past, only a few have survived to the present and are in use.

Many of the devices that we use on a daily basis employ a single crystal element as the heart of the mechanism. A good example is the "quartz-crystal" watch that you wear on your wrist. Here, a quartz crystal is made to vibrate under an applied voltage. Its vibrations are coupled to a sensing circuit and translated into seconds, minutes and hours by using and counting the known resonant frequency of vibration of the crystal. The resonant vibrational frequency is determined by the angle of "cut" (or angle in relation to the crystallographic axes) and the size of the crystal. There are many other sensing devices based upon a crystal component such as a device which controls heating, translates force into electrical voltage, modulates light, or regulates motion. We use them as electrical heaters, strain gauges, laser controllers or piezoelectric devices.

Obviously, the major difference in the single-crystal and polycrystalline (crystallite) state is a matter of size. For the single-crystal, the size is large (≥ 10 cm), whereas in the polycrystalline state, the size of the crystals is small (10 μm = 0.001 cm.) The methods for obtaining one or the other differ considerably. They include formation and growth from:

4.2.1-   <u>Single Crystal</u>   <u>Crystallites (Powders)</u>

	Single Crystal	Crystallites (Powders)
	Liquid solvent	Vapor
	Vapor	Melt
	Melt	Flux
	Molten salt	Precipitation from solution

If we are to grow a single crystal, we will need some sort of furnace, which is simply a closed space, heated by electrical elements wherein the internal temperature is controlled.

I. Heating Elements and Crucibles

Before we consider crystal growth in detail, let us examine the **hardware needed to obtain a melt.** Many of the furnace components available have temperature limitations, and many of the crystals we might wish to grow have high melting points, i.e.- >1600 °C. Thus, we may have to choose certain combinations in order to get a furnace which will adequate for the task at hand.

There are many ways to build a furnace. Basically, a furnace consists of a few essential parts, each of which can be varied according to the final operating requirements needed for the furnace. The following shows the essentials required for the proper design of a furnace, as shown in 4.2.2. presented on the next page:

The power source, relay, set-point controller and sensor need not be discussed here. However, the insulation, heating elements, and crucibles involve materials, and need to be examined in more detail.

4.2.2.- Elements of Furnace Design
       a. A power source,
       b. Heating elements
       c. Thermal insulation,
       d. A crucible to hold the melt
       e. A sensor for temperature-feedback
       f. A temperature set-point controller
       g. A relay or other power-control device for the power source.

Heating elements can be metal wires or ceramic rods which become incandescent when an electrical current flows through them. Their compositions are critical and have been developed over many years to optimize their performance as heaters. Insulation is used to retain the heat generated and to disperse the heat uniformly throughout the heated space, namely the internal cavity of the furnace. Table 4-2 summarizes the thermal properties of Insulation and Heating Elements for furnace-construction.

TABLE 4-2

INSULATION	Max. Temp. Usable	HEATING ELEMENTS In Air	Max. Temp.	Power Needed Voltage	Current
Glass Wool	600 °C.	Nichrome Wire	900 °C.	med.	med.
Fiberfrax	1350	"Globars"™(SiC)	1475	low	high
Quartz wool	1100	Kanthal Wire	1300	med.	med.
Fire Brick	1100 to 1650°C	$MoSi_2$	1700	low	high
Alumina(foam& beads)	1850°C	Pt (40% Rh)	1800	low	high
Zirconia (ZrO_2)	2400	ZrO_2 :Y	1900	low	high
Magnesia	2800	$LaCrO_3$	1900	low	high

In furnace insulation, Fiberfrax™ (fibers of aluminum silicate) and "fire-brick" (bricks made from insulating silicate compounds) are the two most commonly used materials. For very high temperature work, Zirconia (= ZrO_2) in the form of beads, "wool", and boards pressed from fibers are used for temperatures above 1600 °C. Alumina, i.e.- Al_2O_3 , is much cheaper, but does not have the thermal shock, or very high temperature capability (i.e.- > 2000 °C.) of Zirconia. Those heating elements given in Table 4-2 are generally used in air atmosphere, up to about 1800 °C. If one needs to produce a melt above 1800 °C, it is necessary to use refractory metals which must be used in an inert atmosphere.   The following lists some of these heating elements:

Table 4-3

Furnace Elements For NON- AIR Usage Only

		Voltage	Current
Mo or W wire	2400 to 2800°C	low	high
Iridium wire	2400	low	high
Graphite	3400	low	high
R.F. Current	2800	low	high
Oxy-hydrogen flame	4000	NA	NA

The temperatures given here are approximate, and are presented solely for comparison purposes.

Wire-wound heating elements are used most frequently. They are usable in air up to about 1200 °C. and consist of a heating wire or coil, wound upon an insulating- preform. They are cheap and will last for considerable lengths of time, especially if they are run at < 1200 °C. (i.e.-Kanthal™ wire). On the other hand, wire made from precious metal, i.e.- 60% Pt - 40% Rh = {Pt(Rh)}, can be operated up to 1800 °C. in air, and will operate continuously at 1700 °C. for long periods. However, it is expensive and special care is required when wrapping a furnace core to form the furnace element. Another idiosyncrasy of this type of heating element is that the power-source needs to be especially designed. The Pt(Rh) wire has a very low resistance at room temperature. As it warms to a few hundred degrees, its resistivity changes considerably. Thus, a **current-limiter** is needed in the power control circuit during start-up, or

the {Pt(Rh)} wire will melt. "Globars"™ (silicon carbide rods) are the next most frequently used furnace heating elements. They will operate continuously at 1450 °C. and intermittently at 1500 °C. A newer type of element, Mo wire coated with silicide, i.e.- $MoSi_2$ (to protect the Mo wire against oxidation), has become common. Such heating elements have appeared as "hairpins" and will operate at 1750 °C. in air on a continuous basis. Another heating element coming into use is the defect conductor, ZrO_2 , doped with Y_2O_3 to make it conducting. However, this type of heating element has special operating characteristics which hinder its wide-spread usage. It does not become conductive until 600 °C. is reached. But, it can be operated continuously at 1800 ° C. in air for long periods of time, with a maximum of 1900 °C. It does require very high currents and low voltage to operate satisfactorily. Still another type of heating element is that of lanthanum chromite, $LaCrO_3$. These are available in the form of rods and will operate at 1800 °C. in air for long periods.

Most heating elements fail due to the development of internal flaws in their structure. For example, a "globar" is formed by compressing SiC particles to form a rod, and then sintering it. During operation (especially if it is operated at the upper end of its temperature-operating range), it develops "hot-spots". These are due to oxidation and formation of localized resistive areas within the rod (due to diffusion and collection of vacancies at grain boundaries). These areas dissipate power locally, so that the rod eventually fails, i.e.- melts locally at the hot-spot and becomes non-conductive. Wire-wound elements fail in a similar manner, except that formation of vacancies within the metallic structure is the most prevalent mechanism which causes "hot-spots" in the operational heating coil. These vacancies also migrate to grain boundaries. The grain-boundary-junction decreases in conductivity, due to vacancy defect formation. Thereupon, hot-spots form within the wire during operation, causing ultimate failure of the heating element.

Among those heating elements which require the use of neutral, reducing atmospheres, or vacuum, Mo- wire or W-wire in the form of heating coils, graphite in the form of rods or semi-cylinders, are most often used. Iridium wire is also used but it is very expensive. Both Mo and W wire are usable up to 2800 °C while Ir can be used only to 2400 °C. Graphite heating elements can be used above 3000 °C.

We have also included R.F. (radio-frequency) current as a heating element, although it is only a heating **method** when employed with a suitable succeptor. Finally, one other method is listed for the sake of completeness, that of the oxy-hydrogen flame. It generates combustion products (H_2O) but the RF- method can be used in any atmosphere including vacuum.

The most critical element in melt-growth of single crystals is the container, or crucible. The first requirement for selection of a suitable crucible is that the crucible does not react with the melt. The second is that it be thermally shock- resistant. A third is that of operational-temperature capability. A fourth is that it be stable in the chosen atmosphere. These requirements eliminate many potential crucible materials for a given application. For use in air, the silica crucible (SiO_2) has no peer when used with oxide-based materials up to 1200 °C. It is almost non-reactive, thermally shock-resistant, and inexpensive. Mullite, i.e.- aluminum silicate, is more reactive, less shock-resistant but cheaper than silica. It is used in making glass-melts for the most part. Table 4-6, given on the next page, lists the crucible materials most often used and their temperature capabilities.

Most metals can be used as crucible materials, but only the precious metals seem to possess the non-reactivity required for melts at the higher temperatures. Alumina is less shock- resistant than silica but can be used at higher temperatures. Shock-resistance relates to how fast one can heat the crucible and its contents up to the melting point of the material without cracking the crucible. ZrO_2 and BN are most useful for metal melts. Pt is used for melting in air and remains chemically inert to most melts.

TABLE 4- 4

COMPOSITIONS SUITABLE FOR CRUCIBLES

FOR USE IN AIR   **FOR USE IN NON-OXIDIZING ATMOSPHERE**

MATERIAL	MAX. TEMP.	MATERIAL	MAX. TEMP.
Silica	1200 °C.	Iridium	2400 °C.
Alumina	1700	Zirconia*	2400
Platinum	1750	Magnesia*	2800
Mullite	1400	Carbon*	2800
Boron	1400	Platinum*	1700
Nitride		* Not in hydrogen	

For preparation of melts in inert atmosphere, Ir stands alone. It is non-reactive, has a very high temperature limit, and is not subject to thermal shock or stress. Pt can also be used but at lower temperatures. Ir is about 4 times as expensive as Pt. A good rule of thumb is to use a **metal** crucible for oxide melts and an **oxide** crucible for metal melts, whenever possible.

II. Growth of Single Crystals From the Melt

For the most part, single crystals are grown from a melt of the compound, provided that melting the compound does not cause it to decompose. The following parameters are very important when considering growth of a single crystal from a melt:

4.2.3.-   Melting point
                Partial vapor pressure of melt
                Thermal stability of solid phase
                Reactivity of liquid or gaseous phase
                Thermal conductivity of both liquid and solid phases.

Materials tend to grow polycrystalline. This behavior is related to the 2nd Law of Thermodynamics and the Entropy of the system. What happens is that a large number of nuclei begin to appear as the melt temperature approaches the freezing point (but before it freezes). All of these nuclei grow at about the same rate, and at the freezing point produce **a large number** of small crystals. If we wish to restrict the growth to just one crystal, we would like a single nucleus to grow preferentially at the expense of the others. But usually, it does not. However, if we use a "seed" crystal and set up conditions so that it will grow, then we can obtain our desired single crystal. A number of techniques have been developed including the Kyropoulous, the Czochralski Method and the Bridgeman (Stockbarger) methods. However, they are dependent upon obtaining a stable "seed" upon which to start and grow the single crystal.

a. Necking In a Seed

The problems of obtaining a seed-crystal are not simple. We can freeze the melt to a polycrystalline state. When cool, we examine the boule (after first removing the crucible) to try to find a single crystal large enough for a seed (≈ 3 - 6 mm.). We could also cast the melt into a mold and then look for seeds. We could also freeze a polycrystalline rod by pulling it vertically from the melt. We would use a small loop of Pt wire to catch part of the melt by surface tension. By rotating the wire loop while pulling vertically, we find that a polycrystalline rod of small dimension builds up. Once we have the polycrystalline rod, we can reheat it next to the melt surface so that it remelts and "necks-in" as shown in the following diagram:

4.2.4.-

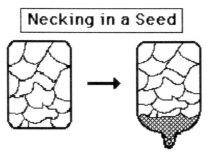

Fused Single Crystal Seed Tip

What happens is that the crystallites melt and fuse into a small tip. If we do this carefully, we will have our "seed". The tip's small size limits regrowth of the remelted part to that of a single crystal. Then, returning to the melt, we can initiate the growth of a single crystal, provided that growth-conditions are suitable.

Another way to obtain a seed is to dip a capillary tube into the melt. Surface tension causes the tube to fill and when it freezes inside, a small seed results. The difficulty with this method is that it is difficult to obtain a tube of proper diameter, made of the proper composition. Glass softens at too low a temperature and quartz melts around 1400-1500 °C. Usually, we are restricted to metals and even then, we must be able to cut the tube to obtain the seed, since it is confined within the tube. Once in a while, we can use the tube directly and obtain growth directly upon the seed, even though it has remained within the tube.

There is one other method that can be employed to obtain a seed. We use a metal crucible having a small tip and cause a melt to form as shown in 4.2.5., presented on the next page.

The tip acts in the same way to form the seed. Nevertheless, we have the same problem as when we use the metal capillary. We need to obtain the seed free from its holder. Thus, the crucible must be sacrificed, or else the whole mass must be extracted from the crucible with the seed in the tip intact.

4.2.5.- <u>Crucible for Use in Obtaining A Seed from a Melt</u>

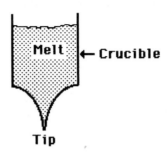

If there is a sufficient difference in contraction between the mass and the crucible, perhaps then we can obtain the whole mass intact, including our seed. But this is rarely the case.

It should be pointed out that once we have grown at least one single crystal of a given composition, we then can save a small piece to use as a seed. It is only when we try to grow a single crystal from a material which has never been attempted and accomplished before that we incur the problems which we have outlined above for obtaining a seed. Nonetheless, once a seed is available, then we have the problem of getting the crystal to grow in a proper manner.

b. Melt Growth Using The Czochralski Method

The most common method for growing single crystals, whether they be a metal or a complex mixture of oxides, is to pull a single crystal from a melt. This is the so-called **Czochralski Method**, using the technique invented in 1918 by the Polish scientist Jan Czochralski (which he called crystal pulling). Large crystals can be grown from the liquid formed by melting any given material (providing that the material does not decompose upon melting). One attaches a seed crystal to the bottom of a vertical arm such that the seed is barely in contact with the material at the surface of the melt and allows the crystal to slowly form as the arm is lifted from the melt. Let us examine this method in more detail to illustrate its versatility in crystal growth:

4.2.6.-

Design for a Czochralski Crystal Growth Apparatus

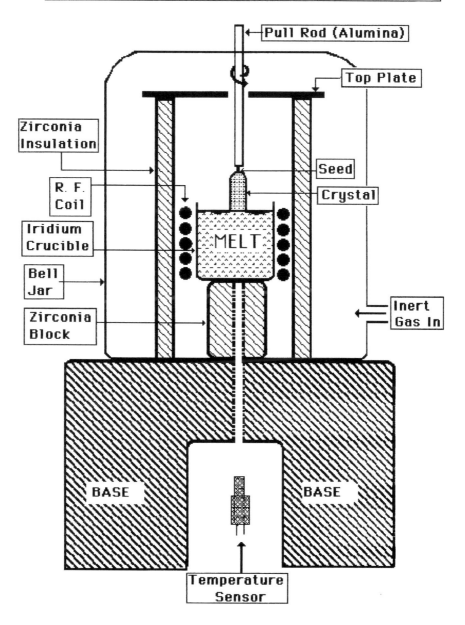

Pull Rod (Alumina)

Top Plate

Zirconia Insulation

Seed

R. F. Coil

Crystal

Iridium Crucible

MELT

Bell Jar

Zirconia Block

Inert Gas In

BASE

BASE

Temperature Sensor

As an example, we will grow a crystal from a melt of oxides whose melting points exceed 1800 C. Automatically, we are limited to use of an Ir crucible and we will use an R.F.- generator for the power source.

In 4.2.6. (see above), we illustrate the typical setup for the crucible and melt in the Czochralski apparatus. We place the Ir crucible containing the mixture of oxides on a circular ZrO_2 platform. This acts as a thermal barrier for the bottom of the apparatus. We may then place a larger ZrO_2 cylinder around the crucible for further insulation (Note that the choice of a thermal insulator is not critical except that it must be able to withstand the anticipated temperatures to be used). An R.F. coil is placed around the outside of the insulation in a position where it can electrically couple with the metal crucible. Finally, an outside wall of insulation is put into place and a top cover plate is put into position. At 2000 °C., the outer wall thickness of ZrO_2 needs to be at least 2.5-3.0 cm. The whole is then covered by a bell-jar having a hole at the top. The dome and its contents are flushed for several minutes with an inert gas before the R.F.-generator is turned on. Gas flow is maintained throughout the entire operation because we are using an iridium crucible. As the crucible heats ups, it's entire contents degas. As the materials heat up and melt, they contract so that it may be necessary to add more material to the melt until the crucible is full before actual crystal growth is attempted. This usually takes 3-4 crucible volumes of material. Thus, the top-hole needs to be large enough to accommodate this step.

1. Growing a Single Crystal

Pulling a good single crystal is not easy. There are many factors involved, each of which imposes restraints upon the others. Some depend upon the crystal pulling system design, while others depend upon the nature of the material being grown. These factors, for a given system, become **parameters**. That is, they are inter-related.

The parameters which control CZOCHRALSKI GROWTH are listed in the order of their importance in 4.2.7., given on the next page. All of these

4.2.14.-Defects Produced in a Crystal as a Function of Growth Conditions

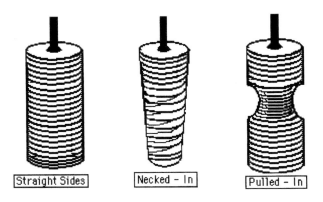

If the sides of the crystal are kept straight during growth, the lattice planes remain intact. However, if the crystal is "necked-in" or "pulled-in" during growth, line-defects appear in the lattice structure. If this is a laser-crystal that we are growing, the line defects have a localized effect upon the refractive index of the crystal, making it useless as a laser crystal.

3. Design of a Crystal Pulling Apparatus

To grow a single crystal, a proper design consists of a heavy base to minimize effects of external vibrations on the surface of the melt, a precision screw driven by a controlled reversing-motor (so as to control rate of pulling precisely), and a precision motor controlling rate of rotation. Such a design is shown as 4.2.14., presented on the next page.

Note that we have not shown the R.F. generator. However, we have shown one design of the pulling apparatus (there are many). Here, we have shown the mechanical parts of the apparatus in addition to the arrangement of the crucible and melt already shown in 4.2.6. A reversing motor is used to turn a precision screw to move the pull rod away from the melt in a precise manner. The function of the "pull-motor" is to turn the crystal so that it grows in a regular manner. You will note that we did not show a temperature sensor at the bottom of the crucible. This control is optional and can be used to monitor the bottom temperature of the

4.2.14- One Design of a Czochralski Crystal Pulling Apparatus

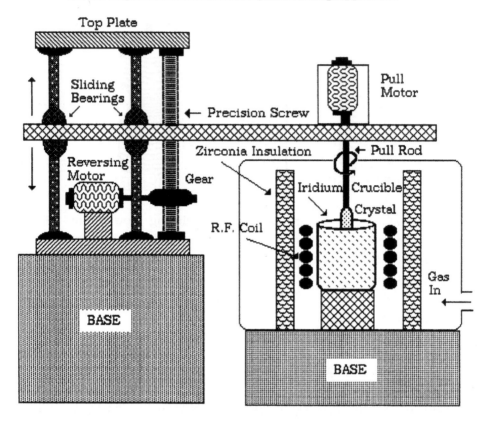

melt. Not shown is a fixed temperature sensor which monitors the temperature of the top of the melt. This sensor must be used since temperature adjustments are based upon its readings. There are other refinements to this apparatus that can be made. We have only shown the basic components needed to grow a single crystal.

The next most important parameters in Czochralski growth of crystals are: the **heat flow** and **heat losses** in the system. Actually, all of the parameters (with the possible exception of #2 and #9) are strongly affected by the **heat flow** within the crystal-pulling system. The next section addresses this factor.

4. Heat Flows in the Czrochralski System

A typical heat-flow pattern in a Czochralski system involves both the crucible and the melt, as shown in the following diagram:

4.2.15.-

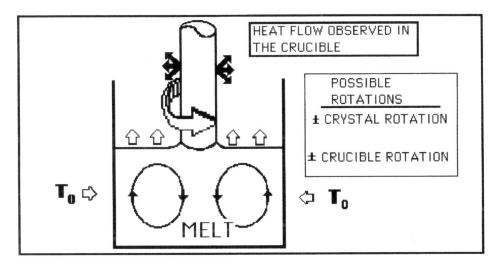

In this case, we see that heat radiates from both the sides and the top of the melt even though the crucible is being heated by the eddy-currents from the RF-generator coupling. Note also that the crystal being pulled also radiates heat. However, this is for a static crucible. The circular currents set up in the melt are the results of the crystal being turned as it grows. The circular heat flow pattern causes the surface to radiate heat.

The crystal also absorbs heat and reradiates it further up on the stem. If the crystal is kept stationary, we see that the heat flow pattern is uneven. Thus, it is mandatory that the crystal be rotated so as to control crystal diameter, so as to obtain a defect-free crystal. The above description applies to the system where only the growing crystal is rotated.

There is at least one other way to "stir" the melt so as to control heat flow. This is illustrated as follows:

6.4.11.-   ± crystal rotation = ± crucible rotation
             ± crystal rotation ≠ ± crucible rotation

We can rotate the crucible by itself, or in conjunction with the crystal. But another complexity arises, namely what direction of rotation and what relative speed of rotation should we use for both, or either?

If we rotate the crystal clockwise and the crucible counter-clockwise. then the heat flow patterns become complex indeed. These complexities have been studied in detail. A simplified version of the effects of rotation upon stirring patterns in the melt are given in the following diagram:

4.2.16.- Stirring Patterns Observed for a Static and Rotating Crucible

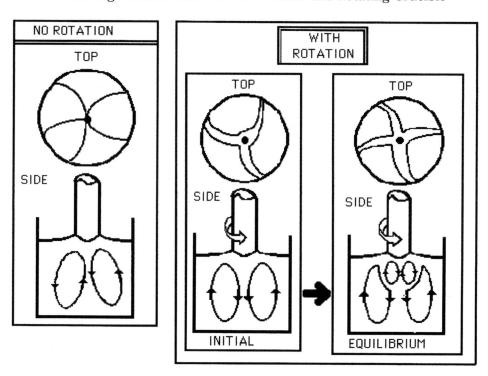

Here, we show some of the changes that occur in heat flow patterns as the crystal grows. These patterns are indicative of the heat-flow patterns and

thermal gradients that exist in the melt. However, if we turn both the crucible and the crystal, we have at least 8 combinations from which we can make a choice. The effect on stirring patterns is shown as 4.2.17. on the following page.

In this diagram,, we have shown clockwise rotation as ⊞ , while counter-clockwise rotation is given as ⊟ . the speed of rotation has not been defined except as "slow" and "fast". If we rotate in the same direction, but at a slow rate for the crystal and turn the crucible much faster, we get the stirring pattern shown in #1.

Note that a completely different stirring pattern occurs if crystal and crucible are turned in opposite directions (compare #1 to #6). We have indicated the cases where the crucible rotation is dominant and those where the crystal rotation dominates.

The best compromise seems to be fast- rotation for the crystal and slow or no rotation for the crucible. Of all the possible methods of stirring the melt, the static-crucible method seems to be the best, and this is the method used by most **crystal-growers.**

The next best method seems to be rotating the crucible at a slow rate, counter to the direction of the crystal rotation. It is clear that crystal-rotation needs to dominate the stirring pattern so that mixing of the melt continues while the crystal is growing.

However, we have not shown all of the complexities that can occur while the crystal is growing. To do that would require more space than we have at the moment. You will note, however, that the problem of building a Czrochralski crystal pulling apparatus is not easy and requires a number of factors and controls if the apparatus is to be capable of growing several different kinds of single crystals, i.e.- from low temperature melting compounds to high melting compounds. It is for this reason that several commercial varieties of Czrochralski crystal pulling machines are now available. These machines incorporate many of the improvements that we have been discussing.

4.2.17.-

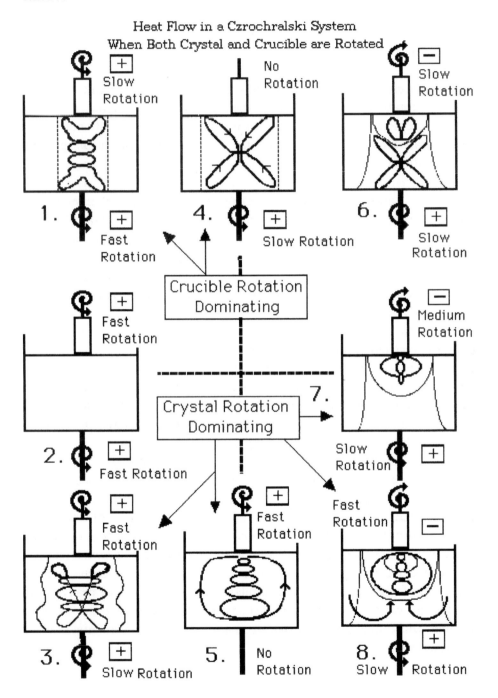

Heat Flow in a Czrochralski System
When Both Crystal and Crucible are Rotated

5. An Automated Czrochralski System

The Czochralski Method has remained the most frequently employed method for obtaining single crystals. The seemingly formidable complexities involved have been conquered by the use of computer-control of the crystal-growing Parameters. One such system, available commercially, is shown in 4.2.18., presented on the next page.

Here, we show only a bare outline of the individual components in the overall system. This SYSTEM is capable of operation in inert atmosphere or vacuum. A melt/seed contact monitor is provided as well as a CCTV camera for observing and controlling the crystal diameter as it grows. Note that both the crucible and crystal rotation can be controlled. In order to control the heat-convection patterns which normally appear in the melt, an external cryomagnet is supplied. Its magnetic field controls heat losses, plus it maintains a better control of the crystal growth. A slave micro-processor controls both crystal diameter and meniscus-contact of the growing crystal.

As we have stated, this is most important if we wish to obtain a crystal essentially free from ingrown defects, i.e.- line defects. This takes the form of a melt/seed contact monitor and a separate monitor is provided for the operator as a CCTV camera for observing the crystal diameter as it grows. There is also a crystal annealing furnace to remove any crystal strain that may have been induced by the crystal-growing conditions. A base heater helps to maintain a uniform temperature gradient in the melt-crucible during crystal growth.

Note that both the crucible and crystal rotation can be controlled. In order to control the heat-convection patterns which normally appear in. For the melt, an external cryomagnet can be supplied. Its magnetic field helps to control heat loss to maintain a better control of the crystal growth. This has not been shown but includes a source of liquid nitrogen for the cryomagnet as well as a controller to maintain the correct magnetic field needed for defect free growth of the crystal.

4.2.18.-

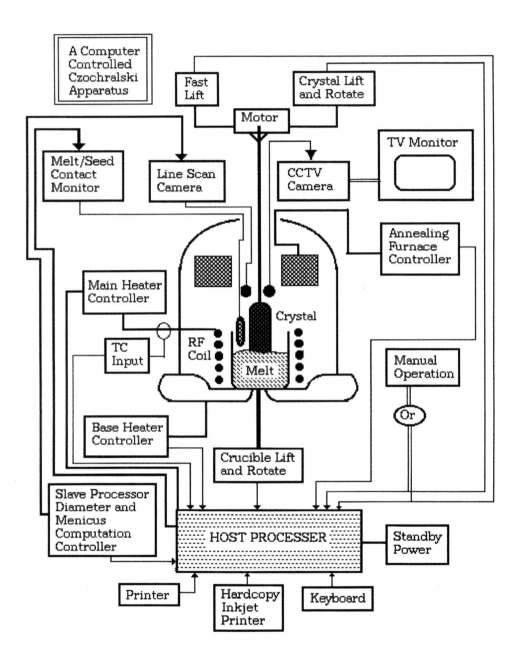

A Computer Controlled Czochralski Apparatus

Fast Lift

Crystal Lift and Rotate

Motor

TV Monitor

Melt/Seed Contact Monitor

Line Scan Camera

CCTV Camera

Annealing Furnace Controller

Main Heater Controller

Crystal

TC Input

RF Coil

Manual Operation

Melt

Or

Base Heater Controller

Crucible Lift and Rotate

Slave Processor Diameter and Menicus Computation Controller

HOST PROCESSER

Standby Power

Printer

Hardcopy Inkjet Printer

Keyboard

4. The induction heating coil is a helix so that the RF field is not symmetrical. Thus, one side of the iridium crucible will be hotter than the other side.

5. Another source of variation is the temperature of the cooling water. The power delivered to the Ir crucible is: $W = I^2R$, where r is made up of two components, i.e.- $R_D + R_I$, where the former is the resistance of the crucible (and its contents) and the latter is the resistance of the coil itself. It has been found that variation in temperature of the cooling water itself has a definitive effect of the power output of the RF heating source.

At this point you might think that the problems associated with growing laser crystals are nearly impossible to overcome. However, this was the point of the diagram given in 4.2.18. By automating the crystal-growing system, we find that thermal variations at the liquidus interface can be diminished and nearly abated. Now, we can show the parts of the overall crystal-pulling system of 4.2.18., not shown but mentioned above. This is shown in 4.2.20., given on the next page.

Note that the crystal is being weighed in real time so that a crystal diameter can be calculated. This assumes that the crystal remains round and no voids occur during growth. By keeping track of the diameter as the crystal grows, avoidance of ingrown defects can be avoided as well as being able to remain as close as possible to the critical pulling rate, R_C. The power to the RF coil is thereby controlled precisely by a separate feed-back loop.

Not shown is a further improvement consisting of a water-temperature control for the cooling water in the RF coils. It has been found that even changes of a few degrees (° C) is mirrored in changes of crystal diameter in the growing crystal. It is these changes that have made possible an increase in pulling rate of most laser crystals, including that of YAG:Nd^{3+}, without exceeding the critical growth rate mandated by the individual phase diagrams of the solids. In this way, critical defects in the crystals have also been avoided.

4.2.20.- Control Loops Used to Control Crystal Diameter

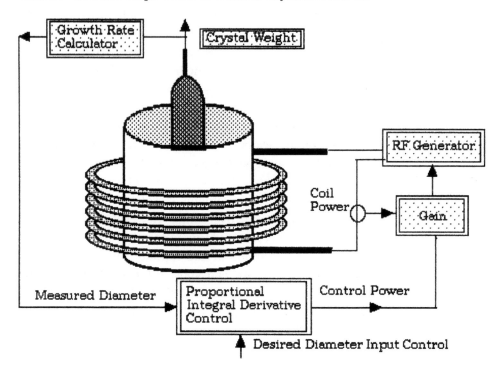

In order to show you what is meant here by critical defects, consider the following diagram, which shows growth defects in a crystal before processing into laser rods.

4.2.21.- Growth Defects in a Crystal Before Processing into Laser Rods.

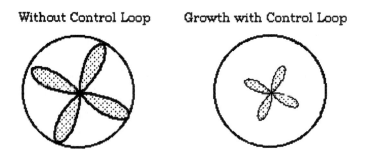

In many cases the constitutional defects (which are only seen under polarized light) are nearly abated and in some cases completely eliminated. In the cases shown above, four rods of selected diameter can be cored from the pulled crystal rod on the left. The rod on the right allows much larger diameter rods to be cored, or if desired, several more rods of smaller diameter to be obtained.

A final comment on the Czochralski Method: GaAs has become important in construction of integrated circuits for computers because of the promise of speed of response and density of components, compared to silicon wafers. Both As and Ga tend to oxidize in air as they approach the melt stage and As_2O_3 sublimes. Both elements are toxic to man. In the melt stage, they have high partial pressures as well, so that the use of an inert atmosphere is not sufficient to allow growth of a single crystal.

One method that has been used for this case has employed a "liquid encapsulating agent". As shown in the following diagram, this consists of a lower melting agent used to form a liquid barrier, floating on the surface of the melt which serves as the source for the crystal (in this case, Ga-As).

4.2.22.-

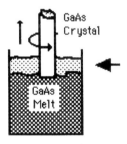

The arrow indicates the liquid barrier layer. This use of a barrier melt illustrates that there are several ways to grow crystals which would be difficult to obtain under "ordinary" means of crystal growth, i.e.- prevention of oxidation and evaporation of GaAs during crystal growth.

c. Melt Growth Using the Bridgeman-Stockbarger Method

Whereas the Czochralski method requires rather elaborate equipment to obtain single crystals of good quality, the Bridgeman (Stockbarger) method uses a fairly **simple** apparatus. This is shown in the following diagram:

4.2.23.-

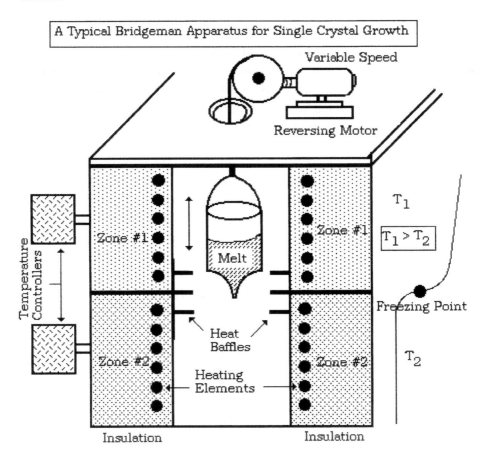

Note that we are using a shaped crucible in order to induce a seed to make the crystal grow. Under optimum conditions, one can grow quite good crystals by this method. However, one must experiment before the final conditions become apparent for the particular crystal being grown.

The major components needed to grow single crystals by this method are:

* A two (2) zone furnace
* Two (2) set point temperature- controllers
* A crucible with "seed" tip
* A constant-speed elevating and lowering device for the crucible.

The procedure involves obtaining a full crucible of melt whose composition mirrors that of the crystal we wish to grow. We then lower the crucible containing the melt through a baffled zone within the furnace at a slow rate. The baffles within the furnace produces a uniform temperature decrease between the two zones, resulting in a temperature profile like that given on the right in the Figure. The temperature set-point of the upper zone is just above the melting point of the material to be grown, and the lower zone is set just below the melting point of the material. This results in a smooth temperature profile, above and below that of the crystal fusion point. Thus, if we can induce a seed to form in the tip of the crucible, then, as the crucible is lowered through the freezing part of the temperature zone, the rest of the crucible will freeze as a single crystal. The parameters for the **Bridgeman Method** are:

4.2.24.- PARAMETERS INVOLVED IN THE BRIDGEMAN METHOD

1. Temperature Profile within Furnace
2. Rate of Crucible Raising or Lowering
3. Melt Temperature
4. Supercooling of Melt
5. Annealing Temperature
6. Thermal Conductivity of Melt and Crystal
7. Thermal Expansion of Crystal
8. Crucible Material Used
9. Chemical Stability of Material Being Grown

Parameter #1 is a matter of furnace design. There should be provision for several baffles so as to adjust the temperature- gradient distance. Then, it is a matter of adjusting the two set-point temperatures to achieve the proper gradient. For some crystals, the gradient needs to be sharp, whereas for others it can be more gradual. The two set-point parameter depends primarily upon the degree of supercooling experienced in the melt, prior to nucleation. This in turn may be affected by purity of the constituents. The raising and lowering mechanism can be simple, that of a gear-driven motor with a counterweight for the crucible-melt mass. It is convenient to use the lower set- point as the annealing temperature for the crystal. But, for very high melting points, this may not be possible. Then, we need to add a lower annealing furnace. As the crystal freezes, it may not do so uniformly. If so, internal strain results (a polariscope will reveal this). Furthermore, if the expansion coefficients of the crystal and crucible are too disparate, then external strain on the crystal will generate internal strain. The crystal will than have to be annealed to relieve this strain. The **Bridgeman Method** is useful for many types of crystals. The key to getting this method to work is to induce a seed to form in the "seed-tip". If the crucible "seed- tip" is not properly designed, the method will not work.

d. Melt Growth Using the Kyropoulos Method

A comparable way to grow crystals is the **Kyropoulos Method,** which uses a type of apparatus similar to the Bridgeman-Stockbarger method. In the Kyropoulos method, we first form a melt and then introduce a seed. By raising the melt through the temperature gradient, a single crystal will grow from the point where the seed engages the melt. The apparatus is shown on the next page as 4.2.25.

Note that we have **reversed** the temperature gradient within the furnace in 4.2.22. and that the top is cooler then the bottom, where the melt is first formed. In other variations of the Kyropoulos method, we raise the melt through the freezing point of the melt, by raising it past the baffles of the furnace. Another possible method is one where we use a single- zone furnace, stabilize the melt, lower the furnace temperature to incipient

4.2.25.-

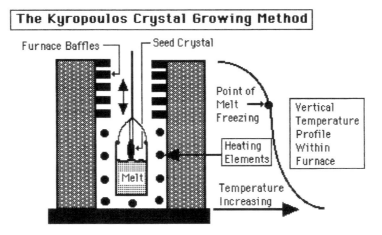

nucleation and then cool to form the crystal. However, this works only for a few systems. The reason is probably supercooling. There are only a few melts where a large degree of supercooling occurs. Addition of a seed will then cause very rapid growth of the single crystal. But, it is usually strained and must be carefully annealed to obtain the mass intact. If the apparatus shown in 4.2.25. is used and the crucible is exactly positioned so that a vertical temperature gradient is maintained **exactly along the crucible height**, the melt, when seeded, will grow into a single crystal more slowly. Then, the degree of supercooling is not as important. As a matter of fact, melts with little or no degree of supercooling can be induced to form single crystals by following this procedure.

e. Crystal Growth Using a Variation of the Kyopolous Method

The problems in getting a "seed" to form in the Bridgeman crucible was, perhaps, the basis for a new and different type of crystal growth based upon a variation of the Kyropolous method. It was noted that if one could get the melt to "super-cool" (that is- not crystallize when the melt temperature was lowered just slightly below the crystallization temperature), one could then induce rapid crystallization by "touching" the surface of the super-cooled melt with a seed of the same material. In this method, the seed is introduced after the melt temperature has been stabilized and then brought to incipient nucleation. This is illustrated in:

4.2.26.

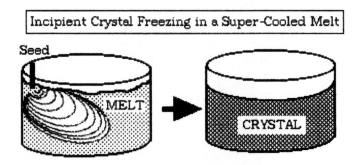

It is this method which is currently being used to manufacture sapphire boules as large as 18 inches in diameter. The boule is then cut into slabs which are polished and used for UV transmitting windows, substrates for various electronic devices and even non-scratching faces on your watch. Note that sapphire is about as hard as diamond and has a heat-transmitting capability almost equal to many metals, while remaining essentially chemically inert and electrically non-conductive.

Sapphire has become the material of choice for many applications, including windows for watches and other devices where other materials like glass would be easily scratched. In addition to hardness, sapphire has high mechanical strength for high pressure and high shock loading. It is also transparent from 1420 Å to 50,000 Å (far infra-red). This makes it a good candidate for optical components in the integrated circuit design field. However, sapphire has a rhombohedral lattice structure which results in anisotropic properties. That is, it refracts light at different angles according to the wavelength of the light. Such behavior is called birefringence. Sapphire can be used if a {0,0,0,1} oriented crystal is used. However, getting the crystal boule to freeze along the (0001) planes is very difficult. Usually, the preferred growth orientation is on the (11$\underline{2}$0) or (10$\underline{1}$0) axis. The growth of a large boule such as shown in 4.2.26. allows one to orient the crystal and cut optical parts along the (0001) axis to obtain high quality sapphire optical elements such as lenses, prisms and shaped optics having little or no birefringence.

Although we have illustrated how the crystal is frozen in place within its melt container, the actual situation is not so simple. It has been only within the past few years that the technique for obtaining a melt at the desired supercooling stage has been realized. The method is called the "heat-exchanger method" (HEM). The following diagram only shows a part of the total design of the HEM apparatus:

4.2.27.-

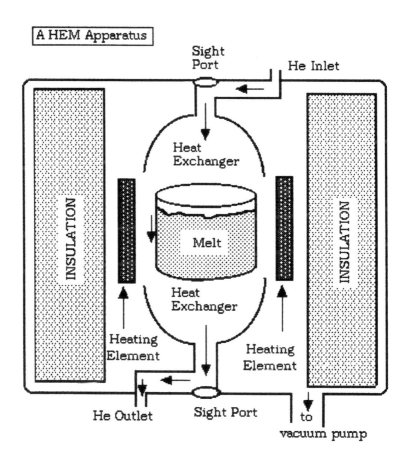

In this case, we have not shown all of the parts of the apparatus, just the essential ones. You will note that the heating elements create the melt by radiative heating in a vacuum. Once the melt is obtained, the vacuum is turned off and helium gas is flowed through the heat exchanger to progressively cool and create a supercooled melt. In this case, there is a temperature gradient formed in the melt from top to bottom. By carefully measuring this gradient (the instruments are not shown), one can achieve a supercooled melt which will freeze upon command into an oriented single crystal. In some cases, a single crystal seed is placed on the bottom of the crucible before the melt is to be formed but care has to be made not to completely melt the seed crystal before the freezing step is initiated. Sapphire single-crystal boules of 33 cm in diameter have been routinely fabricated which weigh up to 65 kg.

The other crystal being grown in this way is calcium fluoride, CaF_2. Such crystals are as large as 30 cm in diameter and 10 cm thick. Because of its low index of refraction and low dispersion, clear, colorless fluorite of optical quality is used for apochromatic lenses. The crystal is cut into pieces and fashioned into optical components. CaF_2 is transparent from 0.1 μm (1000 Å) to 7.0 μM (70,000 Å - mid-infra-red). Lithography is the technology used. This is an optical technique which uses a photo-mask to isolate certain areas of a photo-resist from radiation which polymerizes any part of the photoresist exposed. The un-polymerized resist is then washed off, leaving the desired circuit consisting of patterned lines. This is repeated many times. Integrated circuit manufacturers form line circuits on a single crystal disc of silicon. They have progressed from relatively thick lines to 0.18 μm to 0.13 μm lines.

Line-widths in the integrated circuit alternate with "trenches" so that the metal "lines" connect individual transistors in the circuit, be it a CMOS or MOSFET type. The alternation looks like this: ⎍⎍⎍⎍⎍⎍⎍⎍⎍⎍⎍⎍.
The density of lines for a 0.13 μm (130 nm) spacing is thus about 38,500 lines per centimeter (1/2 of the total space, assuming a linear cross-section). If a tighter spacing is sought, the light source is restricted since its diffraction limit is close to the Franhofer limit and must be changed. Originally, ordinary light sources, i.e.- incandescent or discharge lamps were used but soon met their limitations. Even monochromatic line-

sources, i.e.- lasers, have approached their diffraction limits using quartz optics. Although Excimer, i.e.- rare gas, lasers are available, quartz absorbs too much of the optical power, i.e.- does not transmit the ultraviolet light, to be of any use. Currently, excimer lasers operating at 193 nm. are being used with CaF_2 optics. This allows lines to be formed as small as 100 nm. When the 157 nm laser becomes the workhorse of the microlithography process to form IC lines, it is expected that lines as small as 50-70 nm will become possible.

f. Edge Defined Crystal Growth From the Melt

Many times, the form of the crystal obtained is not suited for the end use contemplated. For example, we have already shown how a flat plate of α – Al_2O_3 (sapphire) is manufactured, using the HEM method. This material is often used as a base for integrated circuits (IC's) because of its high thermal conductivity and its low electrical conductivity. Si is vapor deposited on the surface of the alumina plate, and through various photographic techniques and selective etching (with suitable additives diffused into the silicon layer), an integrated circuit is built up. The IC relies upon the high thermal conductivity of the alumina base for its long life since "hot-spots" can quickly destroy the IC. In the past, the Czochralski method was used to grow an α – Al_2O_3 crystal. It was then sliced into wafers, polished on both sides and then into smaller flat plates of the required size. However, α–Al_2O_3 (sapphire or corundum) is extremely hard (9.5 on the Mohs scale) and is difficult to work with.

One method that was used was to grow α – Al_2O_3 directly as a flat ribbon. By using **edge defined growth**, one can do this. The apparatus consists of a normal Czochralski melt with an anvil at the surface of the melt, as shown in the following diagram, given as **4.2.28.** on the next page.

Once the crystal has started to grow, we pull it through the anvil, thus defining its size. Once it is in the form of a strip, as shown below, it can be drawn and wound over a large wheel. The strip is later cut directly into plates, with no polishing required. Because of the high melting point of α–Al_2O_3 (M.P. = 1920 °C), an iridium crucible and anvil are needed.

4.2.28.-

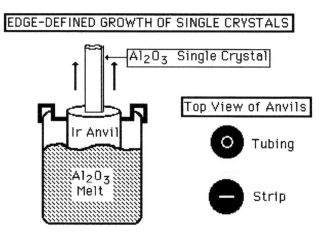

But, the method is more versatile than we might suppose. We can grow crystals in the form of tubing, rods or strip, as shown by the anvils at the right. In fact, we can grow in nearly any configuration we might wish.

Tubing in the form of $\alpha - Al_2O_3$ is used for the construction of the familiar high- pressure sodium- vapor lamps used for street- lighting. End- caps of niobium metal are sealed on the tubing; the capped tubing is evacuated, sodium metal is added and the whole sealed off and mounted. Operation of the lamp occurs at ~ 800 °C and about 15 atmospheres internal pressure of sodium vapor. These operating conditions mandate the use of an optically transparent, chemically- and thermally- stable tubing such as $\alpha - Al_2O_3$. In fact, no other material is known that will successfully withstand these operating conditions.

Edge-defined growth of a single-crystal tubing of $\alpha-Al_2O_3$ for use in high-pressure sodium- vapor lamps was initially tried and found to be successful. However, this method was deemed too expensive since polycrystalline tubing formed by high-pressure sintering at 1700 °C and 100 atmospheres pressure performs as well and at much lower cost per tubing. Additionally, the HEM method has largely supplanted edge-defined growth of $\alpha-Al_2O_3$ strips or wafers for use in integrated circuit design.

Here, we show how the impurities change as a function of the number of passes, given a specific distribution coefficient for the impurities in the melt relative to those in the growing crystal. Note that as the number of passes reaches 10, we obtain a change of some 90 times in the impurity content **at the front of the rod.** But, at the back, i.e.- $L/Z = 8$, we approach an ultimate distribution with 20 passes (the dotted line). Here, we have used:

4.2.36.- $k_0 = 0.5$ and $L / Z = 8$

If we could see the internal distribution of impurities after the rod had been zone-refined by 20 passes, we would see the distribution of impurities as shown in the following:

4.2.37.-

Rod: Zone-Refined by 20 Passes

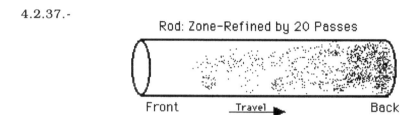

Front   Travel ▶   Back

In this diagram, we show the actual distribution of impurities in the crystal after it has zone refined by 20 passes. The density of dots represents the concentration of impurities. What has happened is that we have moved many of the impurities from the front to the back, but not all. and we finally end up with an impurity distribution we cannot change. To understand this, we must examine the impurity-leveling factor in more detail.

2. The Impurity Leveling Factor

When a melt-zone is moved through a long crystal, impurity concentrations build up in the melt zone due to rejection by the growing crystal. We can also say that the distribution coefficient favors a purification process, i.e.- $k << 1$. Another reason (at least where metals are concerned) is that a solid-solution between impurity and host ions exists. It has been observed that the following situation occurs:

4.2.38.-

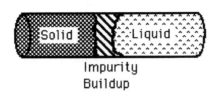

Impurity
Buildup

The impurity, x , builds up at the solid- liquid interface as the zone moves. We can write for the distribution coefficient:

4.2.39.- $\qquad k_x = c_{xL} / c_{xS}$

where xL is the impurity- concentration in the liquid, etc.,

and we estimate k from:

4.2.40.- $\qquad k\ N_I\ /\ N_{MX}\ \exp\ -\ (E_M - E_X)\ /\ kT$

where I is the impurity, MX is the compound under consideration, and E_M, E_X are the activation energies for formation of solid solution of MX · IX . Note that we have assumed that I is a cation impurity in the crystal. We can differentiate between 3 separate cases, as shown in 4.2.41., presented on the next page.

It should be clear that if the impurity freezes into the solid from the melt, there must be a certain amount of solid solution formation, even though the impurity content be less than 0.01%. Case I & II show simple solid solution behavior for k < 1. For $c_{xL} > c_{xS}$, we will get an improvement in purity in the crystal. But for $c_{xL} < c_{xS}$, the opposite effect is seen. Thus, two different impurities could manifest opposite behaviors, depending upon what host they were in.

Nevertheless, simple solid-solution for impurity systems is rarely the norm. The most prevalent case is that of Case III of 4.2.42. Limited solid solution occurs, and we get a two-phase system, as shown in that diagram, also given on the next page.

4.2.41.- Behavior of Impurities as a Function of Type of Solid Solution
Formed

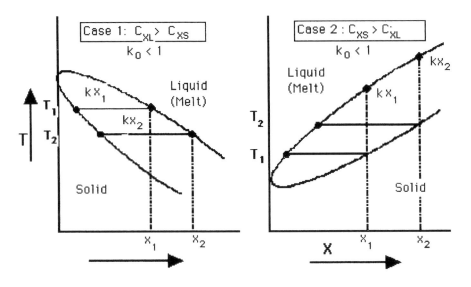

4.2.42.-

Behavior of Impurities as a Function of the Type of Solid Solution Prevalent

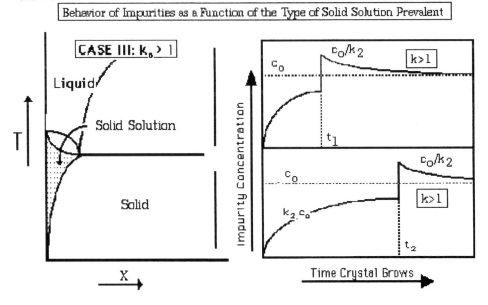

In 4.2.42., we show the phase diagram of the crystal composition and two cases for impurity segregation between melt and crystal as it grows in time (given on the right of the diagram). Note that an initial purification occurs in both cases but the distribution coefficient for the upper case is such that the amount of impurity **actually incorporated** into the crystal, $k_i \cdot c_0$, rapidly approaches the **original** impurity concentration, c_0. At t_1, it actually **increases** over the original concentration at some point within the solid crystal.

The same thing occurs for the case at the bottom of 4.2.42. except that the distribution coefficient is such that the impurity buildup is slower. This illustrates the fact that impurity segregation and purification processes are dependent upon the type of impurity involved and its individual segregation coefficient. As we illustrated above in 4.2.34., the problem is that the impurity is initially rejected from the solid, but its concentration **builds up in front of the growing crystal.** The segregation coefficient, k_i, then operates on that increased concentration and the product, $k_i c_0$, increases.

If we use individual zone lengths, as shown in 4.2.34., we would have:

4.2.43.-　　ZONE　　　　　　IMPURITY CONCENTRATION

1.	$(c_0 - k c_0)$
2.	$(c_0 - k c_0) + k(c_0 - k c_0)$
3.	$(c_0 - k c_0) + k(c_0 - k c_0) + k^2(c_0 - k c_0)$, etc.

This is the reason for the behavior shown in 4.2.38. The impurity front builds up until its concentration surpasses the original concentration (sometimes by manyfold). Thereupon, the impurity concentration levels out (but not necessarily to the same c_0 we have used for illustration in 4.2.38.). It is for these reasons that zone-refining has found limited application, since actual purification by this method does not produce a crystal with complete uniform impurity distribution.

The conclusion we must draw is that we should purify our raw materials

before we attempt to grow our single crystal, not after. Zone refining has been found to be effective only in a few limited cases. It is not applicable to the general case or to high melting crystals. But if we wish to do **zone-leveling of an added impurity,** then we have a very useful technique. A good example would be to add a small amount of Ga to Ge . We can add a small volume of Ga to the front zone and will find that it has become evenly incorporated into the Ge crystal after several passes, using zone-refining.

While the above description might seem quite complex, one needs to know what is happening as one attempts to grow a single crystal, even if the description is somewhat limited in scope, as is the above narrative. It should be clear that the distribution of impurities between the melt and solidifying solid is a function of the phase diagram applicable, i.e.- the phases present, and the nature of the material, both chemically and physically, being used. If you need to know more concerning impurity distribution in crystals, a number of publications dealing solely with this subject can be found in the library.

h. Crystal Growth from the Melt- The Verneuil Method

In our discussion to this point, we have formed a melt and then processed it to grow a single crystal. In the Verneuil method, only a small amount of material is actually melted at one instant and a single crystal is gradually constructed. This method has also been called the "flame-fusion" method of crystal growth. If a powder is blown through a flame, it will melt if the flame is hot enough. The only flame hot enough to do this is the oxy-hydrogen flame. However, a specially designed burner is required so that the crystallites will melt during the short time that they pass through the flame-front. The original work was accomplished by Verneuil (1931) and the apparatus is named after this investigator.

The tip of the torch is important. One design used for such a torch is shown in 4.2.44, given on the next page. The center tube is used for oxygen gas which transports the powder. The burner is designed so that the outer tubes contain only hydrogen gas, with the **interstices** between

4.2.44.-

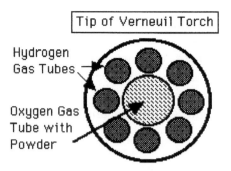

the H_2 - tubes transporting additional oxygen gas. This design will melt Y_2O_3 powder which has a melting point of 2380 °C.

As shown in the following diagram, given as 4.2.45 on the next page, the overall Verneuil apparatus consists of a sealed hopper to contain the powder, the TORCH itself, a refractory pedestal to hold the growing crystal, and an after-furnace to anneal the crystal. Thermal stability is important in this method because the high temperatures reached may be sufficient to cause decomposition of the material.

The Verneuil Method of crystal growth is not generally applicable to all types of crystals. There are serious deficiencies in the method. For example, there is a large temperature drop of hundreds of degrees over a few millimeters within the crystal. This causes a large difference in thermal expansion within a limited space, and consequent strain. Many crystals are not refractory enough to withstand the stress buildup and so crack into many smaller parts. This makes it very difficult to obtain a single crystal of any size. Thus, if a crystal is grown by this technique, it must be annealed carefully in order to obtain it intact.

Generally, the method is restricted to crystals like Al_2O_3 whose thermal conductivity is high and whose refractive nature makes it possible to obtain a crystal. For the most part, one is restricted to growing simple oxides by this methods. Complex oxides such as Ca_2SiO_4 or $Ca_2P_2O_7$ generally cannot be grown easily and the Czochralski Method becomes the method of choice.

4.2.45.-

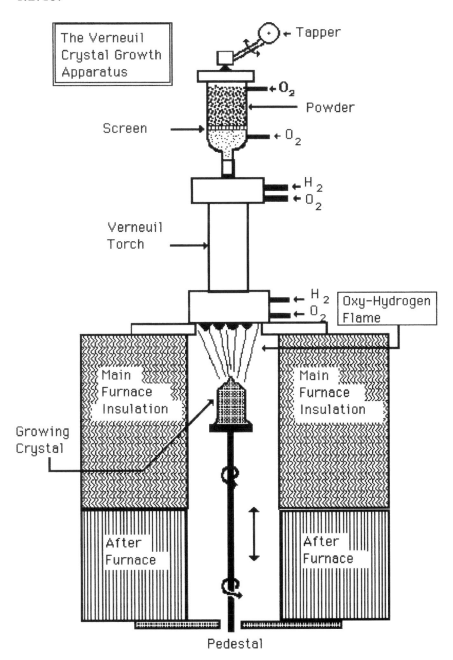

The Verneuil Crystal Growth Apparatus

Tapper

O₂

Powder

Screen

O₂

H₂
O₂

Verneuil Torch

H₂
O₂

Oxy-Hydrogen Flame

Main Furnace Insulation

Main Furnace Insulation

Growing Crystal

After Furnace

After Furnace

Pedestal

The procedure for using the apparatus shown in 4.2.45. is given as follows:

4.2.46.-   Operating Instructions for Verneuil Apparatus

1. Fill powder hopper, close and attach to burner.
2. Light burner and adjust flame to oxidizing.
3. Adjust pedestal to proper height and begin rotation.
4. Slide after-furnace into position.
5. Turn on and adjust oxygen flow in powder hopper.
6. Begin tapping.
7. As melt begins to build up, adjust both height and speed of rotation so that crystal begins to form in a regular manner.
8. When the proper crystal size has been reached, stop powder flow by shutting off the oxygen flow to the hopper. Continue flame operation and pedestal rotation.
9. Lower the crystal into the after- furnace, and anneal crystal, Gradually lower furnace temperature and let cool to room temperature.

The parameters involved in the Verneuil method are:

1. Melt Temperature
2. Thermal Stability at Melt Temperature
3. Temperature Gradient present in Furnace and after-furnace.
4. Sintering Volume Losses
5. Thermal Conductivity of Crystal
6. Annealing Temperature Required
7. Effects of Reducing Flame on Material
8. Chemical Stability of Material being Used

Although this method was used initially to grow very refractive crystals of high melting point (because Czrochralski equipment was not adequate to attain the high temperatures needed), recent improvements in

achieve planarity and smoothness of the individual discs are extremely important. It has been found that any crystal defects which appear in the surface of the silicon disc are mirrored in the integrated circuit being formed on the surface of the disc. Thus, many "dice", cut from the disc to make operating transistors do not function properly and must be discarded.

Under conventional crystal growth conditions of silicon boules, vacancies agglomerate to form small, low-density octahedral **voids**, commonly called "D-defects". Interstitials can form distributed **dislocation** clusters. Both have serious effects upon the process of forming integrated circuits. The void defects have been clearly associated with dielectric breakdown failures in integrated circuits, while the dislocation defects have been associated with controlling the rate of the electrical defect producing reactions due to integrating the management of defects throughout the entire growth process. In other words, the dislocation defects tend to cause the formation of defects in the integrated circuit at the point of the dislocation defect. Additionally, such a defect may not appear until the IC is being operated. Such defects are known to be related to certain classes of leakage current failures. The void defects have recently become a particular concern throughout the industry in spite of their extremely low density.

In the growth of the single-crystal silicon, the crystal boule is pulled from the melt and is allowed to cool as it is being pulled. By annealing the upper part of the boule and finally subjecting it to a controlled cooling rate, this "zone engineering" ensures that the reactions that produce either the void or dislocation cluster defect are completely suppressed. The result is crystals that are completely free of both void and dislocation cluster defects. Eliminating one or the other can be readily achieved. Crystals have been made that are completely free of both classes of defects and have earned this material the name "perfect silicon,".

At present, defect-free silicon crystals have been achieved at only at diameters of 200 mm. Comparisons of crystal quality were made among three techniques: a typical conventional Czrochralski crystal growth technique, a slow-cooled controlled reaction and the "perfect silicon" process. The quality levels achieved in D-defect levels of the material is mirrored in the gate oxide integrity of these materials.

j. Actual Imperfections of Crystals Grown from a Melt

Now we can summarize **all** of the imperfections likely to appear in crystals grown from a melt. Some of these, particularly **stacking- faults**, were discovered only when single crystals were grown large enough so that the deviation from long range order became apparent, as we discussed for the case of silicon.

The intrinsic defects found in crystals include vacancies, interstitials, impurities and impurity compensations. reverse order, and combinations such as V- S and I- S, etc. Their numbers are well described by:

4.2.50.- $N_i = N_0 \exp - \Delta E_i / kT$

where i refers to the intrinsic defect. In addition, we have dislocations, both edge (line) and spiral (screw), the latter being three- dimensional. Their numbers are well described by:

4.2.51.- $\Delta G_i = n \Delta H_i - T (\Delta S_{config} + n \Delta S_{vib})$

This equation arises because both of these **extrinsic** defects affect the energy of the crystal. We can also have grain boundaries which may be: clustering of line defects or mosaic blocks. The latter may be regarded as very large grains in a crystallite.

Another imperfection in crystals is called "twinning". This usually happens when **enantiomorphs** are present, or possible. A good example is quartz, i.e.-

4.2.52.- α– quartz $\rightleftharpoons$ β– quartz

573 °C

In this case, **two** crystals grow and are joined at a given plane, each being a mirror image of the other.

The other crystal imperfection we have not covered is "stacking faults". A good example is SiC. Here, we have **two sublattices,** one based on Si and the other on C, each of which is hexagonal. In stacking **alternate** layers of identical atoms, we first stack Si and then C. If we refer to "A" as the 1st layer, "B" as the 2nd layer, and "C" as the 3rd layer, then the normal stacking sequence for the hexagonal lattice is:

4.2.53.-   **ABC - ABC - ABC**

where we have shown three sequences. For SiC, "A" in the first sequence is Si, "B" is C, and "C" is Si. We still have the same packing since only "A" atoms are over "A" atoms, as shown in the following diagram:

4.2.54.-

STACKING SEQUENCES IN THE HEXAGONAL LATTICE

LAYERS

A	──	A	A	A	A	A	──	A
C	──		C	C	C	C	──	C
B	──	B	B	B	B		──	B
A	──	A	A	A	A	A	──	A
C	──		C	C	C	C	──	C
B	──	B	B	B	B		──	B
A	──	c_0 A	A	A	A	A	──	A

a_0

Here, we have arranged the layers on a two-dimensional structure, even though the layers are arranged in three dimensional order. Note that only two crystallographic axes are indicated, We call this the natural stacking sequence because of the nature of the hexagonal close- packed lattice.

This behavior is more common than one might think, particularly those having the hexagonal crystal structure. Thus, in addition to point defect deficiencies in the lattice, we also may have stacking defects, i.e.- line defects and spiral defects, i.e.- growth around a line defect. However, SiC also exhibits other stacking sequences, as shown in the following:

4.2.55. β - SiC   =   ABC ABC

SiC - 4H   =   ABCA ABCA

SIC - 6H   =   ABCACB ABCACB

SiC - 15R   =   ABCBACABACBCACB   ABCBACABACBCACB

These arranged layers are called "polytypes" and are prevalent where simple compounds such as SiC and SiN are involved. In many cases, the properties of such compounds depend, to a large extent, upon the specific stacked layers obtained during formation. For SiC, we can also have "polytypes" where two stacking sequences like 4H - 6H can combine to form a unit. Another "polymorph" is 4H -15R. This phenomenon has been thoroughly studied and polymorphs of   87R and 270R have been reported.

Another type of stacking fault is called **"polystructure"**. A good example is ZnS, which is dimorphic (has two forms). The cubic form of ZnS is called sphalerite, whereas the hexagonal form is called wurtzite (These are mineral names, after the first geologist who discovered them). The stacking sequence for sphalerite is: AB or ABBA; that for wurtzite is: ABC. A polystructure sometimes results when sphalerite is converted to wurtzite:

4.2.55.- ZnS $_{cubic}$ $\overset{\sim~1200~°C.}{=}$ ZnS $_{hexagonal}$

The polystructure sequence is: ABC ······ AB······ABC·····AB······ABC. This may be regarded as cubic- close- packing within a hexagonal- close- packed structure. We may also note that ZnS is subject to the same "polytype" stacking faults as those given for SiC above.

Thus, the stacking fault patterns noted are a function of the type of lattice involved, not on the chemical composition of the material. For the most part, these stacking faults are found only in the high symmetry lattices such as hexagonal close-packed and cubic close-packed structures.

k. Crystal Growth from a Liquid Phase

You will note that we have formed a liquid phase for growing a single crystal by melting the component(s). There are other ways of growing single crystals which use a liquid phase which does not have the same composition as that of the crystal we wish to obtain.

1. Growth of Single Crystals From a Molten Flux

This method is related to single crystal growth from a melt in that it employs a molten flux which dissolves the material and redeposits it upon a selected substrate. That is, the molten flux acts as a transport medium. The temperature of the flux can be varied to suit the material and to promote high solubility of the solute material in the molten solvent. However, the method is limited in that one is restricted by the temperature at which one can use the molten flux. One example is "YIG", yttrium iron garnet, i.e.- $Y_3Fe_5O_{12}$. This material is used in the Electronics Industry as single crystals for microwave generating devices. It can be grown via the molten flux method. A typical molten-flux apparatus is shown in 4.2.56., as follows on the next page.

The apparatus is simple and consists of a furnace whose temperature can be accurately controlled, a crucible to hold the molten flux, and a method of rotating the crucible to stir the flux containing the dissolved material. Single crystals can be induced to grow along the sides of the crucible by controlling the temperature and supercooling of the molten flux.

The steps involved in using this method are as follows:

1. A flux such as lead borate is melted. PbB_2O_4 is useful in this method because it will undergo supercooling rather easily.

2. The material which is to form single crystals is dissolved in the molten flux to near saturation (Note that this requires prior knowledge regarding solubility of compound in molten flux).

3. The crucible, which is usually platinum, is rotated to obtain a uniform temperature distribution within the melt- compound solution.

4.2.56.- A Molten-Flux Apparatus

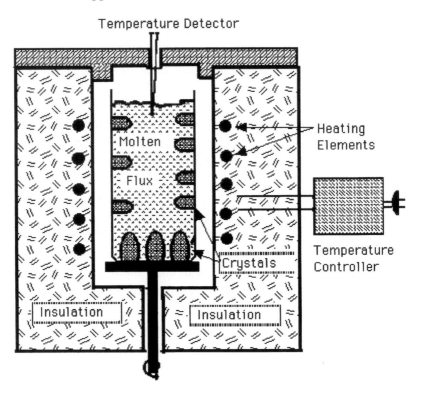

4. The solution temperature is gradually lowered to incipient nucleation. At this point, because of the physical arrangement of the heating elements, a temperature gradient will exist along the length of the crucible, from top to bottom.

5. Single crystals will begin growing along the bottom of the crucible, and sometime later along the edges.

6. The crucible is kept rotating to maintain a uniform mixing and heat flow while the crystals are growing.

The crystal-growing parameters for this method are given as follows:

4.2.57.- Molten Flux Growth Parameters

1. Flux melt temperature
2. Solubility of MX in BX_2
3. Degree of supercooling
4. Temperature gradient achieved
5. Rate of rotation used

This method has serious deficiencies for use as a general method. We find that if we dissolve MX in a BX_2 flux and grow an MX crystal, it is likely to be contaminated with B, or even BX_2.

Naturally, the flux to be used depends upon the crystal to be grown. Many times, it is a matter of trial and error to establish the proper flux needed. The two most important parameters are: the **degree of supercooling** and the **temperature gradient** achieved. If these are not within the correct range, one does not obtain single crystal growth. This method has been used in the past only because of its relative simplicity of apparatus and materials. However, the quality of crystals so-produced has been rather poor. Crystals produced by this method are suitable for structure determinations, but are poor in optical quality and are not suited for electronic applications.

Nonetheless, there is one area where molten flux growth has been very successful. That is epitaxial growth of a single crystal film on a substrate, such as that used for production of bubble memories. This is shown in the following, given as 4.2.58. on the next page.

In this case, we use molten lead tetraborate (PbB_4O_7), which melts at about 960 °C. The melt is raised to 980 °C. and the requisite oxides are dissolved therein The temperature is lowered to near 960 °C. and supercooling begins. It should be obvious that the degree of supercooling that the melt will undergo before the dissolved oxides precipitate out depends upon the amount of oxides dissolved. **This is generally determined by trial and error.**

4.2.58.- Epitaxial Growth on a GGG Crystal Wafer

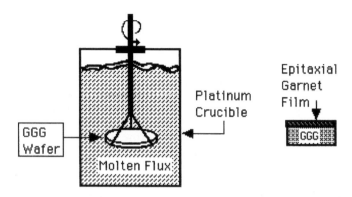

We lower the temperature to just above that point. A substrate slice of single- crystal GGG (gadolinium gallium garnet = $Gd_3Ga_5 O_{12}$) is then submerged in the molten flux, while being rotated.

Since the substrate is cooler, a garnet film will grow on its surface under these conditions. The surface of the GGG slice will have been carefully polished to minimize surface defects. It takes about 6 minutes to grow a single-crystal film about 50 - 100 μ thick. The film grows in an epitaxial manner, that is- it builds up on the crystallographic planes of the substrate itself.

In this case, purity is another parameter for epitaxial growth in addition to those given above. The purity of the film grown, and its intrinsic defects, affects the operation of the bubble memory film. Obviously, impurities cannot be tolerated.

2. Hydrothermal Growth of Crystals

Single crystal growth by hydrothermal methods utilizes water as the material-transport medium. The method is most often used for growing single crystal quartz. Quartz (SiO_2) is not very soluble in water, but its solubility increases considerably at higher temperatures. Thus, growth is accomplished at high pressures in a sealed autoclave. These quartz crystals are used as resonant frequency "tuning forks" for timing applications in wrist watches. As shown in the following, seed crystals are hung within an autoclave on a revolving hanger.

4.2.59.-

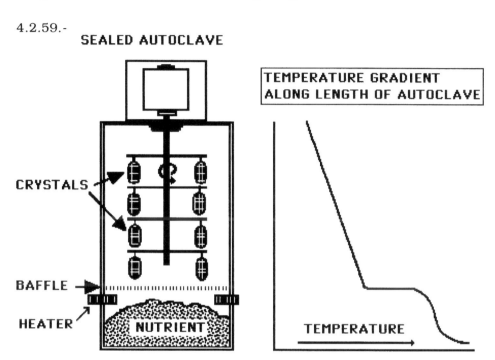

Nutrient (high purity sand or natural crystal quartz) is contained at the bottom and dissolves as the autoclave is heated. A temperature gradient is most often used so that the nutrient dissolves and is transported to the cooler area where it is redeposited as single crystal material. Note the temperature gradient at the baffle.

The solubility of the solute (in this case, quartz) is a function of both pressure and temperature. **Pressure** could be in theory be used as the controlling parameter rather than temperature. However, it is difficult to design an apparatus with a **pressure gradient**, whereas obtaining a temperature gradient is fairly easy.

Actually, what is controlled is the critical volume from the equilibrium:

4.2.60.- $H_2O_{(g)}$ ⇔ $H_2O_{(l)}$

where l and g refer to the liquid and gaseous state, respectively.

The phase diagram for water is shown at the lower left of the following diagram:

4.2.61.-

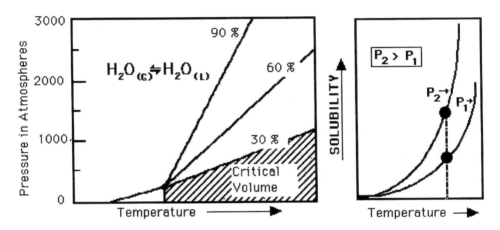

Above the critical temperature, water exists as a gas at all pressures. But, if we fix the volume of both gas and liquid (as in an autoclave), then no liquid is possible below the critical volume (~ 30 % by volume). By filling the available volume greater than 30% with water, we can go to very high pressures (i.e.- > 3000 atmospheres) and still maintain a liquid volume.

In hydrothermal growth, the materials usually grown as single crystals are those classified as insoluble at standard temperature and pressure (STP). For quartz, growth is usually accomplished at 85% fill and 2000 atmospheres at 350 °C. The parameters for hydrothermal growth of single crystals are given on as follows:

4.2.62.- Parameters Controlling Hydrothermal Growth of Single Crystals.

    1. Operating Pressure
    2. Operating Temperature
    3. Solubility in super-critical water
    4. Degree of supersaturation
    5. Degree of supercooling
    6. Purity of Nutrient

The last three factors require some further consideration. We will discuss these in light of the growth of quartz, since it is this crystal for which the most experience has been gained. The rate of crystal growth is a function of seed-crystal orientation. The rate of growth, r_{hkl} , may vary several orders of magnitude, depending upon the {hkl} plane orientation. However, a certain degree of supersaturation is mandatory as the nutrient dissolves, passes through the baffle, and moves to the seed area. Otherwise, the nutrient would precipitate before reaching the seed crystals. It has been determined that the rate of crystal growth is a direct function of the degree of supersaturation, Δ Sat., as follows:

4.2.63.- $\qquad r_{hkl} = \alpha \cdot k_{hkl} \cdot \Delta\,Sat$

where k_{hkl} is the seed orientation factor and α is a constant dependent upon the crystal system. However, it has been found that the degree-of-supercooling factor is contra-indicative to the growth rate. To understand this, imagine the face of a growing crystal. Rejection of impurities at the growing interface causes a localized increase of impurity concentration.

This is shown in the following diagram, given as 4.2.64. on the next page. If the superimposed temperature gradient is T_a, i.e.- dT/dl is a constant, and T_e is the equilibrium curve, i.e.- dC/dT_e , then supercooling will occur. The rate of deposition thereby changes. If it becomes slower, crystal growth slows whereas if it speeds up, then dendritic growth or faceting occurs. Note that the impurity- rejection mechanism, observed for the case of Molten Flux growth, also occurs here.

4.2.64.-

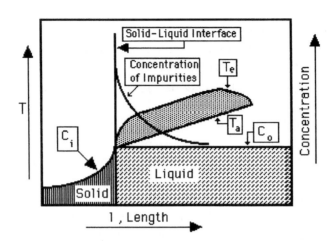

For all these reasons, purity of materials is one of, **if not the**, most important of the parameters controlling hydrothermal growth.

While we have discussed how to produce quartz crystals by hydrothermal growth, we have not addressed the applications in which they are used. Two very important technical fields use quartz crystals. These are: as frequency oscillators and as thin-film monitors. Both of these technologies take advantage of the fact that quartz can be act as a crystal oscillator by applying a controlled voltage.

Piezoelectricity was discovered in 1880 by Pierre and Paul-Jacques Curie, who found that when they compressed certain types of crystals including quartz, tourmaline, and Rochelle salt, along certain axes, a voltage was produced on the surface of the crystal. The next year, they observed the converse effect, the elongation of such crystals upon the application of an electric current.

Some solids, notably certain crystals, have permanent electric polarization. Other crystals become electrically polarized when subjected to stress. In electric polarization, the center of positive charge within an atom, molecule, or crystal lattice element is separated slightly from the center of negative charge. Piezoelectricity (literally "pressure electricity") is observed if a stress is applied to a solid, for example, by bending, twisting, or squeezing it. If a thin slice of quartz is compressed between two electrodes, a potential difference occurs. Conversely, if the quartz crystal is inserted into an electric field, the resulting stress changes its dimensions. This was the basis for the discovery that quartz resonators could control the frequency of an electrical oscillating circuit.

The operating frequency of a particular resonator, usually called a "thickness-shear" resonator, is primarily determined by the thickness of the crystal plate, i.e.- the thinner the plate, the higher is the frequency of operation. But, a thin rectangular plate with large lateral dimensions is subject to breakage. This led to larger and thicker flat plates as large as 1.0 inch on a side. Obviously, the dimension of the original quartz crystal presented a problem, particularly if the crystal was limited in size. The early market was for commercial radio stations and ham operators in the 1920's and 1930's. Crystal control was used by the Military in the 1940's for controlling Radar. As the demand for higher frequencies arose, it was found that round plates could withstand the stresses of high-frequency flexion much better, and these became the standard at that time. With the advent of larger hydrothermally grown quartz crystals with improved lattice properties, i.e.- lack of growth-induced defects, this problem was solved. Nowadays, the quartz resonators, driven by improvements in the ever-expanding market for smaller-sized packages in surface-mount technology for integrated circuits, are cut in highly defined modes from the crystal. The terms "strip resonator", "AT strip resonator" and "BT strip resonator" are commonly used to describe the low profile, surface mount quartz crystals used in oscillator circuits.

To understand this terminology, consider the diagram given on the next page as 4.2.65. Quartz exists in two forms: α-, or low, quartz, which is stable up to 573 °C (1,063 °F), and β-, or high, quartz, stable above 573 °C. The two are closely related with only small movements of their constituent atoms during the alpha-beta transition.

4.2.65-

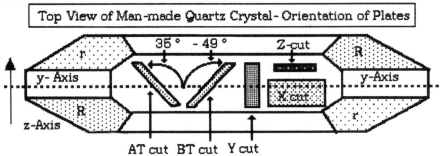

The structure of α-quartz is trigonal, with either a right- or left-handed symmetry group.

The structure of beta-quartz is hexagonal with again either a left- or right-handed symmetry group, equally populated in crystals. At the transition temperature, the localized tetrahedra, i.e.- SiO_4 , of high-quartz twist, resulting in the lower symmetry of low-quartz. That is, atoms move from special space group positions to more general positions.

What is shown in the above diagram of 4.2.65. is an alpha-quartz crystal, oriented along the y-axis. The z-axis is at 90° while the x-axis is perpendicular to the field of view. Y-cut plates are cut from the crystal in a way so that the thickness dimension is parallel to the y-axis. In this way, the resonant frequency of a Y-cut plate is primarily determined by its thickness. However, Y-cut plates exhibit some undesirable traits, such as temperature instability and fragility, that make them unsuitable for commercial use. So most plates today are cut at specific angles to the z-axis (the y-axis is the major mechanical direction in the trigonal lattice) as shown for the AT and BT cuts. This results in plates having more desirable traits of frequency stability versus temperature. The difference in AT and BT cuts are nearly opposite in crystal lattice orientation and results in substantial differences in their operating characteristics. The term "resonator" simply means a crystal plate sandwiched between two electrodes. Most of today's demand for surface-mount resonators is for these AT and BT crystals because they can resonate at very high frequencies.

The following diagram shows the main types of vibrational modes that can be induced in a crystal resonator plate:

4.2.66.-

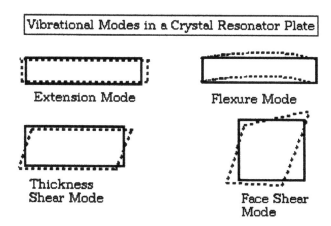

AT and BT plates are made to vibrate primarily in the thickness shear mode. However, it is important to realize that quartz crystals can be made to vibrate in any one or combination of these modes. The vibrational modes can be induced electrically, acoustically, thermally or by some combination of all of these factors. The thickness shear frequency response of an AT plate can be described in terms of the fundamental frequency. For a rectangular AT cut plate, the equation for calculating the approximate frequency of vibration is:

4.2.67.- $f_{m,n,p} = 0.5 \ (1/\rho)^{1/2} \{C_{66} \ m^2/T^2 + C_{11} \ m^2/L^2 + C_{55} \ m^2/W^2 \}^{1/2}$

where m, n and p are integers, ρ is the density of quartz (2.65 gm/cc), T is the thickness in mm, W is the width in mm, and C_{66} , C_{11} and C_{55} are the elastic constants of quartz in the x, y and z directions. If m = n = p =1, we have the fundamental frequency. However, it is easy to obtain unwanted "overtones" such as third or fifth harmonics, i.e.- (3,1,1) or (5,1,1) overtones. Overtones are suppressed by careful sizing and cuts of the crystal plate.

The operation of a crystal quartz resonator can be explained as follows:

4.2.68.- Equivalent Circuit for a Resonating AT-cut Quartz Plate

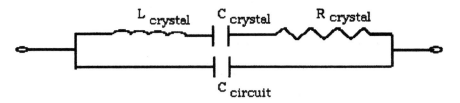

The actual resonator is made by attaching electrodes to opposite sides of the quartz plate. The metal can be evaporated upon the surface or a silver paste attached. Application of a voltage and current causes the crystal to vibrate. The fundamental frequency can be estimated from the induction, capacitance and resistance of the crystal plate in an operating mode, i.e. $L_{CRYSTAL}$, $C_{CRYSTAL}$ AND $R_{CRYSTAL}$. The crystal is assumed to be vibrating at a specific frequency and order of overtone. $C_{CIRCUIT}$ is the capacitance of the circuit to which the resonator is electrically connected. Then:

4.2.69.- $f = 1 / 2\pi (L_{CRYSTAL} \cdot C_{CRYSTAL})^{1/2}$

where L and C are the motional inductance and capacitance during vibration as a resonator. L is in millihenries, C is in picofarads and f is in mega-hertz.

The other application of quartz resonators is the quartz-microbalance. These devices monitor layers of films deposited by changing frequency as the layer adds thickness to the quartz plate. Since the resonant frequency entirely depends upon the thickness, i.e.- mass, of the plate, the quartz resonator is uniquely suited for this application. Although an anti-reflection coating of MgF_2 was sufficient 20 years ago, nowadays current designs call may call for a 24-layer stack of films with alternating refractive indices. With high speed optical communication systems, the required stack may consist of up to 256 layers, all of which must be monitored as to thicknesses deposited sequentially.

Quartz resonant sensors measure film thickness by monitoring change in fundamental frequency of the quartz microbalance. The sensitivity is remarkable. A uniform coating of as little as 10 Å of aluminum will cause a change of 20 mega-hertz in frequency, easily measured by today's electronics. The useful life of this quartz sensor depends upon the thickness and type of coating monitored. If a low stress metal such as Al is deposited, layers as thick as 1,000,000 Å have been measured. At the other extreme, high stress dielectric films such as MgF_2 will reach less than 2000 Å before the crystal malfunctions. What this means is that the quartz microbalance sensor is incorporated within the thin-film coating chamber, as shown in the following diagram, and is subjected to repeated coatings until the weight builds up to the point where it cannot no longer vibrate at its fundamental or overtone frequencies:

4.2.70.-

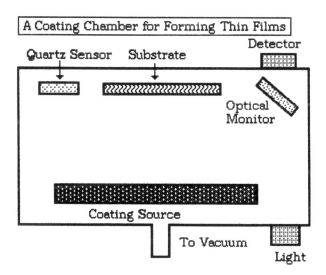

You will note that we have presented only a bare outline of the components in this diagram. We will address these techniques in more detail in a later chapter on thin film technology and its ramifications to the advancement of our "Information Age".

The quartz microbalance has some short comings in its use as a thin-film monitor. Frequency shifts can be positive or negative, depending upon the film being deposited. When coatings are added to the crystal surface, the resonant frequency decreases linearly. If the coating is removed, the resonant frequency increases. The following are a summary of the factors which affect the resonant frequency of the microbalance when used as a thin-film monitor:

4.2.71.- Causes of Resonant Frequency Changes in a Quartz Sensor

> 1. Real time additions of a thin film from the coating source
> 2. Stresses induced from deformation by deposited high stress films like MgF_2 and SiO_2.
> 3. Vibrations induced through the mounting hardware.
> 4. Variations in the voltage used to oscillate the crystal.
> 5. Chemical changes in the film being monitored as it is being deposited
> 6. Adhesion failure of the monitored coating on the quartz electrodes.
> 7. Radiofrequency interference in the monitoring circuit.
> 8. Temperature changes in the crystal when high temperature coatings are being deposited.
> 9. Splatters of material from a faulty coating source as it is being ablated by the various means used to do so.
> 10. Ablation of the crystal surface by high-energy plasmas used to clean the substrate before film deposition.

All of these contribute to early crystal failure. In the case of thin-film stacks of 100 or more, the coating chamber cannot be opened to the air before the coating is completed. If the crystal fails, then the process is ruined and will have to be started again with a new crystal sensor. The latest efforts to prevent early crystal failure are beginning to bear fruit, but we will not address these here at this point

1. Vapor Methods Used for Single Crystal Growth

If the material whose single crystal we want is volatile or sublimable, then we may choose a vapor-method of crystal growth. These methods have been used for a variety of crystals including ZnS and CdS. In this method, a carrier- gas is most often used for material transport and for the sulfides, H_2S is the gas of choice. The following diagram, given as 4.2.72. on the next page, shows a simple apparatus to grow crystals by the vapor transport method:

4.2.72.- A Furnace for Growing Crystals from the Vapor Phase

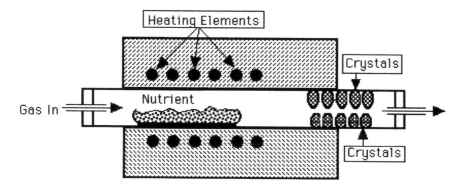

In this method, we start the gas flow and then begin heating. It is important to heat the material to just below the point of sublimation (volatilization) and let the system come to equilibrium. Gas flow need not be rapid but needs to be sufficient to carry the volatiles to the cooler part of the furnace. The tube used in the furnace is most often a silica-tube, although metal tubes have sometimes been used. The choice depends upon the nature of the material being sublimed and crystallized.

For ZnS and CdS, it is important to exclude all traces of oxygen since these materials are easily oxidized, namely-

4.2.73.- $\quad 2\ ZnS + 3\ O_2\ =\ 2\ ZnO + SO_2 \uparrow$

The operating parameters for the vapor phase method of growing crystals are shown in 4.2.74., presented on the next page.

Obviously, whether the material has a low (high vapor pressure) or a high temperature of sublimation is important because of furnace- construction material considerations.

4.2.74. - Operating Parameters for Vapor Phase Growth of Crystals

1. Temperature of sublimation (volatilization) of material
2. Amount of gas flow used
3. Degree of furnace- temperature set- point **above** material sublimation temperature.
4. Temperature gradient at crystal-growing junction of tube.

In general, we cannot use this method for materials which are volatile above about 1200 °C. because most materials of construction cannot withstand the corrosive nature of the vapors at this high temperature. If we set the temperature of the furnace **above** the sublimation point of the material, it will volatize **all at once**. This would necessitate setting the gas-flow such that all of the material is transported by the gas. It would be better to set the temperature just below the vaporization (sublimation) point. There will be enough material transported to begin growth of single crystals. In fact, it has been determined that the slower the growth, the better are the crystals obtained.

There is another method that has been sometimes employed in the vapor phase growth of crystals. This method uses an **evacuated** capsule as shown:

4.2.63- Another Method Sometimes Used for Vapor Phase Growth

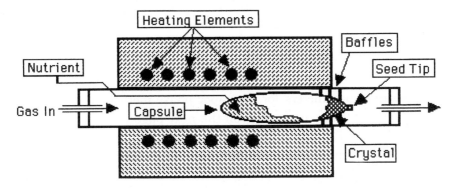

The capsule is generally made from quartz, although platinum is sometimes used. **The capsule needs to be evacuated to remove any residual gas before heating is started.** Otherwise, the internal pressure would build until the capsule would explode. Even though it is evacuated, capsules have been known to explode because the quartz (metal) walls could not contain the internal vapor pressure of the material being grown as single crystal. Care must be exercised **not to handle the capsule hot** before and after crystal growth.

In addition to sublimation or vaporization, we can also use chemical transport as a method of single crystal growth. For example, we could use the apparatus of 4.2.60. to grow crystal of $ZnCl_2$ by the following reactions:

4.2.64.- $ZnO + 3\ HI\ =\ ZnI_2 \Uparrow\ +\ H_2O$

$Zn\ +\ I_2\quad =\ ZnI_2 \Uparrow$

Metal carbonyls are also convenient for growth of certain **METAL** single crystals, as shown in the following:

4.2.65.- $M_{(s)}\ +\ x\ CO_{(g)}\quad =\quad M(CO_{(g)})_x$

$M(CO_{(g)})_x\qquad =\quad M_{(s)}\ +\ x\ CO_{(g)}$

In these cases, one is limited to the growth of single crystals of the transition metals and those metals which can form volatile carbonyl compounds. The alkali metals, alkaline earth metals and certain of those which are allotropic in nature are not at all suited for this type of crystal growth.

Suggested Reading

1. R.R. Irani and Clayton F. Callis, "Particle Size- Measurement, Interpretation and Application" - J. Wiley & sons, New York (1963).
2. John Wulff et al, "The Structure and Properties of Materials - Volumes I, II, III, & IV" - J. Wiley & Sons, New York (1964).
3. Polakowski and Ripling, "Strength and Structure of Engineering Materials" - Prentice-Hall, Englewood Cliffs, NJ (1966).

4. R.E. Newnham, "Structure-Property Relations" - Springer-Verlag, New York (1975).

5. "Optical Microscopy"- General Test (776), Vol. USP 24, pp. 1965-67, Publ. By The United States Pharmacopoeia Society, Rockville, Md. (2000)

6. Robert E. Reed-Hill, "Physical Metallurgy Principles"- Van Nostrand, Princeton, New Jersey (1964).

7. A.J. Dekker, "Solid State Physics" - Prentice-Hall, Englewood Cliffs, New Jersey (1958).

8. Perelomova & Tagieva, "Problems in Crystal Physics"- MIR Publ., Moscow, Eng. Transl. - (1983).

9. W.W. Wendlandt, *Thermal Methods of Analysis*, Interscience-Wiley, New York (1964).

10. P.D. Garn, *Thermoanalytical Methods of Investigation*, Academic Press, New York (1965).

11. W.J. Smothers and Y Chiang, *Handbook of Differential Thermal Analysis*, Chemical Publishing Co., New York (1966).

12. E.M. Barral and J.F. Johnson, in *Techniques and Methods of Polymer Evaluation*, P. Slade & L. Jenkins- Ed., Dekker, New York (1966).

13. W.W. Wendlandt, *Thermochim. Acta* **1** , (1970)

14. D.T.Y. Chen, *J. Thermal Anal.*, **6** 109 (1974) - Part I

15. Chen & Fong, loc. cit. **7** 295 (1975) - Part II.

16. Fong & Chen, loc. cit., **8** 305 (1975) - Part III

17. C. Duval, *Inorganic Thermogravimetric Analysis*, 2nd Ed., Elseveir, Amsterdam (1963).

18. C. Keattch, *An Introduction to Thermogravimetry*, Heydon, London (1969).

19. "Vapor Pressure Determination using DSC", A. Brozens, R.B. Cassel, C.W. Schaumann & R. Seyler, *Proceeding. N. Am. Therm. Analysis Soc.* **22nd Conf.** Denver, Colo. (1993).

CHAPTER 5

Optical and Electronic Properties of Solids

In order to fully understand the nature of phosphors, it is necessary to consider light and the optical behavior of solids, i.e.- the interaction of solids with electromagnetic radiation, including the resonant transfer of absorbed (excitation) energy from one site to another in a solid. We will emphasize phosphors and discuss theories of electronic and vibrational properties of solids as well as solid state luminescence mechanisms in detail. This will establish the groundwork for a specific discussion concerning the design of phosphors and how phosphors are prepared and used in the next chapter. We will then enumerate the many phosphors that have been developed for myriad purposes in the following chapter, particularly those in use since 1991, the date of publication of the first printing of this treatise.

5-1 THE NATURE OF LIGHT

In the early 17th century, the nature of light was not well understood. Nowadays, we know that light is comprised of "photons", which are quantized waves having some of the properties of particles. However, this concept was not made clear until the advent of quantum electrodynamics in the 1930's, following Einstein's concept of relativity in the 1920's.

Such was not the case originally. Beginning with Kepler's *Paralipomena* in 1604, the study of optics had been a central activity of the scientific revolution in Europe. Descartes' statement, in 1634, of the sine law of refraction, relating the angles of incidence and emergence at the interfaces of the material through which light passes, had added a new mathematical understanding to the science of light. Descartes had also made light central to a description of the mechanized theory of nature.

The concept of photons with wave properties has its roots in the study of optics and optical phenomena. Until the middle of the 17th century, light was generally thought to consist of a stream of some sort of particles or *corpuscles* emanating from light sources. Newton and many other scientists of his day supported the idea of the corpuscular theory of light. It was Newton in 1664 demonstrated that sunlight could be separated into colors. He also showed in 1703 that "ordinary" light, such as that from a flame, could be dispersed into its constituent colors by a prism, but the phenomenon was not clearly grasped at that time.

About the same time, the idea that light might be a *wave* phenomenon was proposed by Huygens and others. Huygens demonstrated that every point on a wavefront may be regarded as a source of spherical wavelets, the envelope of which accounts for the position of the wavefront at a later time. Huygens was thus able to explain rectilinear propagation and those involving the laws of reflection and refraction. Fresnel added to this hypothesis by showing that the wavelets can interfere, and this led to a theory of diffraction. However, Newton still regarded light as a stream of particles which were either refracted or reflected.

It remained for Lord Rayleigh in 1887 to explain the wave-nature of light by considering such subjects as electromagnetism, color, acoustics, and diffraction gratings. His most significant early work was his theory in 1871 explaining the blue color of the sky as the result of scattering of sunlight by small particles in the atmosphere. The Rayleigh scattering law, which evolved from this theory, has since become classic in the study of all kinds of wave propagation. Indeed, diffraction effects that are now known to be associated with the wave nature of light were observed by Grimaldi as early as 1665, but the significance of Grimaldi's observations were not understood at the time.

It was thus that many scientists described light as a beam of particles whereas others regarded it as a series of waves. This disagreement remained until the 19th century because various early investigators who had experimented with light had shown that light could be reflected or diffracted by optical means. Some interpreted their observed behavior as that expected from a beam of particles while others felt that their data could only be explained by assuming wavelike behavior for a beam of light. Those who made the most significant contributions were Newton,

Descartes, Huygens and Rayleigh. Although other scientists made additions to the total body of knowledge of optics as well, these four established the major principles regarding propagation and interaction of electromagnetic radiation with matter. Later contributors to our overall understanding of the nature of photons included:

1) Fresnel and Thomas Young (1815) on interference and diffraction respectively

2) Maxwell in 1873 who postulated that an oscillating electrical circuit should radiate electromagnetic waves

3) Heinrich Hertz in 1887 who used an oscillating circuit of small dimensions to produce electromagnetic waves which had all of the properties of light waves

4) Einstein in 1905 who explained the photoelectric effect (He did so by extending an idea proposed by Planck five years earlier to postulate that the energy in a light beam was concentrated in "packets" or *photons*.. The wave picture was retained in that a photon was considered to have a frequency and that the energy of a photon was proportional to its frequency).

Experiments by Milliken in 1908 soon confirmed Einstein's predictions. In 1921, A.H. Compton succeeded in determining the motion of a photon and an electron both before and after a collision between them. He found that both behaved like material bodies in that both kinetic energy and momentum were conserved in the collision. The photoelectric effect and the Compton effect, then, seemed to demand a return to the corpuscular theory of light. The reconciliation of these apparently contradictory experiments has been accomplished only since about 1930 with the development of quantum electrodynamics, a comprehensive theory that includes both wave and particle properties of photons. Thus, the theory of light propagation is best described by an electromagnetic wave theory while the interaction of a photon with matter is better described as a corpuscular phenomenon.

All bodies emit some form of electromagnetic radiation, as a result of the thermal motion of their molecules. This radiation, called thermal radiation, is a mixture of wavelengths. At a temperature of 300 °C, the most intense of these waves have a wavelength of 50,000 Å which is in the *far infrared* region of the electromagnetic spectrum. When the temperature is raised to about 800 °C, such a body emits enough visible radiation to be self-luminous and appears "red-hot". However, by far the most energy is still carried by photons having wavelengths in the infra-red region of the spectrum. But at 3150 °C, which is the temperature of a tungsten filament in an incandescent light bulb, the body then appears "white-hot" and a major part of the energy is in the visible region of the spectrum. The following diagram, given as 5.1.1. on the next page, shows the emission of radiation as a function of "black-body" temperatures. A "black-body" can be simulated by a sphere internally coated with a "black" surface and having a small hole for light to escape. One then observes the wavelengths that escape as a function of the internal temperature of the sphere. For the highest temperatures, the curves are derived from a fit of the equations which correspond to lower temperatures.

Note that even bodies at liquid-air temperatures emit photons between 10 and 100 microns in wavelength, i.e.- 100,000 and 10^6 Å in wavelength. The earth itself at a temperature of 300 °K. has an emission between about 20,000 and 300,000 Å in wavelength, i.e.- 2000 nm. and 30,000 nm. What this means is that even you emit infra-red wavelengths as a consequence of being warm. Note that this is the basis of "night-vision" headgear which can detect human bodies even when it is so dark that other objects are invisible to the naked eye.

In vacuum, all electromagnetic radiation travels at the speed of light. This is given by:

5.1.2.- $$c = \{1 / \varepsilon_0 \, \mu_0\}^{1/2}$$

where ε_0 is the permittivity of free space, and μ_0 is the permeability of free space. The former comes from Gauss's Law and the latter from Faraday's Law. However, the speed of light in media *other* than vacuum is always **slower** than in space. This is believed to be due to resonance interactions between the electromagnetic fields of the electrons

5.1.1.-

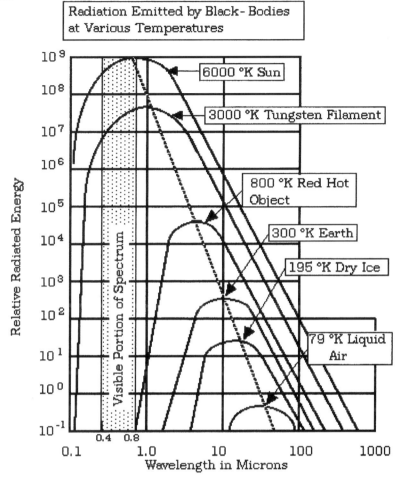

associated with the transparent media and that of the traveling photon. When electrons or other charged particles are accelerated to relativistic speeds, i.e.- a fraction of the speed of light, they emit photons as they travel in a transparent medium, i.e.- Cerenkov effect. Thus, it should be clear that both photons and particles of mass are inexorably interwoven by the matrix of space-time. What this means is that when particles having a given "rest" mass are caused to accelerate near to a limiting speed, i.e.- the speed of light, they are prone to release that excess energy gained through the emission of photons. Einstein was the first to realize this phenomenon, which has since been proven many times over.

Because particles, i.e.- electrons, have wave properties similar to those of photons, we need to differentiate between them. It was de Broglie in 1906 who first postulated the wave nature of particles. In 1927, Davisson and Germer first showed that electrons are reflected from the surface of a solid in the same way that x-rays are reflected. The wave hypothesis clearly required sweeping revisions of our fundamental concepts regarding the nature of matter. The best explanation to date seems to be that a particle must be regarded as an entity not entirely localized in space whereas the photon is a point-source, i.e.- is localized in space. What this means is that an electron is strongly attracted or repelled by electromagnetic fields as it moves through space whereas a photon, being a localized point source, is only weakly affected.

Thus, a photon moves at a constant speed through space while an electron does not. Yet, both have electromagnetic fields thereby associated with each, which are subject to reflection, diffraction etc. The photon-packet thus interacts with the space-time continuum as it moves through space at a constant speed. In contrast, the particle interacts with both the time-space continuum and the electromagnetic fields thereby associated with mass, and its speed is **not** constant, but subject to mass-mass (gravity) interactions as well. The wavelength of the photon (which has no mass) is a function of its internal energy as it moves through space. The particle has a mass which is determined by how much it is spread out in the space-time continuum. Consequently, it has properties we normally associate with "mass".

From the above, it should be clear that a photon is a "force-carrier" while an electron is a "matter-constituent". The notion of particles as mediators of force in nature has provided a framework for testing and developing the **Standard Model** and the associated "Big-Bang" theory of the formation of the Universe. It has also been important for the exploration in depth of many other important questions about the physical world. The most familiar interaction is Electromagnetism. Electromagnetic radiation in its various forms (including radiowaves, microwaves, infra-red-light, visible-light, ultraviolet-light, and x-rays) can be thought of as the exchange of massless photons between electrically charged particles, either quarks or leptons. We can summarize all of the above in that a photon is a quantum of radiation, whereas an electron is composed of matter-quanta.

5.2.- ABSORBANCE, REFLECTIVITY AND TRANSMITTANCE.

When a beam of photons strikes a solid, specific interactions take place which are related to the Quantum Theory. These interactions have been measured in the past and certain formulas have been found to apply. According to HUYGHEN'S principle of electromagnetic radiation **scattering,** when photons come into close contact with a solid, the electric and magnetic field vectors of the incident photons couple with those of the electrons associated with the atoms comprising the solid. This , of course, depends upon the nature of the solid. The radiation can be reflected, transmitted or absorbed fully. In the case of absorption, the energy of the photon changes the energy of the atom or molecule in the solid, resulting in heating at the site of absorption. In transmission of the photon through the solid (assuming it is transparent to the photonic wavelength), no interaction occurs. In reflection (scattering), the photon can undergo either an elastic or inelastic collision with the atoms of a solid. In the former mechanism, the wavelength is unchanged, whereas the inelastic collision changes the wavelength of the photons. That is, part of the energy of the photon is absorbed at the site of atom or molecule of the solid, resulting in an "excited" state in which an electron is promoted into an upper energy level. If the wavelength of the **emitted** photon is **not** changed, then the photon is "scattered" and the reflection is an elastic collision.

This is the basis for the so-called Rayleigh scattering in which blue-light is scattered while all of the other colors of visible light result in inelastic collisions in which the wavelength is changed. Rayleigh showed that the intensity of the scattered light was proportional to the wavelength. We now know that this is due to the Raman scattering effect. That is, the atom or molecule absorbs part of the energy via vibrational (phonon) interaction and reëmits the photon with an altered wavelength. These interactions have been measured in the past and certain equations have been found to apply. The following gives some of these formulas applicable to concepts given later in this Chapter.

5.2.1.- Formulas Applicable to Optical Properties of Solids

    a. Absorbance : $A \equiv \log 1/T = \log I_0 / I$

    b. Transmittance: $T \equiv I / I_0$ (I = measured intensity; I_0 is original)

    c. Absorptivity: $A \equiv A / bc$ (A is measured absorption; b is the optical path-length; c is the molar concentration)

    d.Reflection: Reflectivity $\equiv$ specular reflection, i.e.- at specific angles

    Reflectance $\equiv$ diffuse reflection, i.e.-scattered radiation

    e. Intensity: I is defined as the energy / unit area of a beam of photons, i.e.-electromagnetic radiation.

According to Huygen's principle of electromagnetic radiation **scattering,** when photons come into close contact with a solid, the electric and magnetic field vectors of the incident photons couple with those of the electrons associated with the atoms comprising the solid. This interaction leads to at least four (4) components, namely: R- the radiation **reflected,** A- the radiation **absorbed,** T- the radiation **transmitted,** and S - the radiation **scattered.** This mechanism is illustrated in 5.2.2., given on the next page, as follows.

The original intensity of the radiation is defined as I_0. A part of the intensity is absorbed, another part is transmitted, still another part is scattered, and a part of the total intensity is reflected. The components, S and T, are processes which are **independent** of the wavelength (frequency) of the incident photons, whereas R and A are mostly wavelength dependent.

The **exact amount** of energy extracted from I_0 by each process is a complex set of variables depending upon the type and **arrangement** of atoms composing the solid. The processes, A & S, are related in that both involve absorption of energy, but S is a resonance process.

5.2.2.- Photon Interactions within a Solid

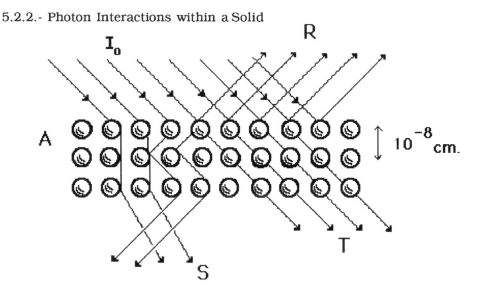

That is, the same amount of energy absorbed in the scattering process is reëmitted by the atom(s). Contrast this to absorption, where a part of the energy is changed into vibrational energy within the solid. We can summarize these properties as:

5.2.3.- Interaction Parameters in Terms of Radiation Properties

$$S = f(\mathbf{L} \cdot \mathbf{S}) \cdot f(I)$$
$$A = f(\lambda) \cdot f(\mathbf{L} \cdot \mathbf{S}) \cdot f(I)$$
$$R = f(\lambda) \cdot f(\theta, \delta) \cdot f(I)$$

where λ is wavelength, $(\mathbf{L} \cdot \mathbf{S})$ relates to electric field vectors of the electrons within the atoms comprising the solid (i.e.- angular momentum and spin vectors), I is the intensity (number of photons impinging per second), θ is the angle between the beam and the atomic lattice, and δ is the atom density. Once you have studied these equations, it becomes obvious that in order to predict and calculate the optical properties of a solid, we need to know more about the discrete factors which control the optical properties of that solid.

Consider an optically homogeneous thin film. By optically homogeneous, we mean one that is thin enough so that no scattering can occur. If a beam of photons is incident to the surface at a given angle (but less than that where all of the beam is transmitted- the so-called Brewster angle), part of the beam will be **reflected** and part will be **absorbed**. This is shown as follows:

5.2.4.- Single Reflection in a Thin Film

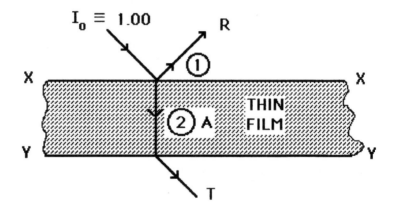

The reflectance, R, is a consequence of the difference in refractive indices of the two media, (1) - air, and (2)- the semi- transparent thin film. The amount of absorption is a function of the nature of the solid. In this case, A, the **absorbance,** is defined as:

5.2.5.- $A \equiv (1 - R) - T$

It is easy to see how this equation is derived from the diagram since T, the transmittance is equal to (1- R) - A. Then, the original beam intensity, $I_0 \equiv 1.00$, is found to be diminished according to the **Beer-Lambert law:**

5.2.6.- $A = \ln I_0 / I = \varepsilon c l$

where l is the pathlength (width of film traversed), ε is a **molar extinction coefficient** of the absorbing species, and c is the concentration of the absorbing species. Although we have shown but one reflection at X, another is equally likely to occur at Y as well. This problem of multiple reflections was worked out by Bode (1954) and we will use it to prove the validity of the equation of 5.1.4. for the case of the homogeneous thin film. First, we define total values in terms of intrinsic, or individual, values for a homogeneous medium, as shown in the following:

5.2.7.- Intrinsic Values of R , A & T

$$R = \Sigma r \quad T = \Sigma t = \ln I_0 / I \quad A = \Sigma a$$

For the case of multiple reflections, **Fresnel's equation** may be used:

5.2.8.- $$r = \{ (\eta\text{-}1)/(\eta + 2)\}^2$$

where η is the refractive index and r is the intrinsic reflection at a single plane. For multiple reflections, we have the following situation:

5.2.9.- Multiple Reflections in a Thin Film

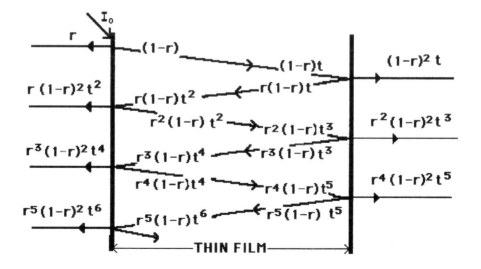

We can show that r is reflected at each interface, and that only (1- r) enters the medium, i.e.- the thin film, at each reflection. The **total** transmission is a series function:

5.2.10.- $\Sigma t = (1-r)^2 t + r^2 (1-r)^2 t^3 + r^4 (1-r)^2 t^5 + r^6 (1-r)^2 t^7 + ..$

and: $\Sigma r = r + r(1-r)^2 t^2 + r^3 (1-r)^2 t^4 + r^5 (1-r)^2 t^6 + ...$

We can rearrange these equations to obtain:

5.2.11.- $T = (1-r)^2 / (1 - r^2 t^2)$
                 $R = r + \{r (1-r)^2 t^2 \} / \{1 - r^2 t^2\}$
                 $A = \{(1-r)(1-t)\} / (1-rt) = (1-R) - T$

which is the result we want. This exercise also shows the relation between A , R, & T in greater detail. However, this derivation only applies to an optically homogeneous thin film. The derivation of scattering coefficients to obtain absorbance from reflectance values of a **powder** was accomplished by **Kubela and Monk** in 1930. Several other approaches to the particle scattering problem have also been taken, but we will not address any of these at the moment.

Consider the energy interactions which occur when a **photon** strikes a solid such as a phosphor. The timeframe of **interaction** is about 10^{-18} seconds, as is easily seen:

5.2.12.-   Speed of photon = $v = 10^{10}$ cm./sec.
                 Distance between lattice planes = d = 10^{-8} cm. = ~ 1 Å
                 Time for photon to traverse lattice = d/v = ~ 10^{-18} seconds

The fact that the photon does traverse the lattice planes does not mean that the photon will be absorbed or even scattered by the solid. The reflectance of the photon is a function of the nature of the compositional surface, whereas absorption depends upon the interior composition of the solid. A "resonance" condition must exist before the photon can transfer energy to the solid (absorption of the photon). In the following, we show in general terms the resonance condition of both R & A, and how this relates to the perception of light as we see it.

5.2.13.- Energy Transfer to a Solid By a Photon

R	A	ENERGY TRANSFER	EXAMPLES
High	High	Moderate	"Colored" solid
Low	High	Very High	"Black" solid
High	Low	Nil	"White" solid
Low	Low	Nil	"Transparent"

Note the difference between a "colored" solid and a "black" solid. In the first case, only parts of the visible spectrum are absorbed or reflected. In the latter case, all parts of the visible spectrum are absorbed and little or none are reflected. In the case of the "white" solid, all visible wavelengths are reflected.

As an example of controlled absorption, consider the case of a pigment. It is quite common to add controlled amounts of a transition metal to a transparent solid to form an inorganic pigment. In one such case, we add ~ 1% of chromium oxide to aluminum oxide to obtain a pink solid, i.e.- "ruby". The Cr^{3+} ion in the Al_2O_3 lattice absorbs blue and green light and reflects mostly the red wavelengths. Thus, both processes, R & A, are wavelength dependent.

Since we have discussed colors, we need to further define the photons that embody them in terms of wavelength, λ, and other symbols commonly used in optics and spectroscopy. This is shown as follows in 5.2.14. given on the next page. Note that the visible region of the spectrum lies between about 1.75 and 3.00 ev. Thus, in a "colored" solid where absorption occurs from a ground state to an upper absorbing state,

5.2.14.- Part of the Electromagnetic Spectrum

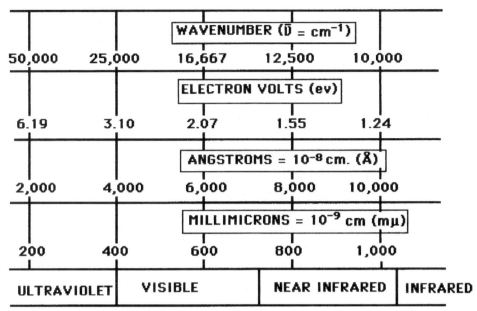

		WAVENUMBER ($\bar{\nu}$ = cm^{-1})		
50,000	25,000	16,667	12,500	10,000
		ELECTRON VOLTS (ev)		
6.19	3.10	2.07	1.55	1.24
		ANGSTROMS = 10^{-8} cm. (Å)		
2,000	4,000	6,000	8,000	10,000
		MILLIMICRONS = 10^{-9} cm (mμ)		
200	400	600	800	1,000
ULTRAVIOLET	VISIBLE		NEAR INFRARED	INFRARED

blue,green,yellow,orange,red

the separation of the upper energy state from the ground state cannot be more than about 3.00 ev. Furthermore, in such a solid, resonant absorption of the photon does not occur unless that photon has a minimum energy **between** 1.75 and 3.00 ev. Moreover, if the solid reflects photons having a wide range of energies, we say that we have a "white" solid. We can, as a general rule, divide inorganic solids into two (2) classes: those which absorb light (pigmented), and those which emit light (phosphors and solid state lasers). That is, we can have pigments (for paints) and phosphors (for TV or lamps). In both cases, we can add controlled amounts of a transition metal ion to the solid in order to control the photon absorption and/or photon emission properties of that solid.

We have now presented a general picture of how photons were discovered and identified as "force" carriers between electrons and the like, i.e.- "leptons". We have also shown, in general terms, how solids interact with

in the solid, results in an **allowed band structure** (or Brillouin Zone) to which the electrons are confined in the solid. This work was first accomplished by Felix Bloch. His equations described the wave-functions of the electrons, as a function of the diffraction conditions imposed by the periodic lattice. The result is called a **BLOCH function:**

$$5.3.4.\text{-} \qquad \Psi_k \ = \ U_k \ (r) \ exp \ [j \ \Bbbk \cdot r \ ]$$

where $\Bbbk$ is a general lattice vector and U_k (r) is a periodic function. The function "r" is a general direction in the lattice, and we require three for the solid. In order to mathematically describe electron energy bands in solids, we find that we must examine the reciprocal lattice. The best way to understand this concept is to examine   actual structures and the Brillouin Zones associated with them, as shown in 5.3.5., given on the next page.

The reciprocal lattice is just what the name says it is. If we have unit cell vectors in the real lattice, the **imaginary** reciprocal lattice vectors are 90° to the real ones. These imaginary vectors define the Brillouin Zone for that structure.

In 5.3.5., we show the allowed electron energy zones in two different cubic crystals. At the top of this figure, we have a face-centered cubic lattice (fcc) with the 1st Brillouin Zone (cell) drawn in. Note that the cells are in reciprocal space and the **reciprocal lattice** is body-centered. The cell-faces of the Brillouin Zone are at right angles to the points of the real lattice. It is a truncated dodecahedron and is derived by cutting off the corners of the cube on the {111} plane to derive the Zone faces at 90 ° to the points of the cubic lattice. At the bottom of the Figure is the Brillouin Zone for the body-centered cubic lattice (bcc). Note the difference between the two reciprocal lattices. In the bcc reciprocal lattice, the Brillouin Zone impinges on the faces of the unit-cell cube, whereas in the fcc reciprocal lattice, it is flat at these points. The faces of the zone represent energy gaps in velocity space, and the surface is sometimes called "**k**" space. Although Brillouin Zones define the extent of k-space, the maximum energy of the electrons (as a function of temperature)

5.3.5.-

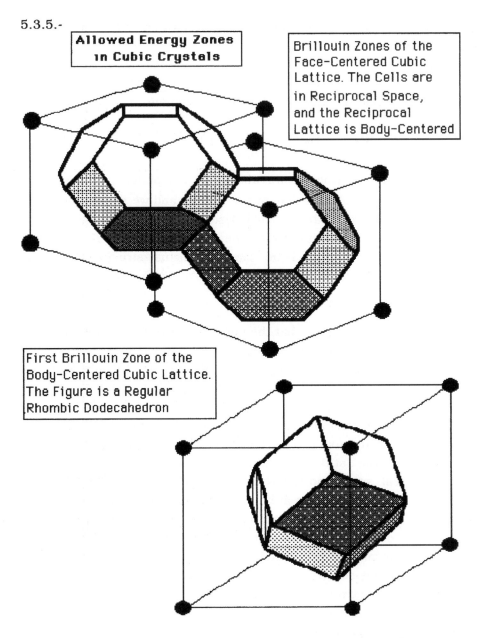

**Allowed Energy Zones
in Cubic Crystals**

Brillouin Zones of the
Face-Centered Cubic
Lattice. The Cells are
in Reciprocal Space,
and the Reciprocal
Lattice is Body-Centered

First Brillouin Zone of the
Body-Centered Cubic Lattice.
The Figure is a Regular
Rhombic Dodecahedron

defines the Fermi energy. Thus, the **Fermi Surface** of the "sea of electrons" in a solid is the upper energy of these electrons, confined to reciprocal space by the Brillouin Zone. To better understand how this works, consider a band model for metallic copper, as shown in the following diagram:

5.3.6.-

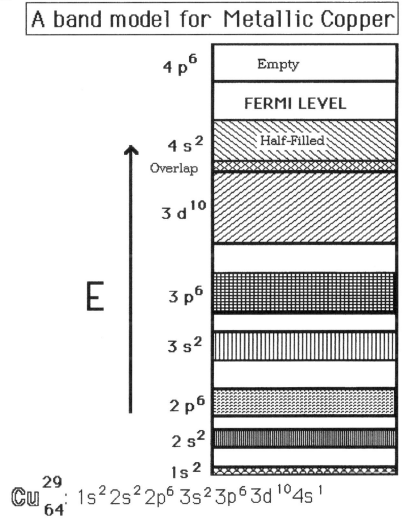

For Cu⁰ , the $4s^1$ electron-subshell is half-filled. This is the top of the filled band, or Fermi energy of the sea-of-electrons in the solid. Copper has a fcc lattice structure. This diagram illustrates the same band structure of copper, but from a different aspect, i.e.- that of energy. Thus, the Brillouin band model shows the placement of the allowed energy zones in the crystal whereas the energy band model of 5.3.6. shows the relative energy of the electron bands, including those of the bonding electrons and "free" electrons in the metal solid.

We are usually more interested in the energy levels of the solid and tend to use the band model when dealing with phosphors rather than the Brillouin band model. To illustrate what we are discussing, consider the model given as 5.3.7. as follows on the next page. Compare it to the band model given above as 5.3.6.

In this diagram, we see the experimentally-determined Fermi surface of the copper electrons within the Brillouin Zone. The "necks" actually define the positions of individual atoms in the real fcc lattice. Compare the actual Fermi Surface to the Brillouin Zone geometry. Also shown is the Fermi Surface of the element, germanium- Ge^{4+} , heavily doped with arsenic, As^{5+} , to make it a n-type semiconductor. It has the same Brillouin Zone as copper but far fewer electrons are available for conduction. This results in "wells", or energy depressions, at the atomic positions, in contrast to the "necks" of copper which extend out to the surface of the real lattice. Thus, the actual occupation of the Brillouin Zone by electrons depends not only upon geometry, but also upon the numbers and energies of the available electrons. However, we conclude that the energy band model is more adaptable to our needs in that it tells us the relative energy of the individual electrons in the solid.

One other example is the compound, TiO. Calculated energy bands are given in the following, presented as 5.3.8 on the following page. TiO has the cubic NaCl structure but is not a dielectric. It is nearly metallic in electronic character. This diagram presents the results of experimental measurements, and the electron bands calculated from that data, presented in terms of energy and direction around the cubic lattice. In this case, lattice intercepts are used rather than Miller Indices. The individual energies, as a function of the electron-band, are labeled at each intercept. It is easy to see why TiO conducts like a metal, since the 3d

5.3.7.- FERMI SURFACES AND BRILLOUIN ZONES IN COPPER

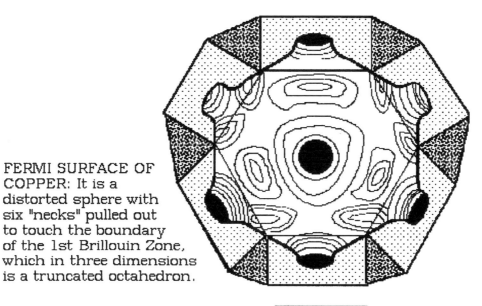

FERMI SURFACE OF
COPPER: It is a
distorted sphere with
six "necks" pulled out
to touch the boundary
of the 1st Brillouin Zone,
which in three dimensions
is a truncated octahedron.

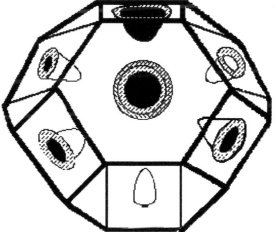

FERMI SURFACE OF
GERMANIUM: It is
doped with Arsenic
to make it an n-type
semi-conductor. The
Fermi surface differs
markedly from that
of copper, showing
far fewer electrons
available for conduction.
Yet, the Brillouin zone
is the same as that for
copper.

and 4s electron bands overlap extensively in the solid. The conduction
band is the 2nd Brillouin Zone in the solid (see 5.3.8. on the next page).
The Fermi Energy rises and falls with temperature and so may be level
with the valence band, or above or below it.

5.3.8.-

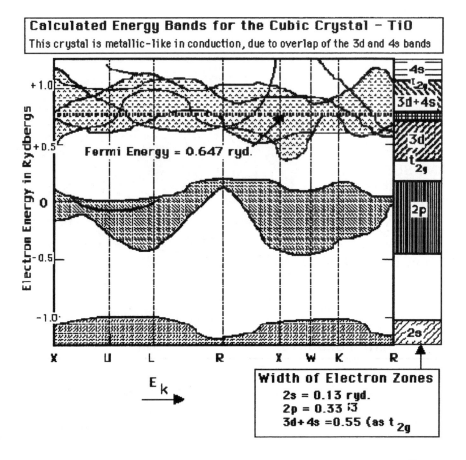

Calculated Energy Bands for the Cubic Crystal – TiO
This crystal is metallic-like in conduction, due to overlap of the 3d and 4s bands

Fermi Energy = 0.647 ryd.

Width of Electron Zones
2s = 0.13 ryd.
2p = 0.33 ⁵3
3d+4s =0.55 (as t$_{2g}$

In a semi-conductor with proper doping, thermal energy is sufficient to bridge the forbidden gap so that conduction results. It is easier to use a two dimensional band model to represent Fermi Levels and Brillouin Zones in the crystal lattice, just as long as we realize that the actual case is three-dimensional in nature.

The ENERGY BAND MODEL has arisen to serve as an easy method of representing electron states in the solid. The Fermi level usually defines the top of the valence band, and is related to the 1st Brillouin Zone in the solid. We generally draw a band model as in 5.3.9.(next page) which shows a comparison of band models for insulators, semi-conductors and conductors.

5.3.9.- BAND MODELS FOR INSULATORS AND CONDUCTORS

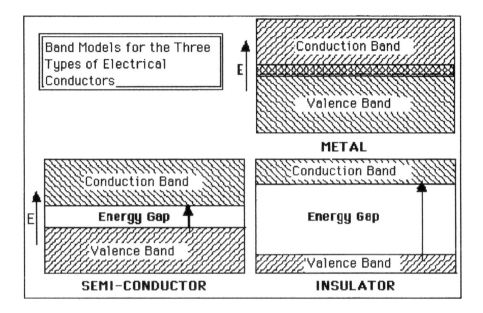

A final comment regarding Brillouin Zones: we generate the faces of these zones at 90 ° to the faces of the real lattice. To do this for the fcc lattice, we cut off the corners of the cube at right angles to the 4-triad axes (see 5.3.5.). If we cut the cube at 90 ° to the 6-diad axes, we generate a rhombic dodecahedron which is the Brillouin Zone for the bcc lattice. Our direction of cutting is 90 ° to the real lattice vectors, but parallel to the reciprocal lattice vectors.

We ought to discuss the mathematics connected with construction of the reciprocal lattice and those of the Brillouin Zone. We will do so in the following Section.

We will start by examining the methods of calculation of energy bands in solids.

II. Calculation of Energy Bands in Crystals

As stated above, the basic equations used to describe electron energy bands in solids are called the *BLOCH FUNCTIONS* . These are a part of quantum mechanics, so we begin by borrowing one of its more famous concepts, that of **"The Particle in a Box"**.

Consider a free electron. It has both particle and wave properties. The general **SCHROEDINGER** equation describing its wave properties is:

5.3.10.- $$H \Psi_e = E \Psi_e$$

where **H** is the *hamiltonian operator* which "operates" upon the wave-function of the electron, Ψ_e, and E is an energy function. There may be more than one solution to the E- function which fits the equation. Indeed, a number of "quantized" solutions to equation 5.3.10. have been shown to exist. But, if we confine the electron to a box having a length, **1** , we find that the number of solutions diminishes considerably. What happens is that a resonance condition arises for the electron when it is confined to the box. Standing waves result whose individual wavelengths are a function of "l". The wavevector is an energy solution for the resonance conditions. Let us define:

5.3.11.- k = any wavevector of the electron

                  Ψ_e = wavefunction of that wavevector

If we set up a Schroedinger equation for the particle in a box, we get:

5.3.12.- $$[- h^2 / 2m \nabla^2] \Psi_k (r) = E \Psi_k (r)$$

where r is any direction in the crystal, $h = h/2\pi$, m is the mass of the electron and ∇^2 is of course the second derivative of the partial equation relating the three directions, x, y, & z in the lattice:

5.3.13.- $$\nabla^2 = (\partial^2 /\partial x^2 + \partial^2 /\partial y^2 + \partial^2 / \partial y^2)$$

If the lattice is cubic, then we have:

5.3.35.- $d = 1 / (h^2 + k^2 + l^2)^{1/2} \cdot (h \mathbf{a} + k \mathbf{b} + l \mathbf{c})$

which is the equation relating lattice planes (Miller indices) to "d", the distance between selected planes.

In the early history of x-ray diffraction, Laue (1905) discovered that by placing a crystal oriented in the proper manner in the path of a "white" x-ray beam, he could obtain a series of spots , arranged as concentric circles, on a photographic film. His geometrical arrangement was:

5.3.36.-

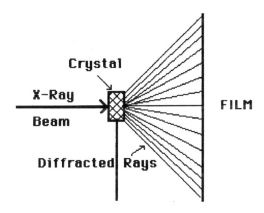

The Laue equations which characterize the x-ray diffraction in this case are:

5.3.37.- $\mathbb{T} = n_1 \mathbf{a} + n_2 \mathbf{b} + n_3 \mathbf{c}$

          $\Delta k = \mathbb{k}_{scatt} - \mathbb{k}_o$

where k_{scatt} is the scattering factor and k_o is the original x-ray beam vector. This leads to:

5.3.38.- $\mathbb{a} \cdot \Delta k = 2 \pi n_1$; $\mathbb{b} \cdot \Delta k = 2 \pi n_2$; $\mathbb{c} \cdot \Delta k = 2 \pi n_3$

We can solve for Δk as follows:

5.3.39.- $\Delta k = 2\pi(n_1/A \, \mathbb{a} + n_2/B \, \mathbb{b} + n_3 C \, \mathbb{c})$

where A , B, and C are reciprocal lattice intercepts. Laue found that by using the reciprocal lattice concept, he could easily label the x-ray points on the film in terms of {h,k,l}. Furthermore, all {0,k,l} , {h,0,l} , or {h,k,0} points lay on separate concentric circles, thus simplifying identification. It remained for Ewald (1908) to explain this phenomenon. Ewald rearranged the Bragg equation to:

5.3.40.- $\sin\theta = n/d \, / \, 1/\lambda$

This equation defines so-called "Ewald Space". Radiation of appropriate wavelength, impinging upon ordered rows of atoms, produces cones of scattered radiation, as follows:

5.3.41.-

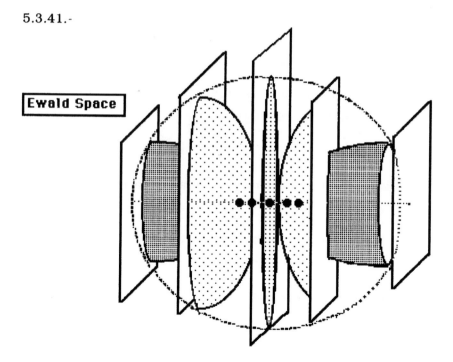

These cones intersect a sphere of radius, $1/\lambda$, and define planes which are separated by $1/d$. θ is the angle between the cones. The distance between the center of a cone and its edge is the reciprocal lattice vector. k_o and k_{scatt} are as defined above. We can set up a scattering equation in terms of the reciprocal lattice vectors, as follows:

5.3.42.- $\Delta k = q\, A + r\, B + s\, C$

where A, B, and C are the reciprocal lattice vectors. Note the similarity to 5.3.36. and 5.3.38.

It is easy to show , using the Ewald construction, that the following combination of vectors, are true:

5.3.43.-
$$A \cdot a = 2\pi \qquad B \cdot a = 0 \qquad C \cdot a = 0$$
$$A \cdot b = 0 \qquad B \cdot b = 2\pi \qquad C \cdot b = 0$$
$$A \cdot c = 0 \qquad B \cdot c = 0 \qquad C \cdot c = 2\pi$$

We can now equate the reciprocal lattice vectors to those of the real lattice vectors:

5.3.44.- $A = 2\pi\, (b \times c\, /\, a \cdot b \times c)$

$B = 2\pi\, (a \times c\, /\, a \cdot b \times c)$

$C = 2\pi\, (a \times b\, /\, a \cdot b \times c)$

where A , etc., is in cm^{-1} and a is in cm. Thus, for the real lattice, the translation vector is :

5.3.45.- $T = n_1 a + n_2 b + n_3 c$

and that for the reciprocal lattice is:

5.3.46.- $G = h A + k B + l C$

where n_1, n_2, n_3 , and h, k, and l are intercepts in the respective lattices. Note that {hkl} are the same Miller Indices referred to in Chapter 1.

It should be recognized that if we multiply $G \cdot T$, we get a scalar product, since we are multiplying reciprocals.

The wave equation for Laue diffraction is:

5.3.47.-　　　　　　$\Delta k = G$

It is also the same for wave propagation, namely:

5.3.48.-　　　　　　$k_0 + G = k_{scatt}$

If we square both sides of this equation, we get:

5.3.49.-　　　　　　$(k_0 + G)^2 = (k_{scatt})^2$

If we have elastic scattering (as we do in x-ray scattering), then $k_0 = k_{scatt}$. Thus, we obtain the WAVE EQUATION for the reciprocal lattice:

5.3.50.-　　　　　　$2 k_0 \cdot G + G^2 = 0$

In this case, G connects points in the reciprocal lattice. We can now proceed to calculate reciprocal lattice vectors for the several figures illustrated above.

BODY-CENTERED CUBIC (refer to 5.3.5.)

Primitive lattice vectors are:

$$a = a/2 \, (x + y - z)$$
$$b = a/2 \, (- x + y + z)$$
$$c = a/2 \, (x - y + z)$$

where a is the unit cell dimension. The unit cell volume is therefore:

5.3.51.-　　　　　　$V = a \cdot b \times c = 1/2 \, a^3$

by a periodic lattice resulted in certain allowed three-dimensional zones within the crystal. These are intrinsic to the structure. We also showed that the electron occupation of the Brillouin Zone may not be complete, and will depend upon the nature and numbers of electrons involved in the structural lattice.

c. Energy Bands in Solids

Finally, we derived the BLOCH FUNCTION to show that these energy bands, in reciprocal space, do have some validity in quantum mechanics. It also gives insight as to the nature of the Fermi level. We also illustrated band models in 5.3.9. What is important to realize that the valence band there is drawn in two dimensions. Actually, it follows the Brillouin Zone or k-space of the crystal lattice in three-dimensions. The Fermi level surface is also affected by both k-space and temperature. It is constrained by reciprocal space, just as the Brillouin Zone is. We use the band model to illustrate certain aspects of each unique crystal. Otherwise, the required model would be quite complex, particularly those crystals with low symmetry. We usually illustrate some specific defect and the band model immediately adjoining it. For a phosphor, this would be the activator (impurity) center. Since we have already (Chapter 2) examined various point defects, let us now illustrate them within the periodic lattice as a function of the energy bands and the Band Model.

The effect of a periodic lattice upon the wavefunctions of an electron is illustrated in the following diagram, given as 5.3.68. on the next page. Note that the positions of the individual atoms within the lattice are shown, i.e.- "lattice sites".

At the left side, we have shown the energy levels of a single electron as the energy, **E** , increases. When lattice sites are imposed, we get a series of bands (shown horizontally) as **E** increases. Note also the unoccupied zones, indicated by the hatched area. Occupation of the allowed bands is a function of the types and numbers of electrons present. At the bottom is the wavefunction associated, as a function of atom-position in the lattice.

5.3.68.-

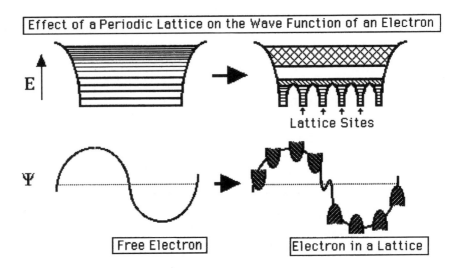

Let us now construct a band model for our semi-conductor, Ge. We illustrate both p-type and n-type defects, as follows:

5.3.69.-

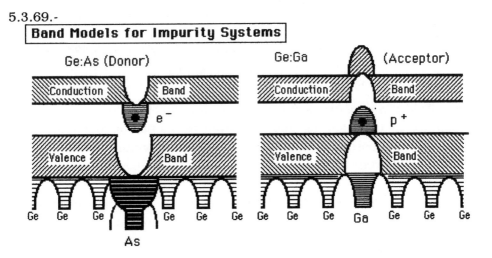

Ga is an acceptor while As is the donor in these diagrams. The individual energy levels of the defects are also shown. Note that they are not the same as those of the lattice atoms themselves. For Ga^{3+}, the positive "hole"

is superimposed upon the energy "hump" caused by the defect, V_M. In the case of As^{5+}, the electron is attached to the conduction band and is available for conduction. The energy required to promote this electron into the conduction band is of the order, $< kT>$, or room temperature. There is an energy well immediately below the electron.

The case for the color center is somewhat different. Here, an electron has been captured by the vacancy, but has its own energy levels, as follows:

5.3.70.-

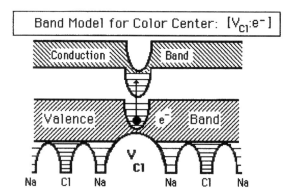

The reason for this difference is that the host crystal is a dielectric or insulator (NaCl), while Ge is a semi-conductor. The vacancy, V_X, is thus seen as a donor, while V_M is an acceptor. Thus, in 5.3.70., we have the example of the defect pair:

5.3.71.- $[V_X , e^-]$

When an interstitial is formed, it becomes a donor upon ionization, as shown in 5.3.72, given on the next page.

We have shown the case for the defect pair: $[M_i^+ , V_M^-]$. Firstly, the V_M^- is seen as a acceptor, while M_i^+ is a donor (But this occurs **after** ionization of the intrinsic defects). They are most likely to be nearest neighbors, so that they tend to cancel each other, as we suspect they ought to do.

5.3.72.-

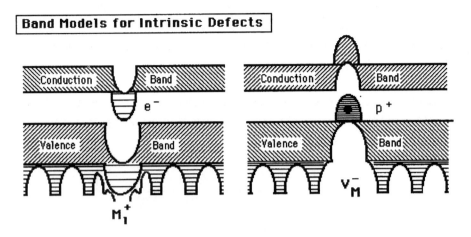

Finally, we show the effect on the energy band of the solid when several defects cluster, as shown in the following:

5.3.73.-

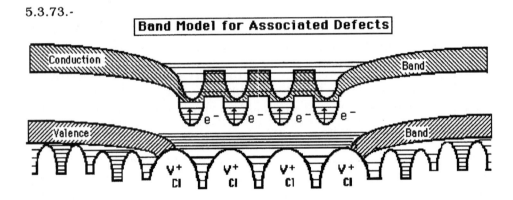

Here, we have used the defect pair:

5.3.74. $[V_X^+, e^-]$

which is an anion vacancy which has ionized. Note how this differs from that of 5.3.70. which is the band model for the color-center. Whereas, in the color-center, the vacancy accepts an electron from an external source,

5.4.9.-   VIBRATIONS IN A CUBIC MONATOMIC LATTICE

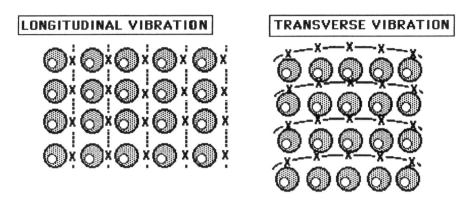

From 5.4.7., we have:

5.4.11.- $M \, du_x / dt = \sum F_i \, (u_{p+x} - u_x)$

where M is a mass. For a traveling wave, we get:

5.4.12.- $u_{p+x} = u_0 \, \exp [i(p + x) \, a \, \vec{k} - i\omega t]$
and:

$$- \omega^2 \, M \, u_0 \, \exp [i \, x \, \vec{k} \, a] = \sum F_i \, (\exp [i\omega t])$$

where a is the unit lattice vector. What we have done is to combine 5.4.8. and 5.4.11. By suitable manipulation, we can get the equations shown as follows:

5.4.13.- $\omega^2 \, M = - \sum F_i \, \{(\exp [i \, p \, a \, \vec{k}]) - \exp (i \, p \, a \, \vec{k}') - 2\}$ or:

$$\omega^2 = 2/M \sum F_i \, (1 - \cos p \, a \, \vec{k})$$

What this means is that a standing wave will occur for nearest neighbor planes of the reciprocal lattice. Since all atoms in this example are equivalent, we can replace $\sum F_i$ by a constant, C., as follows::

5.4.14.- $\omega^2 = 2 \, C/M \, (1 - \cos p \, a \, \vec{k})$

Note that we are still using $\vec{k'}$ and $\vec{k}$ as the interaction vectors of the photon and electrons of the lattice. A plot of the 5.4.14. equation in terms of a traveling wave in a lattice gives:

5.4.15.- A Standing Wave in the Real Lattice

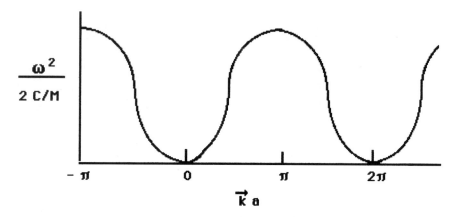

We can also plot this wave in terms of the critical points of the reciprocal lattice, as shown in the following:

5.4.16.- A Standing Wave in the Reciprocal Lattice

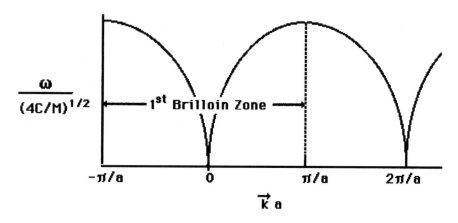

It is obvious that a standing wave exists between $-\pi/a < \vec{k} < \pi/a$. Furthermore, these waves are quantized in energy, as we have said.

Now consider a diatomic lattice. The situation is quite similar and we can use the same approach. In this case, two different atoms are present in the unit cell. An example would be NaCl. If we proceed through the mathematical analysis, using m for one mass and M as the other, we arrive at the result shown in the following:

5.4.17.-

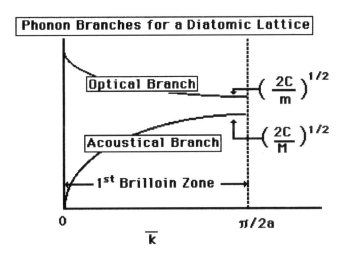

Now we have two (2) phonon dispersion curves, a so-called optical branch and a lower energy acoustical branch. The standing waves are better understood in terms of the actual displacement the atoms undergo:

5.4.18.-

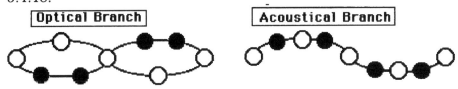

It is easily seen that, for the acoustical modes of lattice phonon vibration, both types of atoms vibrate together (the displacement is in the same direction - which may account for the lower energy required). In contrast, in the higher energy optical modes, each type of atom vibrates together, but in opposite direction to the other, i.e.- the displacement is opposite for each type of atom, the two types being denoted by black or white balls). Thus, we can conclude that in lattices having more than one type of atom, there are both **optical and acoustical** modes of both **longitudinal and transverse** vibration. We label these phonon vibrational modes in a given lattice as:

5.4.19.- QUANTIZED PHONON VIBRATIONAL MODES IN A LATTICE
           <u>Longitudinal</u>   <u>Transverse</u>

               L O   T O   - O = Optical

               L A   T A   - A = Acoustical

This brings us to the general case of phonon modes in a lattice. If there are "y" atoms **per primitive** cell, i.e.- y different kinds of vibrating **groups,** then there are:

5.4.20.-   <u>NUMBER OF BRANCHES OF PHONON DISPERSION</u>

           For y - atoms: Acoustical = 3

                             Optical   = 3 y - 3

Thus, for the case of three atoms present in the lattice, we would get the following number of phonon branches, as shown in 5.4.21. on the following page.

It becomes apparent that we can predict, and calculate the expected number of phonon modes present in a given lattice, by knowing the types and numbers of atoms present. We have now shown how phonons have been defined as quantized lattice vibrations. It should be clear that phonon energy is completely defined by Quantum Mechanics and that if we wish to add or subtract energy during a photon interaction (such as an excitation process in a phosphor), the amount that can be added is a direct function of the type of atoms (ions) comprising the solid and its structure.

5.4.21.-

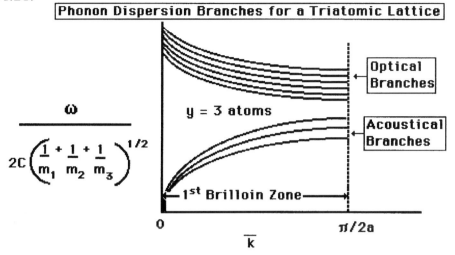

II. The Case of the Impurity Activator Center in a Phosphor

We can now show, in view of our discussion of phonon modes given above, how the phonon frequency around the activator center **differs** from that of the general phonon modes of the lattice. We already know that the activator center in an inorganic phosphor is a substitutional defect, located at a cation site in the lattice. In this case, we will have localized phonon modes because of the induced perturbation by the presence of the defect on the host lattice. What is not so apparent is that the **local** phonon modes around the activator site differ from those of the lattice phonon waves. This arises because the mass of the activator cation differs from those of the cations of the lattice. We use the same mathematical approach as given above, as follows:

5.4.22.- LOCAL PHONON MODES AROUND THE ACTIVATOR CENTER

		Coordinates	
lattice cation :	M	u_1 ,	u_2
activator :	M_0	u_0	

Equations :

$$M_0 \, d^2 u_0 / dt^2 = c \, [u_1 + u_{-1} - 2 \, u_0]$$
$$M \, d^2 u_1 / dt^2 = c \, [u_2 + u_0 - 2 \, u_{-1}]$$

Combining these, we get:

5.4.23.- $\omega^2 M(o) = \omega^2_{max} (M^2 / 2MM_0 - M_0^2)$

where ω_{max} is a cutoff frequency. In other words, there is a set of local phonon modes **surrounding** the impurity activator site which has an upper cutoff frequency.

5.5.- ELECTRONIC ASPECTS OF PHOSPHORS

We can, as a general rule, divide inorganic solids into 2 classes: those that absorb light (pigmented) and those that emit light (phosphors and solid state lasers). In both cases, we add controlled amounts of a transition metal ion to control the absorption and/or emission properties of that solid. At this point, it should be apparent that emission refers to:

1. Absorption of energy (whether by absorption of a photon **or other means such as energy from an electron**)
2. Reëmission of a photon (usually of lower energy, particularly if the original source of energy was a photon).

Thus, one view of a phosphor is that it is a "photon-converter". An inorganic phosphor consists of two parts:

(1) The HOST or inorganic compound
(2) The ACTIVATOR or the added transition metal cation

In general, the host needs to be transparent, or non- absorbing, to the radiation source used for the **"excitation"** process. The activator does just what the name implies, i.e.- it "activates" the host. This combination has several advantages, the most important being that **both the type and the amount** of activator can be precisely controlled. The details concerning those elements that can be used to form the host, and those which can be used as activators will be delayed until we reach the next chapter.

There are three terms we need to become familiar with, as follows:

5.5.1.- Terms Relating to Phosphors and Phosphor Emission

LUMINESCENCE - Absorption of energy with subsequent emission of light

FLUORESCENCE- Like luminescence, but visible emission

PHOSPHORESCENCE- Luminescence or fluorescence, but with delayed
emission of light

Luminescence is the general case in which a higher energy photon is absorbed and a lower energy photon is emitted (such a process is called a **Stokes** process). In this case, the excess energy is absorbed by the solid and appears as lattice vibrational (heat) energy. We shall use fluorescence and luminescence interchangeably, even though, strictly speaking, they are not equivalent. Phosphorescence involves a process where photon absorption occurs but the reëmission process is **delayed**. This delay may be a function of the type of transition metal employed, or a function of the action of solid state defects, including vacancies and the like which "trap" the energy for a time. This occurs most often in the sulfide phosphors.

Thus, we need to define a characteristic **time of decay** associated with luminescence and phosphorescence. In general, we find that the rate of decay is an exponential process. The types of decay that we find in phosphors includes the following:

5.5.2.- Decay Times Associated with Luminescent Processes

RANGE OF DECAY TIMES $(t = 1/e)$

Fluorescence	10^{-9} sec. to 10^{-3} sec.
Phosphorescence	10^{-3} sec. to 100.0 sec.

The decay time, t, is defined by convention as the time for **steady state** fluorescent intensity to decay to 0.3679 $(1/e)$ of its original value.

Note that there are some materials which glow for hours after being excited. This is not due to phosphorescence but to the presence of traps, i.e.- lattice defects, put there deliberately during their manufacture. Such phosphors, notably CaS activated by various cations including Bi^{3+}, are sometimes called "daylight" phosphors since they can be excited by visible light. Once activated, they will glow for hours.

We can now define a phosphor as a solid state material which emits in the **visible part** of the electromagnetic spectrum (this is not strictly true since some phosphors emit ultraviolet light and some emit infra-red light). Let us now examine the configurational aspects of phosphors and then the energy transfer mechanisms which occur once the phosphor is excited. Later in the following chapter, we will examine the solid state chemistry of phosphors.

I. Energy Processes in a Phosphor

Phosphors may be classified as organic (fluorescent organic dyes, as used in dye-lasers) and inorganic (as used in television or fluorescent lamps). We restrict ourselves to the latter for a very practical reason. Although organic fluorescent dyes have found wide spread usage, it remains indisputable that the specifics of the "activator" center have not yet been described, except in general terms of electronic structure of the organic compound. It will become obvious, in light of our later discussions, that the actual activator center in organic dyes must, of necessity, involve an electron or electrons, **isolated** from the vibrational- electronic (vibronic) structure of the organic molecule. Yet, the exact electronic structure of the center in this class of luminescent materials has not yet been established, even at this late date. We will address the approach used to describe organic phosphor excitation-emission mechanisms later in this discussion.

A phosphor is an energy converter. If a photon of 4.95 ev. (2500 Å or 40,000 cm^{-1} = ultraviolet light) is absorbed by a phosphor, then the phosphor might emit a photon of say 2.48 ev. (5000 Å = 20,000 cm^{-1} = green light). This is a so-called Stokes process where less energy is emitted than is absorbed. Note that there are also anti-Stokes processes which will be discussed in a following chapter. (The "extra" energy comes from lattice vibrations). The specific energy processes occurring in a phosphor will, of course, depend upon the exact composition of the phosphor.

The sequence of energy processes that occur within a phosphor include:

5.5.3.- <u>Energy Dissipation in a Phosphor</u>

    a. Absorption of energy (from a variety of sources)
    b. Excitation within the activator center to form an excited state
    c. Relaxation of the excited state (it is here that energy is lost to the vibrational states of the lattice)
    d. Emission of a lower energy photon from the excited state, and relaxation to the original (ground) state

In a phosphor, absorption of energy may occur in the host, or directly in the activator center (the added transition metal ion). But it is the activator center (or site) which becomes excited, whether by transfer of energy from the host, or not. That is, the activator center absorbs the energy and changes its electronic energy state from a ground state to an excited state.

There may be several excited states possible, but only **the lowest** excited state is involved in photon emission. If more than one excited state does occur, each excited state relaxes to the next lower energy excited state until the final excited state is reached, whereupon photon emission occurs in due time. Whether upper excited states are involved depends upon the original energy of the exciting photon. This energy may exceed the minimum energy needed for excitation and the excess energy causes the upper excitation states to become populated or even ionized. It is common to represent the energy states of a phosphor in terms of a potential energy diagram containing only the ground state and the lowest energy excited state, as follows:

5.5.4.-

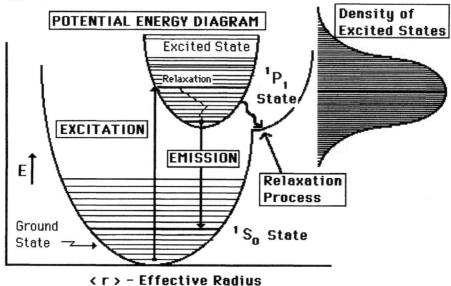

This is a representation for a **single activator center.** Herein, we use spectroscopic notation for the ground and lowest energy excited states (We will explain this later). Thus, the electronic energy (involving a one electron transition) is:

$$^1S_0 \Rightarrow {}^1P_1$$

Let us now interpret the above diagram so that we can understand the electronic processes which occur in a phosphor.

According to the Franck-Condon principle, a change in electronic state takes place in a time nearly instantaneous, as compared to other processes within the activator center, i.e.- 10^{-18} sec. (see 5.2.12. given above). In our diagram of 5.5.4., we have shown an isolated activator center within a given host lattice. Take note that there are several ground state energies, of sharply defined radius values (the horizontal lines, i.e.- <r>). The lowest ground state energy is the lowest energy **occupied state.** If we heat up the lattice, i.e.- increase its vibrational energy, then the other ground state levels will become occupied. Note that **their** limits of

<r> are wider than the lowest energy state. Each of these horizontal lines denotes a separate, so-called, Stark state. When sufficient energy is absorbed, then the center will become excited. Again, the values of <r> are sharply defined in the excited state, but **are different** than those of the ground state. This effect of vibrational modes of the host lattice upon the electronic states of the activator center is called **vibronic coupling,** and is caused by phonon wave propagation throughout the host lattice, and interaction at the activator center. (Phonon waves are **quantized** lattice vibrational waves present in all solids). Remember that unless the host lattice is at 0 °K, then vibrational waves are always present. Once the activator center has become excited, then it relaxes through the set of Stark states until it reaches the lowest energy excited-state from which emission occurs. Note that there is a range of excited states and that the final excited state can also relax, under certain conditions, directly to the ground state **without** emission of a photon. Then, all of the excitation energy is dissipated by vibrational processes throughout the lattice.

Probably the most important point to realize is that this diagram is a static one, whereas the reality is a dynamic one in which there is a large plurality of centers in various energy states which vibrate according to the effective values of <r> for that particular energy state each center finds itself in. This gives rise to a density of states whose occupation centers around the original excitation energy transition, as shown above, but whose occupation density is affected, in **the excited states,** by phonon (vibrational) waves of the host lattice. This variation in <r> is a consequence of host lattice vibrations which affect the relative positioning of the activator center. It is **essential** to realize that the variation in <r> arises from a random interaction of phonons **with the excited center.** That is, the density of states is Gaussian because the interaction perturbation (vibrational + electronic = vibronic coupling) is a random process.

Since the activator center is usually a substitutional defect in the lattice, its vibrational modes will differ markedly from those of the host(as shown below). If both host and activator center were cooled to 0 °K, then the electronic transition would be between a single Stark state of the ground state to a single Stark state of the excited state. This would result in essentially a single energy line (and is called the "zero-phonon" mode). However, vibronic coupling broadens the energy states both in the transition from the ground(absorption) and in the excited state (as shown by the variations which can occur in <r> at each Stark state in the diagram). This action results in a **broad** band both for excitation and for emission. We emphasize that the 4f emitting states of the rare earths are shielded from the perturbation effects of the lattice by the outer valence electron shells, so that **line** emission occurs rather than broad bands.

Note that in a broad band, each photon absorbed, or emitted, has an incremental difference of energy from the other photons, depending upon the vibrations, or phonon modes present during the **instantaneous moment** of electronic change. It is this instantaneous change, as secondarily affected by the slower phonon modes, which cause broad bands of absorption (and/or excitation) and those of emission. It is because phonon energies are quantized that they can add and subtract from the electronic energy process.

From the above description, it should be evident that the electronic excitation-emission transition is a dynamic process which is perturbed by vibronic coupling of the phonon spectrum present in the host lattice. Thus, the host is just as important as the activator center. Another way to describe the overall process is to state that the electronic transition in the activator center involves the zero-phonon line, broadened above absolute zero temperatures by quantized phonon interactions to form a band of **permissible** excitation and emission energies.

Now, what we have not said is that to have an efficient phosphor, we **must** have a luminescence center that is isolated from vibronic coupling from the host lattice. We will address this factor in some detail in the next chapter.

We can represent the energy transition in the activator center of 5.5.4. as:

5.5.5.- $\Delta E_{\text{excitation}} = \{ |^1P_1| \Rightarrow |^1S_0| \} \pm n \, \mathbf{h} \, \omega_1$

where $\mathbf{h} = h/2\pi$ (h is Planck's constant) , $|^1P_1|$ represents the virtual energy of the excited electronic state in the crystal lattice and $h\omega_1$ is the energy of a single phonon state (remember, there is a plurality of sharply defined phonon energies present, i.e.- the phonon spectrum of the lattice). Note that a relaxation occurs in the excited state. This results in a change of $<r>$ (actually how far the excited center can change in effective radius $<r>$), as compared to that of the ground state). This shift is a virtual one and is called the "Stokes Shift". Thus the effective position of the excited state has changed relative in energy to that of the ground state and results in a broad band for absorption (excitation in the case of phosphors) and a broad band for emission, as shown in the following diagram:

5.5.6.- Energy Bands Associated with Phosphors

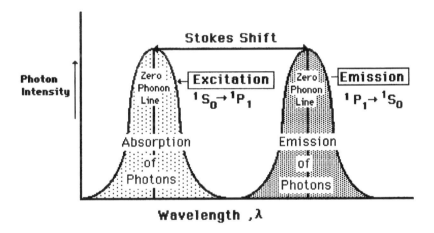

In this diagram, we have shown the absorption band and the emission band as being completely separated from each other for clarity. Usually, they are not. The zero phonon line is the same as the electronic transitions shown in 5.5.4., i.e.- the vertical arrows. Thus, the Stokes shift arises because of a change in $<r>$, as we have said, with the added $\mathbf{h}\omega_1$ phonon energy perturbations (vibronic coupling) which cause a broad band of absorption (excitation) and emission. Note that the rare-earth activators do not exhibit this behavior because the $4f^n$ emitting energy

Depending upon the energy of the electron, the penetration depth depends somewhat upon the constituent atoms of the host. Each electron recoils from collisions with atoms (ions) in the lattice, slowly losing energy with each ricochet until it can no longer lose further energy. Heavy atoms are ionized easier than the lighter elements. At the left, "A" shows how the electron depth is distributed when the energy of the electron is low. As the electron energy increases, the depth also increases until finally at greater depths, ionization of the constituent atoms begins to take place. This mechanism is shown on the right hand side of the diagram. The first excitations of the lattice are caused by the creation of optical phonons. These phonons then travel to an activator site and transfer their energy to the emitting center. At higher electron energies, plasmons (a collective excitation of valence electrons of the host lattice) and/or excitons (which are electron-hole pairs) are created.

Note that these mechanisms apply mostly to semi-conducting phosphors like ZnS:Ag and the like. The plasmon soon decays ($\sim 10^{-15}$ sec.) to an excited state electron in the conduction band. If ionization occurs, large numbers of secondary electrons appear in the conduction band. This accounts for the high cathode-ray efficiency of the sulfides in general. Most lamp phosphors are insulators and do not respond well to electron excitation.

When lamp phosphors are excited by ultraviolet **photon** energy, the photon energy may be dissipated by the lattice or directly in the center. Additionally, **phonon** energy may be **the sole source of excitation** in some phosphors. Actually, some phosphors are capable of energy storage, as by absorption of visible light photons, which is subsequently released by phonon energy. However, this description is only a general depiction of energy dissipation in phosphors and we need to lay a firmer basis of associative properties before we can proceed further.

II. Properties Associated with Phosphors

The following are specific aspects of phosphors we need to discuss before we can delve further into energy dissipation processes.

a. Notation

A language has grown up around the science of phosphors. Most phosphors consist of a host composition plus the activator, added in carefully controlled quantities. The activator itself is a **substitutional DEFECT** and is subject to lattice phonon perturbations. Therefore, it is essential that the charge on the substitutional cation is **equal** to that of the host lattice cations. **Otherwise, an efficient phosphor does not result**.

We denote a phosphor as: $M_aYO_b : N_x$, where M is the cation, YO_b is the anion (assuming we have an oxygen-dominated phosphor and not a fluoride or sulfide), and N is the activator. It is understood that the N-cation is in solid solution in the host-matrix, and the above formula is actually: $[(1-x)M_a)YO_b) \bullet xNYO_b]$. Thus for a tin-activated strontium pyrophosphate phosphor, we would write:

5.5.10.- $Sr_2P_2O_7 : Sn_{0.02}$ $= 0.99 \, Sr_2P_2O_7 \bullet 0.01 \, Sn_2P_2O_7$

In this case, the host cation is Sr^{2+} and the activator cation is Sn^{2+} . Thus, the first part is a short-hand notation for the latter.

However, suppose we have the phosphor: $Sr_2P_2O_7: Sb^{3+}_{0.02}$. In this case, the actual formula is considerably different than the short-hand formula. From our study of defect reactions (Chapter 2), we know that if a trivalent ion substitutes on a divalent ion, it must do so in conjunction with a substitutional defect, usually a vacancy. Thus the actual formula is:

5.5.11.- $0.99 \, Sr_2P_2O_7 \cdot 0.01 \, \{(Sb^{3+}, V^+)_2 \, P_2O_7$.

where both Sb^{3+} and V^+ are on Sr^{2+} lattice sites. If we added Na^+ for charge compensation, we would have:

5.5.12.- $0.99 \, Sr_2P_2O_7 \cdot 0.01 \, \{(Sb^{3+}, Na^+)_2 \, P_2O_7$

or $Sr_2P_2O_7 : Sb^{3+} : Na^+$.

However, it is well to note that Na^+ is **not** an activator in the strictest sense but is a member of the lattice structure, placed there to avoid formation of a lattice vacancy. If a vacancy is formed, it is likely that the resulting "phosphor" may be non-luminescent or will exhibit a low degree of luminescent efficiency. Although you will find that many phosphors like $ZnS:Ag^-$, Al^{3+} : K^+ are listed as requiring the K^+ activator, in reality this cation is present for charge compensation in the lattice. This fact has not been recognized by many prior workers in the field. Furthermore, some zinc (cadmium) sulfides have several listed "activators", part of which are undoubtedly there for charge compensation. Examples include:

$$ZnS: Ag^- :Al^{3+}$$
$$ZnS:Zn^0 :Al^{3+} :Cl^-.$$

b. Quantum Efficiency

We have already stated that proper choice of host and activator is essential to obtain an efficient phosphor. Now consider the case where one-hundred (100) photons are incident upon the phosphor. From our discussion given above, we know that a few are reflected, some are transmitted, and, if the phosphor is an efficient combination of host and activator, most of the quanta are absorbed. But not all the absorbed quanta will result in an activated center. and, once these centers become activated, not all will emit a subsequent photon. Some become deactivated via relaxation processes. To determine just how efficient a phosphor may be, we measure what is termed the "quantum efficiency", i.e.- QE. The quantum efficiency is defined as:

5.5.13.- $QE \equiv$ photons emitted/ photons absorbed

To obtain specific values, we measure total energy of emission and the total energy absorbed. It is easier to measure intensity of photons emitted as a function of wavelength. This gives us:

5.5.14.- $QE = \{(I\ d\lambda)_{emission}/ (Id\lambda)_{absorption}\}\{(1 - R)_{absorbed}$
$$(1 - R)_{reflected}\}$$

where I dλ is the integrated energy of the emission band or that of the absorption band and (1-R) is the integrated energy of the absorption or reflectance band. In this equation, we correct for reflection as well. Generally, phosphors which have QE's of 80%, or greater , are considered to be efficient phosphors. This subject will be discussed at greater length in the next chapter.

c. Decay Times

We have already briefly discussed luminescence decay times. Reiterating, the decay time of a phosphor has been defined as the time for the **steady state** luminescence intensity to decay to $1/e$, or 0.368, of its original intensity. It has been found that the intensity of photon emission builds up in the order of microseconds. i.e.- ~ 10^{-6} sec. to a specific value, i.e.- the excitation process takes only a few microseconds. Since the intensity also decays in microseconds (if the excitation source is removed), there is an equilibrium value attained in the presence of the excitation source, which is a combination of both excitation time and decay time. This so-called steady state is called I_0 , and is promulgated by the population of emitting sites in the phosphor which are simultaneously being excited and decaying back to the ground state (with emission occurring during the process).

Two types of decay are normally encountered. We have already mentioned exponential decay. The other is a logarithmic decay. It is essential to realize that both types of decay arise from a Gaussian array of emitters. That is, the energy states of the excited centers **and** the energy spectrum of the emission band are the consequence of a random process of phonon perturbations of the excited state of the activator center. Hence a random distribution of excited states results. The distribution is Gaussian and the decay times of this Gaussian array of **potential** emitters can be described via the binomial derivation, as shown in 5.5.15. on the next page:

This equation presents the change in numbers of excited states with time in relation to the decay time, τ , and is a Gaussian expression for a

5.5.15.- EXPONENTIAL DECAY

$$I(t) = -dn/dt = n / \tau \quad \text{and } n = n_0 \exp(-t/\tau)$$

population of emitters where n is the number of emitters, τ is the decay time and t is a specific time. I(t) is the intensity of photons emitted per second at that specific time. We can also make the substitution:

5.5.16.- $I(t) = [n_0 / \tau] \exp(-t/\tau)$

which may be more useful in some cases.

The other type of decay encountered is a logarithmic decay:

5.5.17.- $I(t) = -dn/dt = \alpha n^2 \quad \text{and } I(t) \approx 1/t^2$

The coefficient, α, has been determined to be a fractional constant. By combining as above, we get:

5.5.18.- $n(t) = n_0 / (n_0 \alpha t + 1) \quad \text{and } I(t) = n^2 / (n \alpha t + 1)$

Fortunately, we do not encounter logarithmic decays very often and sometimes it is difficult to differentiate between the two types. The only way we can really differentiate between the two is to obtain the decay curve and to subject it to mathematical analysis.

However, there is a very useful reason for determining the type of luminescence decay curve present. Confirming the presence of an exponential decay curve means that only **one type of emitter is present.** If a logarithmic decay process is found, it usually means that more than one type of emitting center is present, or that **two or more decay processes** are operative. While this does not occur very often, it is useful to know if such is present. This phenomenon occurs more in cathode-ray phosphors than in lamp phosphors, i.e.- sulfides vs: oxide- hosts.

d. Band Shapes

As we have said, the excitation process results in a density of states arising from a random process of phonon perturbation of the excited state, both before it relaxes and afterwards as well. It is this random formation of Gaussian energy states that give rise to a broad band in excitation, and to a broad band in emission. The zero-phonon line arises from the nature of the electronic transition taking place which is broadened by the vibronic coupling process. As the temperature rises, increased phonon branching results and the emission band is even further broadened, as follows:

5.5.19.-

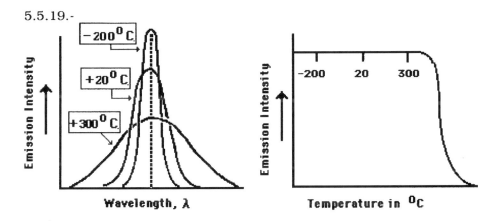

In the above diagram, the emission intensities of the three bands remain constant. It is the peak intensity which changes as the band broadens. To this point, we have accepted the fact that vibronic coupling leads to broad band excitation and emission in a phosphor. Take note that the above diagram is the result of experimental measurements **which prove** that as the temperature increases, the phonon spectrum becomes broadened, thereby leading to broadening of the bands. Thus, at - 200 °C., the number of phonon vibrations is restricted and a rather sharp emission band is seen. As the temperature rises, the number of separate phonon branches increases (the empty phonon levels become occupied) and the emission (excitation) band is further broadened. Note that at some temperature above 300 °C. (in this example), the phonon vibrations increase to the done as follows. The individual angular momentae are **vectorially** summed together, as are the individual spin momentae. Then, the sums are combined to give an overall value, namely:

5.5.22.- $L = \sum l$; $S = \sum s$; $J = L \pm S$

Russell-Saunders coupling is used because it agrees with experimental values **most of the time.** However, there are cases where it has been found better to sum the individual "j" values and then sum these to give the final value.

In Russell-Saunders coupling, sometimes called L-S coupling, the individual the spectroscopic terms are based on the L - values obtained (which are small whole numbers). The spin term is a resultant of 2S + 1. The central term is thus defined like this:

$$^{2S+1}L_J$$

L- values and terms are given in the following:

5.5.23.-   RUSSELL-SAUNDERS TERMS

L - Values	Term
0	S
1	P
2	D
3	F
4	G
etc.	

Actually, these terms originated with Balmer (1890) who first organized the spectra of the ionized hydrogen atom. He named each succeeding energy series as:

S = singlet; P = principle ; D = diffuse and F = fine.

5.6.1.-   Multipolar interaction
                 Exchange interactions
                 Phonon assisted energy transfer
                 Excitation diffusion
                 Sensitization of luminescence

We will address all of these in the following discussion.

We can summarize the material in Section 5.5. as follows. Essentially, we have treated a phosphor as a one-electron process. The sequences for changes in electronic states are:

5.6.2.-   <u>ELECTRONIC CHANGES</u>

         <u>ABSORPTION</u>   <u>EXCITATION</u>   <u>EMISSION</u>

For Sn^{2+}: hv_1 $^1S_0 \Rightarrow {}^1P_1$ $^1P_1 \Rightarrow {}^1S_0 + hv_2$

The energy dissipation changes associated with these electronic changes in the Sn^{2+} center are shown in the following:

5.6.3.- ENERGY DISSIPATION PROCESSES

$$E_{absorbed} = E_{original} - E_{reflected}$$

$$E_{excitation} = E_{absorbed} - E_{lattice\ absorption}$$

$$E_{lattice\ absorption} = E_{excitation\ relaxation} + E_{emission\ relaxation} + E_{excited\text{-}center\ quenching\ relaxation}$$

And the total energy involved is the sum total of these:

5.6.4.- $E_{original} = E_{emission} + E_{excitation} + E_{lattice\ absorption} + E_{reflected}$

where lattice absorption is all the relaxation processes which result in loss of energy. However, this equation does not include vibronic coupling where energy flow from the lattice to the center occurs. Now, we can

Einstein then proceeded to define an intensity of absorption as:

5.6.8.- $I_{absorption} = N_n \ B_{nm} \ \mathbf{h} \ \omega \ \Delta l \ p(\omega) \ d\omega$

where: $E = \mathbf{h}n$; Δl = pathlength ; ω = frequency of radiation;
         $p(\omega)$ = frequency density, and:

B_{nm} = Einstein coefficient of Absorption
B_{mn} = Einstein coefficient of Induced Emission
A_{mn} = Einstein coefficient of Spontaneous Emission
(from one pole to the other in the order given)

Thus, n is the ground state here while m is the upper energy state. Einstein noted **two types of emission** and had to define two (2) types of emission coefficients. These were delineated as two types of emission processes:

1) spontaneous
2) induced.

The former occurs in phosphors while the latter is the basis for **laser action.** From Beer's Law, we get:

5.6.9.- $I = I_0 \ \exp(-\varepsilon c l)$

and: $N_n = N_0 \ \exp(-E_n/kT)$;
                $N_m = N_0 \ \exp(-E_m/kT)$.

That is, the numbers of atoms in both the ground state and excited state is an exponential function of the energy of either. Combining these equations gives us:

5.6.10.- $N_n / N_m = \exp[-\{E_n - E_m\}/kT]$

As in 5.6.8., Einstein's definition of emission intensity for spontaneous emission is:

5.6.11.- $I_{emission} = N_n A_{mn} \hbar \omega c$

where c is a constant, i.e.- the velocity of light and $\hbar$ is $h/2\pi$. Using this, we can show that:

5.6.12.- $N_n B_{nm} p(\omega) = N_m B_{mn} p(\omega)$ - $N_n A_{mn}$

at a steady state condition of equilibrium. Note that these Einstein equations are in terms of a frequency density, $p(\omega)$. We now redefine the equations in terms of $p(v)$ which is a radiation density. This gives us:

5.6.13.- $p(v) = (N_n A_{mn})/[B_{nm}(N_m - N_n)]$

$= A_{mn}/B_{nm} \{\{exp[(-E_m + E_n / kT) -1]^{-1}\}$

Note that these two equations are equal to $p(v)$. They state that the radiation density $(N_n A_{mn})$ divided by the difference in numbers of excited and ground states is equal to the ratio of the emission coefficient divided by the absorption coefficient times the exponential value of the reciprocal of the differences of energy of the emission and excitation. In comparison, one form of Planck's Law is:

5.6.14.- $p(v) = (8\pi hv^3)/ c^3 \cdot 1/[exp[(\hbar v/kT) -1]$

Making this substitution gives us:

5.6.15.- $A_{mn} = (8\pi hv^3)/ c^3) B_{nm} = 8\pi \hbar \omega^2 B_{nm}$

Einstein's equations give us a way of obtaining values for the two types of coefficients which we can use to further develop other concepts regarding energy transfer in solids, particularly those processes associated with phosphors.

c. Electronic Transition Moments

Electronic transition moments are defined as the probability for a given excitation energy transition to take place. It should be evident that the transition moment depends upon the spin-orbit coupling of the electrons in both the ground and excited states. This mandates the use of wave-equation methods and quantum mechanics. At first, we shall restrict our discussion, while attempting to be lucid and concise. We will then present the details of the required calculations to be complete. You may wish to skip the latter part now but will have it readily available for future use.

1. Concise Hamiltonian Calculation of Transition Moments

By use of the Schroedinger equation, i.e.

5.6.16.- $H \Psi = E \Psi$

where H is the so-called Hamiltonian **operator** and E is the eigen (energy) states, we can solve for the wavefunctions, Ψ, in terms of specific time-dependent coefficients, a_k (t), as:

5.6.17.- $\Psi = \Sigma \, a_k \, (t) \, \Psi^0_k$

This allows us to define a **probability of emission** in terms of these coefficients as: $a_m^* \cdot a_m$, where a_m^* is a conjugate (see the details given in the next section). By suitable manipulation, where we solve for the energy of the eigenstates, $\mathcal{E}$:

5.6.18.- $\mathcal{E} = \int (\Psi^*_k \, |E_k| \, \Psi_k \, dT \,) \, / \, \int (\Psi_k{}^* \, \Psi_k \, dT \,) =$
$\qquad\qquad (a^*_k \, a_k \, E_k) / \, \Sigma \, (a^*_k \, a_k)$

This is the so-called Hartree-Fök method used in quantum mechanics, where T is a characteristic time as differentiated from t, an undefined time. By defining a dipole moment as $\mathbb{m} = \Sigma \, e_i \, \mathbb{x}_i$, where $\mathbb{x}_i$ is a unit vector in the x-direction, and using the electric and magnetic vectors, $\vec{F}$ and $\vec{M}$, already defined, we can get an equation relating B_{mn} and $\mathbb{m}$, as follows:.-

5.6.19.- $4\pi^2 \, / \, h^2c \, (\mathbb{m}_x)^2{}_{mn} \, (\vec{F}^0_x(\omega))^2 \, t \, = B_{mn} \, [6/4\pi \, (\vec{F}^0_x \, (\omega))^2 \, t$
and: $B_{mn} = 8\pi^2/3h^2 \, c \, (\mathbb{m}_x)^2{}_{mn}$

This gives us:

5.6.20- $A_{mn} = 8\pi\,\mathbf{h}c\omega^2\,B_{mn} = 64\pi^4\,w^3\,/3\mathbf{h}\,(\mathbf{m}_X)^2\,_{mn}$

Note that we now have **both** Einstein coefficients in terms of the **dipole moment**. If we can obtain a value for either, we can then calculate the dipole moment, or vice-versa. However, the most important part is the dipole transition moment which can also be written as an integral:

5.6.21.- $(\mathbf{m}_X)_{mn} = <\Psi^*_m\ |\ \Sigma\,e_i\ x_i\ |\Psi_n\ dT >$

Sometimes, this form is more useful.

<u>2. Detailed Calculation of Electronic Transition Moments</u>

We start with the one-dimensional, time-dependent, Schröedinger equation:

5.6.22.- $H\,\Psi = -(h/2\pi\,i)\,\partial\Psi/\partial t$

A "ket" (half of a total integral) for this Hamiltonian is $|L\ S\ J\ J'_z >$. It is well known that the Hamiltonian actually consists of two parts, the unperturbed part and the time-perturbation part, i.e.-

5.6.23.- $H^0\,\Psi + H'(t)\,\Psi = -(h/2\pi\,i)\,\partial\Psi/\partial t$

To solve for the wavefunction, Ψ, we obtain values in terms of specific coefficients, a_k , i.e.-

5.6.24.- $\Psi = \Sigma\,a_k\,(t)\,\Psi^0_k$

This involves the so-called Hartree-Fök, or Slater determinant, methods, named after the originators of the calculations. (Actually, Hartree-Fök methods use wavefunction integrals whereas Slater shortened the method of approximation by use of determinants). For the time-perturbed Hamiltonian, we get:

5.6.25.- $\Sigma \, a_k(t) \, \Psi^o_k \, H'(t) \, \Psi^o_k = -h/2\pi i \cdot \Sigma(d[a_k(t)]/dt) \, \Psi^o_k$

One of the rules in quantum mechanical calculations is that we must always multiply by the **conjugate** in order to maintain symmetry. In this case, the conjugate is: $Y^o_k{}^*$. We also define the **probability of emission** as: $a_m \cdot a_M{}^*$ (where the conjugate is also used). By combining and separating the equations, we get:

5.6.26.- $- d[a_k(t)]/dt = -2\pi \, i/\mathbf{h} \exp 2\pi \, i \, (E_m - E_n) \, (t/h) \int \Psi^*_m \, H'(t)\Psi_n \, dT$

where we have introduced the subscripts, m & n, as defined above, along with T , a characteristic time. As t approaches T, we get the following eigenstate solutions:

5.6.27.- $\mathcal{E} = [\int \Psi^*_k \, |E| \, \Psi_k \, dT \,]/ \, [\Psi^*_m \, \Psi_n \, dT \,] = \Sigma a^*_k \, a_k \, E_k \, / \, \Sigma \, a^*_k \, a_k$

We now define a dipole moment as:

5.6.28.- $\mathbf{m} = \Sigma \, e_i \, \mathbb{X}_i$

for a uni-dimensional operator, with $\mathbb{X}_i$ as a unit vector in the x- direction. Then:

5.6.29.- $\vec{F}$ = electric field vector = Σe_i

           $\vec{M}$ = magnetic field vector

such that the Hamiltonian:

5.6.30.- $H'(t) = \vec{F} \, (w, t) \, \mathbf{m}_x$ or: $(\mathbf{m}_x)_{kk} = \int <\Psi^*_k \, | \, \mathbf{m}_x \, | \, \Psi_k \, dT>$

In this case, $\mathbf{m}_x$ is a hermetian operator. We thus get:

5.6.31.- $p(\omega) \, d\omega = 1/8\pi \, [\, \{\vec{F}(\omega, t)\}^2 + \{ \vec{M} \, (\omega, t)^2\} \,] = \{ \vec{F} \, (\omega, t)^2 \} \, /4\pi$

where $\vec{F}$ is defined as:

energy absorption and energy transfer in a solid can be translated to fairly simple terms which we can realize and grasp.

We find that the intensity of emission is a function of the oscillator strength (for spontaneous emission). This leads us to multipolar selection rules, which are a useful way to classify electronic transitions and interactions. We can classify such transitions as: dipole - dipole (dd) and dipole-quadrupole (dq) interactions (see 5.6.6. given above).

The multipolar selection rules are shown in the following:

5.6.50.-   MULTIPOLAR SELECTION RULES

Electric	Rules	Intensity	Magnetic	Rules	Intensity
dd	$\Delta J = \pm 1$	1.0	dd	$\Delta J = \pm 1, 0$	0.1
	$(g \to u)$			$[(u \to u)$ or $(g \to g)]$	
dq		10^{-4}	dq		10^{-7}
qq		10^{-7}			

Note that we have normalized dd- transitions to 1.0, so as to show the relative intensities of other types of transitions. "g" refers to the German "gerade" = even, while "u" is ungerade = uneven states. Thus, the transition $^1S_0 \Rightarrow ^3P_1$ is a ΔJ transition of $0 \to 1$, or $(g \to u)$, which makes it an electric dipole transition. In general, most of the transitions that we encounter are dd or dq in nature. Once in a while, we encounter a magnetic dd transition, particularly in the case of certain rare earth transitions.

5.7. - MECHANISMS OF ENERGY TRANSFER IN SOLIDS

Considerable effort has been made in the past to clarify and quantify how energy is transferred between lattice sites within a solid. The literature is filled with specific examples in the area of non-radiative energy transfer. Sometimes, combinations of two, or more, separate mechanisms operate simultaneously. We are going to classify the various mechanisms by types, according to a general mechanism. You will note that we have already presented a listing of resonant energy transfer mechanisms among excited centers or states in 5.6.1. which include the general types of interactions and exchanges that occur in phosphors. We now will describe these in more detail as follows.

I. Radiative Transfer (Radiation Trapping)

Consider two (2) atoms in a solid, separated by a few unit cell distances and capable of emitting a photon. Let us suppose that one atom, N_1, absorbs enough energy to become excited. It will then emit a photon. This photon may then be captured by the other atom, N_2, which then becomes excited itself. The following diagram shows two possible cases of radiation transfer by trapping.

5.7.1.-

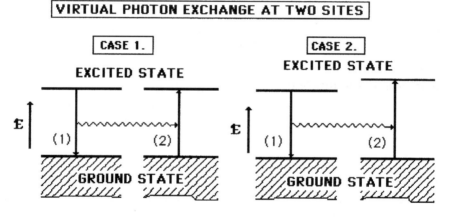

Note that we have used the term "Virtual". Virtual is used here in the context that the photon is ephemeral and only exists momentarily as the exchange is made. On the left side of this diagram, the energy levels are equal while those excited states on the right are not. The excess energy in this exchange is supplied by **phonon absorption** from the lattice. This process, radiation trapping or transfer, is common in a luminescent solid. It might be thought, therefore, that the quantum efficiency in phosphors could exceed 100%. However, it rarely, if ever, does. That is, if one has an assembly of radiators, the excitation energy from a single emitted photon **may** be passed from site to site until it reaches a non-radiative energy sink where the energy is dissipated as phonon energy to the lattice.

It is for this reason that quantum efficiency does not usually exceed 100%. There are, however, some phosphors which absorb a high energy photon and then emits **two photons.** However, the efficiency does not exceed 100%. Only 1 or 2 phosphors have been said to have an efficiency of 100% but in view of the above discussion, there is some doubt as to the accuracy of the measurements.

II. Energy Transfer by Resonance Exchange

By resonance transfer, we mean two radiators which are radiating at the same frequency. This is quite different from the example given above for radiative exchange. Note that we are not speaking of lattice vibrational frequencies but of radiation frequencies. We thus speak of two "coupled" oscillators. To further illustrate this concept, suppose we take two dipoles. there will be an **electric moment**, $\mathbb{E}(\theta)$, which is a vector product whose strength is a function of the angle, θ , between the vector moments of the dipoles:

5.7.2.- $\mathbb{E}(\theta) \;=\; (w^4 \; r^2 \; / \; 4\pi \, R_0 \; c^2) \; \sin^2 \theta$

where R_0 is a critical distance in the lattice. We can show this function to be proportional to the **degree of overlap** between energy states of the two different dipoles. To show how this arises, we define the following as given in 5.7.3. on the next page. Then, from equations developed before:

5.7.4.- $\mathcal{T}_{emission} \;=\; h \, c^3 \; / \; M \, \omega^3$ and: $U = \mathbb{m}_x^2 \; / \; R_0^3$

where $\mathcal{T}_{emission}$ is the characteristic lifetime of emission. If Site 1. becomes excited, and if its $\mathcal{T}_{emission}$ is sufficiently long, there can be a transfer of excitation energy to Site 2., with a characteristic transfer lifetime, $\mathcal{T}_{transfer}$.

5.7.3.- DEFINITIONS FOR THE CASE OF RESONANCE EXCHANGE

$$
\begin{array}{lll}
M_X & \equiv & \text{dipole moment} \\
R_O & \equiv & \text{critical distance} \\
U & \equiv & \text{interaction energy} \\
E & \equiv & \text{total energy} \\
\Omega_{ij} & \equiv & \text{overlap energy} \\
P & \equiv & \text{probability of energy transfer} \\
U_W & = & \text{specific energy transferred} \\
\lambda & = & 2\pi c / \omega \quad (\omega \text{ is a frequency})
\end{array}
$$

This lifetime is related to the other factors as:

5.7.5.- $\qquad \tau_{transfer} \cong E / U_W \cong h / U \cong \mathbf{h} R_O^3 / \mathbb{m}_X^2$

This gives us: $\quad P = U / \mathbf{h} \Omega \cdot [\Omega_1 / \Omega_2]$

We can immediately see that the probability of energy transfer is a direct function of the **overlap** of the energy states between the two sites involved. To further illustrate this phenomenon, consider two optically active sites in a lattice: $[1^* >$ and $[2 >$, where the star (*) indicates the excited site. We have already shown that the optical line (zero phonon transition between spectroscopic ground and excited states) will homogeneously broaden via vibronic coupling to produce an assemblage of excited oscillators within a specific band of energy as defined by:

5.7.6.- $\qquad E_{excitation} = \pm \mathbf{h} \Sigma n \, \omega_1$

where $\Sigma n \, \omega_1$ is a summation of all phonon modes of the lattice. In our case, we have the situation given in 5.7.7. presented on the next page.

This process is a resonance process between matched oscillators and is non-radiative in nature. One important consequence is sometimes called "sensitized luminescence". Note that $|1>$ has a higher energy than $|2>$ and that there is an overlap area of energies caused by vibronic coupling broadening of the zero-phonon transition.

5.7.7.-

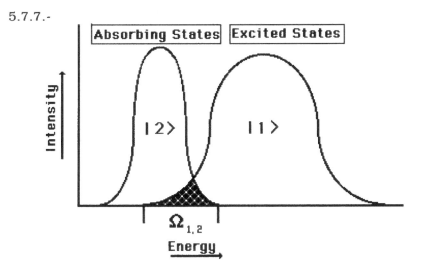

It is this overlap of energies which allows transfer of energy from one oscillator to another, even though they are separated by several lattice plane sites. The overall process may be described by:

5.7.8.- $T_{transfer} \cong (\Omega_{1,2})^2 \, \mathbf{h}^2 / U^2 \, \Omega_1 \cong (\Omega_{1,2})^2 \, \mathbf{h}^2 R_0^6 / \Omega_1 \, \mathbf{m}^4$

where $\Omega_{1,2}$ is the overlap between Ensemble 1 states, i.e.- $|1^* >$, ready to emit, and Ensemble 2 states { $|2 >$} ready to absorb. There is a **critical** radius between any excited state **site** in $|1^* >$, and the receiving site in $| 2 >$ which is described as:

5.7.9.- $R_0 \cong c / \omega \, (\Omega_1 / (\mathcal{T}_{emission} \, \Omega_{1,2})^{1/2} \sim 1 / 2\pi \, (\Omega_1 / (\mathcal{T}_{emission} \, [\Omega_{1,2}]^2)^{1/2}$

Actually, 5.7.7. only applies to dd-coupling processes which are resonant. We must also classify resonant energy transfer processes as defined in 5.7.10., presented on the next page.

We have already illustrated Direct Exchange Coupling. The dd-dq exchange is very similar, except that the intensity is lower because of the oscillator strengths. Virtual phonon exchange involves phonon wave coupling at the two sites where a part of the energy is transmitted

5.7.10.-   Coupling Processes in Phosphors

Electric Multipole Coupling, including dd-dq, i.e.- $Sb^{3+} : Mn^{2+}$

Magnetic Dipole Coupling

Direct Exchange Coupling (between like Poles)

Virtual Phonon Exchange Coupling

through lattice phonon waves. We must further **sub-divide** each type given in 5.7.10. into resonant (matched) and non-resonant (unmatched) energy coupling. The latter sub-division requires **additional energy** which is supplied by the localized phonon modes present at the receiving site. The overall process is therefore a direct (resonant) overlap of multipole fields or an indirect process where part of the excitation energy is supplied by phonon modes at the receiving site.

III. Energy Transfer by a Spatial Process

In this process, the energy exchange occurs over a rather long distance in the lattice, and is executed as an excited energy state carried by a phonon wave to the receiving site. This energy mechanism was first developed by Dexter (1962). If we define $\mathcal{W}$ as a rate of energy transfer, it can be shown that:

5.7.11.- $\mathcal{W} = 2\pi / h^2 \{< 1, 2^* | H_\Omega | | 1^*, 2 >\}^2$

where H_Ω is a Hamiltonian operator describing the overlap. Note that we did not go into detail concerning how the formula was derived. Dexter defined the excited ion (which we have called $| 1^* >$) as a sensitizer (S) and the absorber (A) ion (herein called $| 2 >$) as an activator. Thus, the probability of energy transfer would be, in this terminology, P_{SA} . Using appropriate values substituted into 5.7.11., we can obtain equations of probability of energy transfer, in terms of the type of coupling involved. This is shown in the following:

5.7.12.-

$$P_{SA}\,(dd)=[9\eta^4 c^4\,Q_A\theta\,/\,128\pi\mathbf{h}\;\mathcal{T}_S\;\mathcal{T}_A\;R^6{}_{SA}]\;[(e\,/\,[(k)^{1/2}\;\varepsilon_v]^4\;[\textstyle\int(f_S\,(E)dE\,/\,E^4)]$$

$$P_{SA}\,(dq)=[135\pi\;\eta^9\;c^8/4h^6\;\mathcal{T}_S\;\mathcal{T}_A\;R^8{}_{SA}]\;[g_A{}^*\,/g_A]\;[(e\,/\,[(k)^{1/2}\;\varepsilon]^4\;]$$
$$[\,(\textstyle\int(f_S\,(E)\;F_A\,(E)\;dE\,/\,E^4)\,/\;E^8]$$

These equations look formidable but can be solved rather easily since the variables in the two equations can be obtained experimentally. They are defined as follows:

η = refractive index of the host
k = dielectric constant of the host
e/e_v = electric field strength (host compared to vacuum)
$Q_A = \int \sigma_A\,(E)\;dE$ (σ_A is the optical cross-section of the activator)
$F_A\,(E)$ = a shape factor for the activator band,
where A is: $A\,(E) = Q_A\,F_A\,(E)$ and: $\int F_A(E)\;dE \equiv 1.0$.
q = orientation between poles
E = total energy
c = speed of light
g = magnetic coupling vector of ground state
g^* = magnetic coupling vector for excited state

By using appropriate values, we can reduce the equations in 5.7.12. to:

5.7.13.- $P_{SA}\,(dd) = (27/R_{SA})^6 \cdot 1/\,\mathcal{T}_S$

                 $P_{SA}\,(dq) = (14/R_A)^8 \cdot \;\; 1/\,\mathcal{T}_S$

The important part for us is the fact that we can deduce the **critical radius** beyond which energy transfer **does not occur**. In a normal lattice, this distance is approximately:

5.7.14.- $R_{SA}\,(dd) = 25.0$ Å   and: $R_{SA}\,(dq) = 13.4$ Å

As an example. we can use a simple cubic lattice such as NaCl, where a_0 = 2.814 Å. By defining a reduced concentration in terms of the total sites occupied, compared to total sites available, i.e.-

5.7.15.- $X_S = N_S / N_T$ and: $X_A = N_A / N_T$

we can obtain an estimate of the energy transfer efficiency of the two processes, but only if the average transfer lifetimes are approximately equal, i.e.-

5.7.16.- $\mathcal{T}_{SA}$ (dd)   ≈ $\mathcal{T}_{SA}$ (dq)

Using this assumption gives us:

5.7.17.- Effective Receptor Sites for Energy Transfer from an Excited
         Sensitizer Site

$\mathcal{T}_{SA}$ (dd)   = 0.5 X_S = 1 X_A = 2900 SITES

$\mathcal{T}_{SA}$ (dq)   = 0.5 X_S = 1 X_A = 100 SITES

There are 4 sites per unit cell in NaCl. For a given sensitizer site, there are about 24 nearest neighbor sites. Thus, a sensitizer can affect an activator site as far away as about 120 lattice planes. This is indeed a long distance for resonant energy transfer to occur. It is also evident that the number of activator sites to which energy transfer may take place from a single sensitizer site will vary according to the host, i.e.- the phonon spectrum of the host, the lifetime of the excited states of S and A, and the lifetime of the energy transfer process.

We can also express the same energy transfer mechanisms in terms of spherical harmonics and the oscillator strengths. The actual formulas and derivation have been given at the end of Section 5.6. Using spherical harmonics, we obtain quantities like those of 5.7.12. and 5.7.13.:

Even though there is no overlap of vibronically-coupled energies, energy transfer can occur as (dd), (dq) or $H_{exchange}$. through coupling of each state to the lattice via phonon exchange. This involves emission of a virtual phonon at Site 1 (which is in an excited state) to the lattice which is immediately absorbed at Site 2 , with simultaneous **spatial transfer** of resonant energy by an allowed multipole process.

In other words, **two mismatched multipoles** can readjust their energy states (within certain limits as determined by $\mathbf{h} \Sigma$ n ω_l , the energy of the phonon branches of the lattice) so that a resonant energy transfer process can occur. The mathematics are complicated so that we will describe the process simply as a modification of 5.7.11., wherein $\mathbf{h} \Sigma$ n ω_l been included:

5.7.30- $W = 2\pi/ \mathbf{h}^2 \{< n (\vec{k}, \vec{J}) + (\Delta , 1,2^*)| H_{eff.} | n (\vec{k}, \vec{J})(1^*,2) > \}^2$

$x \{ f_1 (\lambda) f_2 (\lambda) d\lambda \int g_1 (\omega_l) g_2 [\omega_l - \Delta\omega (\vec{k})] d \omega_l$

where $\Delta = \pm 1$, corresponding to the creation and/or destruction of a phonon and the other quantities have already been defined.

In addition, n (which is a function of allowed directions in the lattice), and $\vec{k}$ and $\vec{J}$ (which are the phonon coupling vectors) describe the number of phonons involved in the process. Note that what we have done is to include phonon processes as part of the energy transfer process integral. The actual evaluation of $H_{eff.}$ is very complicated and will not be further elucidated.

Suffice it to say that both $H_{eff.}$ (actually H_{dd} , H_{dq} as the case may be) and $H_{elec.-vibr.}$ (a Hamiltonian describing vibronic coupling) must be included. Nevertheless, this type of energy transfer involving phonon creation and destruction at two sites where the oscillators are not originally coupled has been observed experimentally and occurs in many cases where energy transfer between excited and unexcited centers occurs.

We can now summarize these processes as they relate to phosphors.

5.8.- SUMMARY OF PHONON PROCESSES AS RELATED TO PHOSPHORS

As a summary to this Chapter, let us summarize the phonon coupling processes which can occur during luminescence :

I. <u>Excitation of the Activator Center</u>

Absorption of a photon causes activator center to become excited, i.e.- the ground state is converted to the excited state by absorption of energy.

    a. <u>Phonon process:</u> Homogeneous broadening of zero-phonon line
                                             (a transition between Stark States)

    b. <u>Energy flow:</u>   Lattice to center, with phonon energy causing
                            secondary Stark states to become excited so
                            that a population of energy states exists, each
                            varying by the amount of energy present in
                            the phonon that was absorbed.

II. <u>Excited State Relaxation</u>
    The excited center relaxes to an equilibrium excited energy state:
        a. <u>Phonon process:</u> Description of energy results in a change
                                  of effective radius of the excited center.
        b. <u>Energy flow:</u>   Center to lattice via phonon energy
                                  exchange

III. <u>Center Ready to Emit:</u> Excitation energy can be dissipated in one of three (3) ways.
        a. <u>Direct relaxation</u>
            i. <u>Phonon process:</u> Direct conversion of excitation
                                energy to phonon energy
           ii. <u>Energy flow:</u>   Excited center to lattice
        b. <u>Emission of a photon</u>
            i. <u>Phonon process:</u> Relaxation of individual Stark ground
                                state to equilibrium ground state.
           ii. <u>Energy flow:</u>   Center to lattice

c. <u>Resonant and Non-Resonant Processes :</u>   This includes :
    i. Virtual photon exchange
    ii. Resonant energy exchange by overlap exchange
      between two coupled oscillators,
    iii. Electric multipole coupling,
    iv. Direct exchange coupling,
    v. Coupled multipole radiators,
    vi. Spin coupling,
    vii. Coupled multipole oscillators involving creation and
      destruction of phonons at the coupled sites, allowing the
      transition to take place.

d. <u>Actual Phonon Process:</u>   all of the above cases involve phonon interactions at both sites, including creation and destruction of phonons (the non- resonant case)
    i. <u>Energy Flow:</u>   #I - none;
                 #II - lattice to centers;
                 #III - lattice to centers where energy
                     is carried by phonon waves;
                 #IV - lattice to centers;
                 #V- lattice to centers;
                 #VI - none;
                 #VII- self evident.

What we can conclude is that the luminescence process involving a phosphor is indeed complicated and is a **dynamic process** in which the excited center undergoes many perturbations by the lattice during its excited lifetime.

Suggested Reading

1. F. Williams, Ed.- "Proc. International Conf. on Luminescence", N. Holland Publ., Amsterdam (1970).

2. F.A. Kröger, "Some Aspects of the Luminescence of Solids"- Elsevier Publ., Amsterdam, The Netherlands (1948).

3. "Phosphor Handbook", Edited by: S. Shionoya and W.M. Yen, pp. 921, CRC Press, Boca Raton, Fla. (1999).

CHAPTER 6

Design of Phosphors

In the last chapter, we explored the interaction of an electron or a photon with the lattice of a solid like a phosphor. We concluded that a photon is a "force-carrier" while an electron is a "matter-constituent". We also investigated how excitation energy from these two "forces" was absorbed, i.e.- transmitted to the lattice, and then dissipated. This mechanism involved either reëmission of a photon of lower energy or by phonon modes as heat. The modes by which energy was absorbed and then transmitted within the lattice of the solid included resonant energy transfer between various types of "centers" including "activators" and "energy traps" (the former emitting photons and the latter "heat" as low energy phonons). In this Chapter, we shall investigate the design of phosphors, methods of preparation of phosphors, the electronic devices in which they are used, a list of cathode-ray phosphors and a list of those used in lamps. We shall also explore ancillary areas such as color spaces and the specification of color. In order to elucidate the design of phosphors, particularly new phosphor compositions, we must return to our examination of the luminescent center and the energy processes associated with it.

6.1- THE LUMINESCENT CENTER IN INORGANIC PHOSPHORS

We have already discussed the energy processes which occur at the activator site during absorption, excitation, and emission transformations. Both interaction with a particle (electron) and a photon was explored. Briefly, they include phonon processes with vibronic coupling between electronic states of the activator and the phonon energies of the crystal host lattice. Following absorption and excitation at the activator site, phonon creation and destruction at the site populates adjacent Stark levels in the excited center which are further broadened by vibronic coupling. The isolated activator center can be thought of as interacting with its crystalline environment via the crystal field, which is time-varying because the lattice is vibrating. The effective radius of the excited state wavefunctions also changes as vibronic perturbation by the lattice continues. These two mechanisms account for the homogeneous broadening of the zero-phonon line, as discussed previously. Real time phonon energy flow processes continue to affect the activator center, both prior to, and following the emission of a photon. Since each photon emitted has a very slightly different energy, we end up with a **band** of emission energies. The emitted-photon energy differences are so small that no known instrumentation has sufficient resolution to differentiate between them.

Although we have previously shown in Chapter 5 that the excitation transition for Sn^{2+} is: $^1S_0 \Rightarrow {}^1P_1$, the actual mechanism is, in reality, more complicated than originally presented. The precise process has been found to be:

6.1.1.- $h\nu_1 + {}^1S_0 \Rightarrow {}^1P_1 - (h\,\omega_l\,) \Rightarrow {}^3P_1 \Rightarrow {}^1S_0 + h\nu_2$

where $h\,\omega_l$ is a relaxation involving phonon energy. The ground state for Sn^{2+} is $d^{10}s^2$ ($^1S_0$). When this activator cation absorbs photon energy, the $d^{10}sp$ excited state is split by the crystal field into a singlet state, $^1P_1$, and a triplet state, $^3P_1$, of lower energy. Each excited state has $2J + 1$ Stark states, for a total of seven. The potential energy diagram of the excitation-emission mechanism is shown in 6.1.2., given on the next page.

The excitation transition is shown as: $^1S_0 \Rightarrow {}^1P_1$. It is this state which relaxes to the $^3P_1$ state from which emission occurs. However, if we measure the excitation band for several different Sn^{2+} activated phosphors, we usually will obtain two closely connected bands. There are two possibilities. One is that the excitation transition also involves: $^1S_0 \Rightarrow {}^3P_1$, a transition which is "not allowed" since $\Delta S = 2$. The other is that they are crystal field states.

If we compare the excitation bands of various Sn^{2+}- activated phosphors, i.e. - those having various host crystals, we find that activated phosphors, the excitation bands vary significantly, but the high energy band, i.e.- the $^1S_0 \Rightarrow {}^1P_1$ transition, seems to predominate.

6.1.2.-

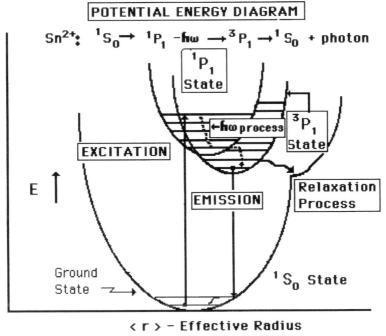

POTENTIAL ENERGY DIAGRAM

Sn^{2+}: $^1S_0 \rightarrow {}^1P_1$ $-\hbar\omega \rightarrow {}^3P_1 \rightarrow {}^1S_0$ + photon

Because the two states are so close in peak energy, one finds it difficult to delineate the actual excitation transition.

Now consider a different type of activator cation such as Cr^{3+}. Suppose we use Al_2O_3: Cr^{3+} as the example. We do this to contrast:

    1). atomic energy level schemes with crystal field splitting

    2). potential energy diagrams

with the intent of comparing the relative merits of these methods of displaying the energy levels involved in absorption, excitation and emission of phosphors.

The energy diagram for Cr^{3+} in Al_2O_3 is given as 6.1.3. on the next page.

6.1.3.-

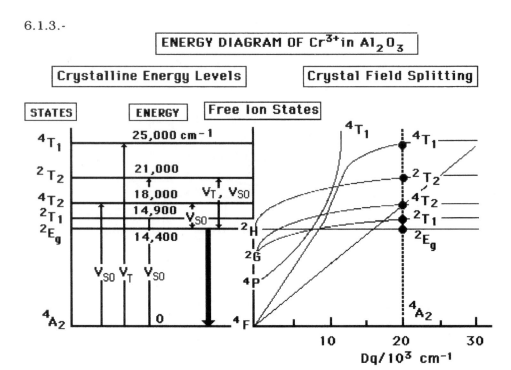

Whereas in Chapter 5, we presented only the energy levels and states of Cr^{3+} in the Al_2O_3 crystal, now we illustrate all of the factors involved, namely the effect of the crystalline field upon the actual states which appear. The electrons of the Cr^{3+} ion are $3d^3$. The free ion states are:

$$^4F, \, ^4P, \, ^2G \text{ and } ^2H.$$

Under the influence of the crystal field, these states split, as shown for a trigonal field, into crystal field components. The crystal field strength of Al_2O_3 is 20,000 cm^{-1} as shown, and this accounts for the exact energies which appear for the excitation transitions observed. We have already discussed these in light of spin-coupling and trigonal field transitions. Cr^{3+} does exhibit an emission transition which is: $^2E_g \Rightarrow {}^4A_{2g}$. However, the QE efficiency is < 5%.

The major point of this discussion is that it is necessary to specify the

$Si_4O_{11}^{5}$, etc. You will note that we have specified oxygen-dominated anions, involving the subgroups:

IIIA, IVA, VA, VIA & VIIA

It is the valence state of the central ion which is important. The actual formula for these anions, that is - the ratio of oxygen atoms to the central ion, will encompass many anionic forms and structures of the final crystalline host. Nevertheless, the valence state of the central ion remains constant.

There is one other class of anions which are important. These are the so-called "self-activated" anions. In this case, if we combine certain of the optically-inert cations of 6.3.1. with those of the following diagram, we obtain a phosphor which does not require an activator cation:

6.3.3.-

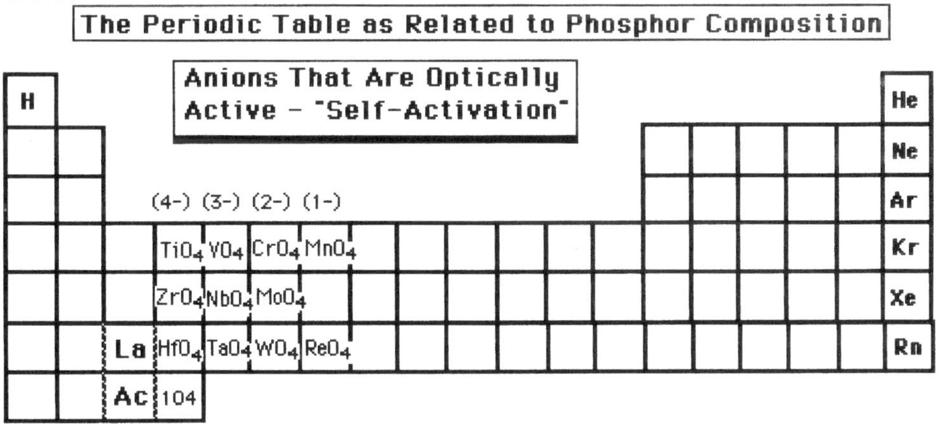

The Periodic Table as Related to Phosphor Composition																	
H		Anions That Are Optically Active – "Self-Activation"															He
																	Ne
		(4-) (3-) (2-) (1-)															Ar
		TiO_4	VO_4	CrO_4	MnO_4												Kr
		ZrO_4	NbO_4	MoO_4													Xe
		La	HfO_4	TaO_4	WO_4	ReO_4											Rn
		Ac	104														

In the above diagram, the **required valence** of the anions is shown at the top of each column. Note that these are all oxygen-dominated anions. The subgroups involved are:

IVB, VB & VIB

In 6.3.3., each of these optically active anions can be combined with the optically-inert cations of 6.3.1. to form the host crystal. The anions of 6.3.3. are optically active because of an allowed transition within **the group orbitals,** called a "charge-transfer" transition. It is akin to molecular orbitals of organic phosphors where the excited state involves a shift in electron-density toward the central cation. Thus, $CaWO_4$ is a phosphor as is YVO_4. We can clarify this point by examining the molecular energy levels of the VO_4^{3-} group as shown in the following diagram:

6.3.4.-

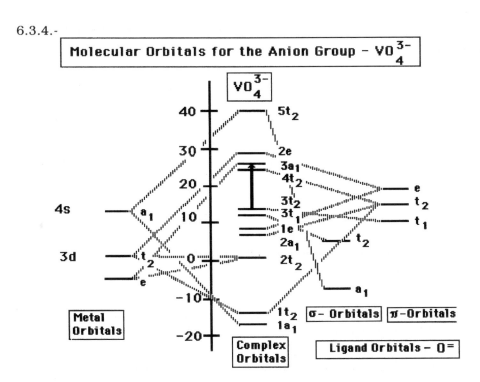

The VO_4^{3-} anion group is tetrahedral and the V^{5+} cation is surrounded by four (4) oxygen atoms which are bound by covalent bonds to the central metal ion. The electronic configuration of the vanadate group is made up of metal orbitals, shown at the left, and ligand orbitals of the oxygens, shown at the right. The excitation transition of the complex grouping occurs from $^3t_2$, an orbital derived primarily from p- electron orbitals of oxygen to $^3a_1$, an orbital derived mostly from 3d electron orbitals of the metal, V^{5+}. Thus, the excited state consists of an increased electron density in the vicinity of the metal ion, along the tetrahedral bonds. For this reason, the mechanism is called a "charge-transfer" transition. This is an allowed transition and the oscillator strength is extremely high, i.e.- in the high thousands, as contrasted to the low hundreds of "ordinary" activators.

Following relaxation (the Stoke's shift), emission occurs, from the transition: $^4t_2 \Rightarrow {}^3t_2$, a dipole-allowed transition. Because these transitions involve molecular orbitals and because vibronic coupling is also present, broad bands of high efficiency are observed both in absorption and emission. This completes the list of possible cations and anions which can be chosen to form a phosphor having high quantum efficiency when combined with a suitable activator.

However, there is also one other class of host which we have not yet addressed. This class comprises those anions which are **not oxygen-dominated.** Thus, we find that phosphors can be formed from:

I-VII compounds, II-VI compounds, & III-IV compounds

The following table lists some of these types of compounds, a phosphor example and physical properties.

Table 6-1

Type	Example & Emission	Physical Properties
I A - VII A	KCl: Tl^+ = near UV	Insulator
II A - VI A	CaS: Mn^{2+} = yellow	Insulator
II B - VI A	ZnS: Mn^{2+} = orange	Semi-conductor
III A- V A	GaP, GaAs, InP	Scmi conductor

Note that when we combine subgroup "A" cations & anions, we obtain a compound which is an insulator (as illustrated by KCl and CaS). But if we change the cation to one from the IIB group, we get a semi-conductor, ZnS. This illustrates the requirement of carefully choosing the proper cation for our phosphor, depending upon the application for which we envisage its usage. It is well to note that the IIIA-VA class of phosphors are not used as crystallite powders like those based on ZnS and the like. They are utilized for many optoelectronic devices such as light-emitting diodes, semi-conducting lasers and photo-diodes in the form of crystalline thin-films. We will address such devices based on these compositions, i.e.- GaP, in more detail in the next chapter.

II. Choice of the Activator(s)

Now we come, in our design of phosphors, to the choice of suitable activators. Because of the limitations imposed by the **ground state perturbation factor**, we are limited to certain cations **in specific valence states.** As we have stated before, this limits us to the electron configuration: $d^{10} s^2$ ($^1 S_0$). The following diagram shows these choices:

6.3.5.-

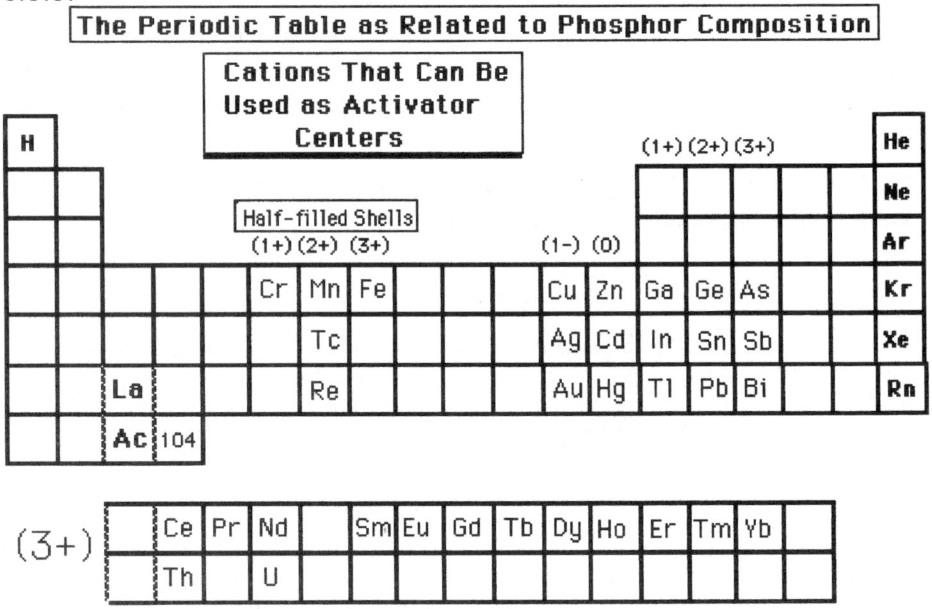

Also shown are the cations having a half-filled shell, and their valence state. We have already pointed out that this type of activator does not show strong absorption, but does show strong and efficient emission once it becomes excited. It is for this reason that some authors have called them "activators" and the others shown at the right-hand side of 6.3.5. "sensitizers". In any case, these "half-filled shell" activators cannot be used singly, but **must** always be employed **in combination with** one of the sensitizer cations. Even though we have listed them, there is considerable doubt that the ions, Tc^{2+} and Re^{2+} can actually be stabilized in a selected host lattice. The same may be said for Cr^{+}. In spite of this, if any of these cations could be stabilized in a host lattice in the valence state shown above, they would function as efficient activators. There are other ions having this same electronic configuration, but have not been presented because of the perceived difficulty of stabilization of the required valence state, regardless of the host crystal. The cations, Mn^{2+} and Fe^{3+}, are well known states, produce efficient luminescence, and have high QE's when incorporated into the proper phosphor host. Note that Fe^{3+} is not used generally as an activator because it is easily reduced to the ferrous state, which is an efficient "killer" of luminescence.

One important point needs to be emphasized. All of the indicated valence states in 6.3.5. are stable states of that element (except as noted). One could probably use Se^{4+} as an activator, **if** one could obtain that state in a crystalline environment. Unfortunately, one cannot. The selenates, as anions, are the stable forms of the selenium ion, and the uncombined form has yet to be observed in a crystal. This brings us to a critical proposition, namely:

IN CHOOSING AN ACTIVATOR, ONE MUST BE ABLE TO FORM AND STABILIZE THE ACTIVATOR IN THE PROPER VALENCE STATE WITHIN THE HOST CRYSTAL IN THE REQUIRED ELECTRONIC CONFIGURATION **IF** EFFICIENT LUMINESCENCE IS TO RESULT.

Let us further illustrate rules for designing phosphors with yet another example. Suppose we choose Mg^{2+} as the cation, SiO_4^{4-} as the anion and Sn^{2+} as the activator. The crystal, Mg_2SiO_4 is well known and could serve as the host composition. In addition, Sn_2SiO_4 is also a known form. Thus, if we attempt to combine them, either in the form of oxides, e.g.- $MgO + SnO + SiO_2$, or directly as the compounds themselves, a question remains as to the success of the combination. We have already said that in order for such combinations of compounds to occur, a solid solution must result.

According to Vegard's Law, in order to form a solid solution, the cations and/or anions of two compounds must have ionic radii **within ± 15%** of each other. The rule is not so rigid if the solid solution content is kept within a limit of no more than 5% of one in the other. That is: In the composition, $(1- x) MSiO_4 \cdot x NSiO_4$, if x is kept below 5% , then a solid state solution will form. Since most activator concentrations are usually no more than 5%, solid solutions can be formed for many of the possible combinations already shown in 6.3.1., 6.3.2. & 6.3.5.

It is for this reason that the phosphor, Mg_2SiO_4: $Sn_{0.02}$ can be obtained, even though the ionic radius of Mg^{2+} = 0.65 Å and that of Sn^{2+} = 1.12 Å . However, we find that the efficiency of this phosphor is rather low and the QE is low as well. The rationale for this is that a cationic radius mismatch has resulted in lattice strain at the activator site and increased loss of excitation energy due to increased vibronic coupling at the site. Conversely, if we choose Sr^{2+} whose cationic radius is 1.13 Å, the so-produced phosphor, Sr_2SiO_4: $Sn_{0.02}$, exhibits high energy efficiency and has a high QE as well.

Therefore, it is apparent that there are other rules for phosphor design that need to be elucidated:

> a. The activator cation and the host cation need to be matched in size so as to obtain maximal efficiency in the produced phosphor. Mismatch creates strain in the lattice, and limits the actual solubility of the activator in the host lattice.

> b. The activator cation needs to be of the same valence state as that of the host cation. Otherwise, the activator will substitute into the lattice with the formation of cation vacancies (for charge-compensation). An alternative is the introduction of a compensating cation for the same purpose.

One other point involves the defect solid state reaction behavior of certain selected hosts. One might wonder how one would stabilize an activator cation in the **metallic** state, e.g.- Zn^0 . If one calcines, i.e.- "fires" at an elevated temperature, the compound, ZnO, in a reducing atmosphere, the following defect reactions take place:

6.3.6.- DEFECT REACTIONS OF ZnO IN A REDUCING ATMOSPHERE

1. $ZnO + x\,H_2 = Zn(O_{1-x}, x\,V_0) + x\,H_2O + (2\,e^-)_x + x\,Zn^{2+}_i$
2. $x\,Zn^{2+}_i + (2\,e^-)_x = x\,Zn^0$
3. $x\,Zn^0 + x\,V_0^= = x\,Zn^0{}_0^=$
4. $ZnO_{1-x} + x\,Zn^0{}_0^= = ZnO: x\,Zn^0{}_0^=$

Note that the defect formation behavior is one where the hydrogen gas reduces the host lattice and forms an oxygen vacancy **plus two free electrons.** These electrons then reduce the **interstitial** Zn^{2+} cation (which became an interstitial cation in order to satisfy the charge compensation mechanism when the oxygen anion was removed by the reducing atmosphere to form water) to the metal valence state. Zn^0 · This atom is too large to remain as an interstitial, and therefore occupies the oxygen vacancy already created. Note that the radius of Zn^0 is 1.48 Å while that of the oxygen vacancy is 1.50 Å. This is one of the few cases where the cation in its metallic form functions as an activator. The semi-conducting efficiency of this phosphor is high, as is its QE.

The hosts that we have discussed to date have been ionic in nature, and dielectric as well (they are non-conductive). There is also another class of phosphor hosts which are covalent and semi-conductive in nature, namely the zinc and cadmium sulfides and/or selenides. The criterion for selection of a semi-conducting host for use as a phosphor includes choice of a composition with an energy band gap of at least 3.00 ev. This mandates the use of an optically inactive cation, combined with sulfide, selenide and possibly telluride. The oxygen-dominated groupings such as phosphate, or silicate or arsenate, etc. are **not** semi-conductive in nature. And, **none** of the other transition metal sulfides have band gaps sufficiently large to be used as phosphors. We have already shown energy band diagrams in Chapter 5.

We find that the ZnS and ZnSe host structures are excellent semi-conductors and exhibit very high efficiencies of photoluminescence and cathodoluminescence, especially when activated by copper, silver and gold. There is a large body of prior literature wherein the valence state of these activators is considered to be: Cu^+ , Ag^+ and Au^+. These are known stable states of these elements. The electron configuration of the elemental state is: $d^{10}s^1$. Note that the d^{10} electron configuration would be optically inert. Therefore, it is likely that the monovalent ion has the ground state of:

6.3.7.- $Ag^+ = d^9 s^1 = {}^2S_{1/2}$

However, in our scheme of activator ground states, this electron configuration would be subject to substantial ground state perturbation, particularly when used in dielectric host crystals. Indeed, they do not function very well as activators when incorporated into such hosts. However, they do function well in the semi-conducting hosts, ZnS (CdS) and ZnSe. The lowest energy excited state for this electron configuration is: ${}^4D_{1/2}$ but the transition is **not dipole allowed.**

A much more plausible explanation considers the silver activator to have the Ag^- electronic configuration. The ground state would then be:

6.3.8.- $Ag^- = d^{10}s^2 = {}^1S_0$

But the Ag^- center would have to substitute on the sulfide site, namely:

6.3.8.- $ZnS: Ag^-_S=$

where the same mechanism as already given for the formation of the ZnO: Zn^0 phosphor applies. But, this does not seem likely due to the mis-match in ionic sizes. Indeed, the ZnS phosphors are "self-activated", and exhibit very high luminescence efficiency under both photoexcitation and cathode-ray excitation. The band gap of ZnS (or fundamental absorption edge as it is sometimes called) is about 3.71 ev (or 3350 Å). Silver introduces an energy level at about 0.31 ev in the band gap (3.40 ev below the conduction level). The "self-activated" phosphor likely has the formula:

6.3.10.- $\qquad$ ZnS : x Zn^0s=

In both of the above cases, the ionic radii of the activator states and the sulfide site are:

6.3.11- $\qquad$ $S^=$ = 1.84 Å
Zn^0 = 1.48 Å
Ag^- = 1.62 Å (estimated)

Thus, the states for these activators as given in 6.3.5. and discussed herein would conform to our rules for functioning efficient activators. However, it is well to note that the mechanism given in 6.3.8. has a **negative monovalent** ion substituting at a negative **divalent** site. This would imply that an additional negative charge is required for charge compensation in the crystal. The emission band of silver occurs at 4500 Å (2.76 ev). The "self-activated" ZnS phosphor emits at about 4450 Å. Therefore, delineation between these two types of sites is nearly impossible. Furthermore, the probability is high that the "self-activated" center forms first in the crystal and that the Ag^- center forms after that. The best explanation for the formation of silver centers seems to be that when a sulfide atom is lost, as in 6.3.6., a corresponding zinc atom becomes interstitial. This would give rise to the following defect reactions, namely-

6.3.12.- Defect Reactions for Formation of the ZnS:Ag Phosphor

1. $ZnS + xH_2 = Zn(S_{1-x} , x\ V_s) + x\ H_2S + (2e^-)_x + Zn^{2+}_i$
2. $Ag^+ + 2e^- = Ag^-$
3. $Zn^{2+}_i + e^-_{lattice} = Zn^+_i$
4. $Zn(S_{1-x} , x\ V_s) = (Zn,Zn^+_i)\ S_{1-x} : Ag^-s$=

However, the above explanation applies to phosphors used in the past in cathode-ray tubes.

The current phosphors being used in color-television tubes are:

$$Blue = ZnS:Ag^-:Al^{3+}$$
$$Green = ZnS:Cu^-:Al^{3+}$$
$$Red = Y_2O_2S:Eu^{3+}$$

Note that Al^{3+} is listed as an "activator". **It is not.** It has been found that if Al^{3+} is incorporated within the ZnS lattice, the resulting phosphor retains its original brightness over longer periods of operation of the television tube and is less subject to degradation. That is, it does not "discolor" over the lifetime of the operation of the display. It should now be obvious that the combination of $Al^{3+}{}_{Zn}{}^{2+}$ and $Ag^-{}_{S=}$ represents a pair of charge-compensated sites which conform to our rules for defect formation in a divalent lattice. It is clear that, after many years of experimentation, the best and most stable ZnS phosphor has been shown to be one where the lattice is free from inherent defects. This was accomplished by substitution of a trivalent cation along with a negative luminescent cation ($Al^{3+}{}_{Zn}{}^{2+}$ and $Ag^-{}_{S=}$) in neighboring divalent sites.

III. Quenchers, or "Killers" of Luminescence

Cations, having unpaired electron spins, function as quenchers of luminescence. Some of these are shown in the following diagram, given as 6.3.13. on the next page.

While this may seem incongruous to some (because we have already cited sub-group IB as an activator-group), it is the actual valence state that is important. Usually, their electronic configuration involves d- electrons as well. Note that certain of these cations are optically active in the **proper valence state.** In the wrong valence state, however, they function as quenchers.

6.3.13.-

The Periodic Table as Related to Phosphor Composition

Cations with Unpaired Spins Which Function as Quenchers of Luminescence

H																	He
																	Ne
			(3+)	(3+)	(3+)		(2+)	(2+)									Ar
			Ti	V	Cr	Fe^{2+}	Co^{2+}	Ni	Cu								Kr
			Zr	Nb	Mo	Ru^{3+}	Rh	Pd									Xe
	La		Hf	Ta	W	Re^{4+}	Os^{4+}	Ir^{4+}	Pt								Rn
	Ac	104															

These encompass the periodic table sub-groups:

IIIB, IVB, VB, VIB, VIIIB & IB

For example, in ZnS:Cu, if the Cu-cation is not stabilized in the Cu⁻ state, it acts as a killer of luminescence. This is particularly true in oxygen-dominated phosphor compositions. This insight has important consequences since the method of preparation thus becomes important. If these cations become stabilized in the wrong valence state, they become energy traps and dissipaters of excitation energy.

These cations are essentially acceptors, even for resonant energy (dq) transfer. Since they cannot undergo an excitation transition because of ground state coupling to the local phonon modes, they function conversely to the luminescence process. Many of them exhibit strong absorption bands at the frequencies of light commonly used for excitation of phosphors. Therefore, excitation energy is dissipated to the lattice by phonon processes, once it has been captured by this type of site.

An important consequence to this observation is given in 6.3.14., presented on the next page.

following description. If one prepares both $CaWO_4 : W$, and $BaWO_4 : W$ (where the W implies self-activation, or emission from the tungstate group), we find that the former composition is a very efficient ("bright") phosphor where the latter is essentially inert (at room temperature).

The reason for this is shown in the following diagram:

6.4.8.-

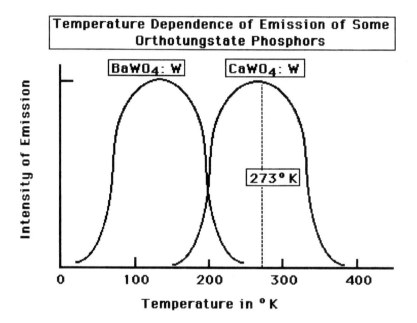

It is easily seen that while the $CaWO_4$:W phosphor has its maximum efficiency of emission close to room temperature, the $BaWO_4$:W is most efficient close to liquid nitrogen temperature. Therefore, the former is nearly inert at the temperature of maximum emission of the latter and vice-versa. This phenomenon is quite common for many phosphor systems, particularly the self-activated ones.

By measuring the temperature dependence of emission, one can estimate E_Q for "temperature quenching". That is, one can measure the temperature at which emission ceases, or is lost entirely, due to vibronic coupling to give an estimate of E_Q , the activation energy for **cross-over** (relaxation and dissipation of energy to the lattice) from excited to ground state.. One of the first cases where this theory was applied by Johnson and Williams (1952) to KCl:Tl, as shown in the following diagram:

6..4.10.-

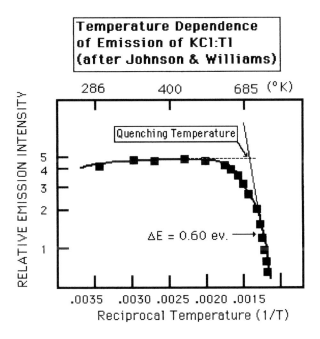

Here, ΔE is our E_Q of 6.4.5. and is obtained from the slope of the line. The so-called "quenching temperature" is obtained from the intersection of the two lines, as shown. It should be clear, then, that temperature quenching of emission will vary upon the nature of the phosphor, the activator and the host lattice. Those phosphors involving "optically-active" cations will be expected to possess lower quenching temperatures whereas those involving "shielded" energy levels like the rare earths will have very high quenching temperatures.

For example, $Y_2O_3 : Eu_{0.06}$ has a maximum brightness output at about 750 - 780 °C (1053 °K) and the emission is clearly visible above the dull-red heat of the furnace. Yet, the phosphor still glows brightly at 1100 °C., even though its emission is effectively masked by the full radiation of the furnace.

By obtaining H and ϕ experimentally, we can then calculate what the absorption intensity (oscillator strength) ought to be.

There are two other factors which affect phosphor efficiencies. One we have already discussed, that of high purity. Reiterating, we must use high purity materials to prepare phosphors because the presence of very small amounts of impurities are killers of luminescent efficiency (energy dissipation traps). The other factor involves "optimum" activator concentrations actually present in the host lattice.

We have already stated that a general formula for a phosphor is:

6.4.18.- $M_{1-x} XO_4 : A_x$

where M is the cation, XO_4 is the anion and A is the activator cation. It has been determined that the "optimum" activator concentration should lie between about:

$$5 \times 10^{-4} < x < 0.20 \text{ mols per mol of phosphor.}$$

There is always an "optimum" activator concentration, but the exact number of sites will vary from host to host. That is, the optimum in one host, being a function of the plurality of localized phonon modes present in the lattice, will not be the same in a different host lattice.

The reason for this is that the emission intensity, $I_{emission}$, is directly proportional to the number of activator centers present, N_A, and is inversely proportional to (an exponential function) the number of quenching centers, N_Q. These quenching centers may be nearest neighbor activator sites which dissipate energy by direct exchange. or, they may be distant sites accessible by resonant exchange mechanisms. What we find is that **if the activator concentration becomes too high**, a loss in energy efficiency (QE as well) is experienced.

We can illustrate this effect as follows in 6.4.18., given on the next page.

6.4.18.-

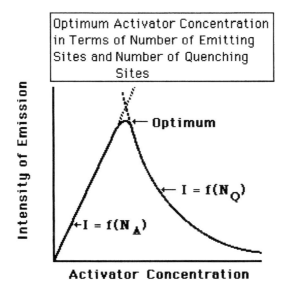

Initially, the emission intensity is linearly proportional to the number of emitting activator sites, N_A , present, and dI/dx is a constant value. At some point, the density of these sites becomes too large and the overall emission intensity falls, according to an exponential rate, as concentration increases. The reason for this is that the original nearest neighbors become quenching sites because of direct exchange coupling, or as given in 5.5.1. of Chapter 5 for the virtual photon exchange mechanism.

Thus, if the number of activator sites increase beyond a specific, but not easily defined, density, then the energy dissipation mechanisms become dominant.

There is a further complication, not yet clarified, which involves the actual "optimum" activator concentration as a function of the host employed. This is illustrated in the following diagram, given on the next page as 6.4.20. The activator concentration peak can be quite narrow and as little as $\pm$ 5 $\times 10^{-3}$ mols per mol of host can change the emission intensity by $\pm$ 50% of the peak value in some extreme cases. The "optimum" activator concentration, being a function of the nature of the activator chosen and the nature of the host is not easily predicted or calculated. It is usually **experimentally determined.**

6.4.20.-

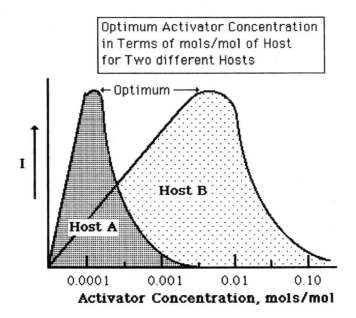

9-5 - PREPARATION OF PHOSPHORS

In this section, we will first describe the preparation of phosphors in terms of the parameters involved, followed by measurements usually required to ascertain the energy efficiency of the phosphors so- produced, and finally how to make such measurements.

I. Phosphor Parameters

Phosphor preparation parameters can be divided into two (2) classes, independent and dependent variables.

a.- Independent Variables

The following lists the independent variables of phosphor preparation:

6.5.1.- Independent Variables

1. Purity of materials	4. Matching of ionic size of activator and host cation
2. Choice of host lattice	5. Charge compensation methods in host
3. Choice of activator	6. Ratio of reacting components

We have already emphasized the importance of purity within the finished phosphor. REITERATING, we must exercise all caution to choose pure raw materials to form the phosphor and prevent further contamination from occurring during processing to form the phosphor.

Let us choose a phosphor composition. For the cation, we choose Ba^{2+}, a silicate as the anion, and an activator such as Pb^{2+}. This could give us a formula such as: $(Ba_{1-x} Pb_x) SiO_4$. We find that Vegard's Law holds for these two cations and they **will** form a solid solution. There is no need for charge-compensation since both cations have the same valence state.

However, by reference to the phase-diagram for BaO and SiO_2, we find that the formation of several compounds is feasible. In this diagram, given as 6.5.2. on the next page, the existence of four(4) compounds is indicated, each of which could serve as our host. It is evident that by changing the ratio of BaO to SiO_2, one can obtain the various compositions. We could prepare the mesosilicate compound by solid state reaction of:

6.5.3.- $$BaCO_3 + 2 SiO_2 = BaSi_2O_5 + CO_2$$

The different compounds which result are:

6.5.4.-

MOLS		PRODUCT
BaO	SiO_2	
2	1	Ba_2SiO_4
1	1	$BaSiO_3$
2	3	$Ba_2Si_3O_8$
1	2	$BaSi_2O_5$

6.5.2.-

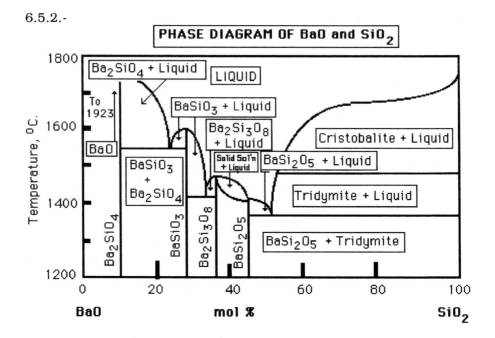

The major question we need to ask is **which** of these compositions will produce the most efficient phosphor. This is a difficult question to answer if we do not already know the answer! It turns out that the $BaSi_2O_5$ composition is the one which works best with the Pb^{2+} activator.

In contrast, Ba_2SiO_4 works better with Sn^{2+}. The exact reason for this **is not known**, but it is suspected that the phonon spectrum of the host and/or the crystal structure is the answer. This brings us to another major rule in phosphor preparation, namely:

6.5.5.-   WHEN CHOOSING A SPECIFIC HOST FOR A SELECTED ACTIVATOR, ONE MUST DO SO BY EXPERIMENTAL MEANS, SINCE THE METHOD OF PROPER COMBINATION OF HOST WITH A GIVEN ACTIVATOR IS NOT COMPLETELY UNDERSTOOD, THAT IS - WHY A PARTICULAR HOST WORKS WELL WITH A PARTICULAR ACTIVATOR IS NOT FULLY UNDERSTOOD AT THIS TIME.

Therefore, in our model phosphor, we would ordinarily prepare all four compositions, and finalize the composition according to our experimental results.

Another independent variable we wish to consider is that of the ratio of reacting components. As we have already shown in 6.5.2., a 1:2 ratio of reacting components produces the $BaSi_2O_5$ composition. However, as we have also shown in Chapter 3, the extent of the solid state reaction between two components (i.e.- the "completeness" of the reaction) depends upon the relative diffusion rate of the reacting species. In our case, this is the Ba^{2+} cation. Thus, if we use a 1.00 : 2.00 ration, we would find that the reaction would proceed to about 98-99% and then nearly stop. The reasons for this have already been discussed in Chapter 3, but can be summarized as follows. The rate of reaction between two solids is an exponential one, rapid in the beginning, but slowing as the components are used up. An asymptote is gradually approached, but the reaction never becomes 100% complete. Therefore, in a solid state reaction to form a given compound, we **always** end up with a small amount of **unreacted components** . Usually, this amount is in the 100 -1000 ppm. range.

Nevertheless, any such "impurity" severely affects the efficiency of the so-produced phosphor. We find that, for the present case of $BaSi_2O_5$, the presence of even very small amounts of BaO has a very strong absorption band in the ultraviolet which detracts form the overall performance of the phosphor when used in a fluorescent lamp. In contrast, an excess SiO_2 is completely transparent in the UV. Even if the phosphor is to be used in a cathode-ray tube, the presence of **excess cation** is deleterious to its successful operation as an efficient phosphor. Our solution is to use an excess of the **non-optically active** anion to promote the completeness of the reaction. In general, the excess required to do so is not more than a few mol%. An example of this is given in 6.5.6., presented on the next page.

If we performed both of these reactions and incorporated Pb^{2+} as an activator, we would find that the phosphor produced by Reaction (1) was much superior in performance to that produced by Reaction (2).

6.5.6.-   <u>Reaction (1):</u>

1.00 $BaCO_3$ + 2.025 SiO_2 = 1.00 $BaSi_2O_5$ + 0.025 SiO_2 + 1.00 CO_2

<u>Reaction (2)</u>

1.025 $BaCO_3$ + 2.00 SiO_2 = 1.00 $BaSi_2O_5$ + 0.025 BaO + 1.025 CO_2

Yet any analysis we might perform to discern the difference between the two phosphors would yield **no** answer because the two phosphor compositions **are the same**, with the exception of the small amount of excess reactant in each case. Therefore, another rule for preparing phosphors is:

6.5.7.- IN FORMULATING A PHOSPHOR COMPOSITION, ONE ALWAYS EMPLOYS A SMALL EXCESS OF THE ANIONIC REACTANT SO AS TO AVOID THE PRESENCE OF STRONGLY ABSORBING CATIONIC SPECIES IN THE END-PRODUCT.

This rule is a general one since all of the anions used (as given in 6.3.2.) are optically transparent. The only case that one will encounter is that where an optically active anion is chosen. Even in that case, one formulates the composition so that only a very slight excess (about 0.005 - 0.01 mol%) of the anion is present in the reacting species to form the final composition.

After we have followed all of these rules, our phosphor formulation turns out to be:

6.5.8.-   0.99 $BaCO_3$
                   0.01 $PbCO_3$
                   2.02 SiO_2

The last factor to consider is that of charge compensation. Suppose we substitute Sb^{3+} for Pb^{2+}. This would require a formulation such as that given in 6.5.9. on the next page.

6.5.9.- Formulation for $BaSi_2O_5:Sb^{3+}:K^+$

0.99 $BaCO_3$

0.005 Sb_2O_3

0.005 K_2CO_3

2.04 SiO_2

The slight increase in silica content is required to maintain a 0.02 mol excess, since the (Sb,K) cations use up 0.02 mol SiO_2 more than the reaction of 6.5.8.

Nevertheless, there are many cases where charge-compensation cannot be effected, in spite of anything we might try to do. A case in point is $CaF_2: Dy^{3+}$. We might try to use Na^+ as a charge-compensating ion, but we will find that it is not effective. Defect reactions of Dy^{3+} with this lattice are:

6.5.10.- $\quad Dy^{3+} \quad \Rightarrow \quad Dy^{3+}_{Ca} + V^{-}_F$

$\quad\quad\quad Dy^{3+} + O^= \quad \Rightarrow \quad Dy^{3+}_{Ca} + O^{-}_F$

Even if **all oxygen** atoms are excluded during the reaction to form the crystal (including that of water), lattice defects will be formed. In fact, the presence of charge-compensating ions do not affect these defect reactions. The phosphor compositions we obtain because of the predominance of the defect reactions are:

6.5.11.- $\quad \{(1\ -x)\ Ca\ ,\ x\ [\ Dy^{3+}_{Ca} + V^{-}_F\ ]\}\ F_2$

$\quad\quad\quad \{(1\ -x)\ Ca\ ,\ x\ [\ Dy^{3+}_{Ca} + O^{-}_F\ ]\}\ F_2$

The phosphor represented in the first formula is a good thermoluminescent (TL) material for measuring exposure to radiation dosage. The vacancies function as traps for energy, and that energy is released by heating the phosphor and measuring the amount of energy trapped (obviously, one has to calibrate the dosimeter before using it experimentally). But, the second is not a good TL phosphor. Thus, we must be very careful to completely exclude the presence of oxygen from any source , if we are to obtain a good thermoluminescent dosimeter material.

Finally, we should reemphasize the effect of impurities on luminescent efficiencies. As we have already said, even parts per million of excess transition metal cations cause unwanted absorption and seriously affect phosphor energy efficiencies. These cations are already present as **Impurities**. However, the inclusion of **excess anion** in the phosphor composition precludes this adverse effect.

b. Dependent Variables

What we mean by dependent variables are those parameters which are interdependent. i.e.- as one is changed, the others are affected as well. They include:

6.5.12.-   Dependent Variables

1. Firing temperature	6. Firing time
2. Size of mass being fired	7. Firing atmosphere
3. Activator concentration	8. Effect of host structure
4. Effect of method of preparation	9. Ratios of reacting components
5. Use of a flux	

It may seem somewhat surprising to see the same parameters listed herein when we have already discussed them as independent variables. However, as we shall see, they are not completely independent parameters and so must be included as dependent parameters as well.

Before we can proceed to an in-depth examination of the dependent variables, we need to examine the steps required to prepare a phosphor. These are shown in 6.5.13., given on the next page.

6.5.13.- STEPS IN PREPARING A PHOSPHOR

    1. Selection of materials

    2. Assay

    3. Weighing

    4. Blending

    5. Firing cycle

    6. Dispersion

    7. Evaluation

SELECTION OF MATERIALS - We require raw materials having an impurity level less than 10 ppm. of total impurity cations. In particular, we need to avoid the "killer" cations given in 6.3.13. It is advantageous to choose a compound which decomposes to form an oxide. A case in point is our chosen phosphor composition where we will choose $BaCO_3$ as a reacting component rather than BaO, because of the reasons stated in Chapter 3 regarding the nature of diffusion reactions and their effect upon the final product. If we chose $Ba(NO_3)_2$, we find that the acidic N_2O_4 produced during decomposition of the nitrate has a deleterious effect on the produced phosphor. This does not occur with the carbonate. In a like manner, if we were to choose Na_2SiO_3 as a reactant, we could do so. However, we would have to use $BaCl_2$ in conjunction with this compound so as to produce NaCl as a final reactant. Then we would remove the NaCl so-produced by washing the product in water. Nevertheless, we would find that very small amount of Na^+ cations had been incorporated into the host lattice, $BaSi_2O_5$, in spite of any precautions we might have taken. For these reasons, we choose certain special compounds, so chosen as to produce innocuous by-products such as gases.

ASSAY: To maintain strict stoichiometry in the end-product, we determine the amount of adsorbed water present on **each of the reacting components.** We then recalculate the exact weight needed to compensate for the amount of water present.

WEIGHING: The exact amounts of components are weighed and blended together.

luminescent state, the following diagram is presented:

6.5.16.-

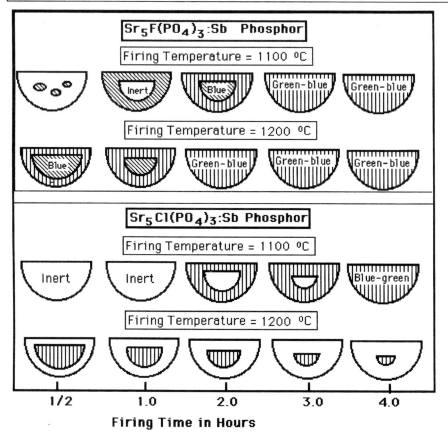

The solid state reactions which occur and the stages of development which occur during phosphor formation are remarkably alike from system to system. These stages are shown for both $Sr_5 F(PO_4)_3$:Sb and $Sr_5Cl(PO_4)_3$:Sb. The emission colors for these phosphors is blue-green.

The stages for the phosphor, $Sr_5F(PO_4)_3$:Sb are as follows:

(1) At 1/2 hour firing time @ 1100 °C., the mass is mostly inert with a few luminescent "spots"

(2) At the 1.0 hour firing time, a blue-emitting phase is obtained, with only a small amount of inert material present

(3) By 2 hours firing time, the blue-emitting phase has shrunk and a green-blue emitting phase has proceeded to grow from the outside; At 3 hours, the transformation is complete to the green-blue emitting phase. When the firing temperature is 1200 °C., the same mechanisms are observed, except that they occur faster.

The stages for the $Sr_5Cl(PO_4)_3$:Sb phosphor are more revealing:

(1) At the 1100 °C. firing temperature, the growth of the blue-green emitting phase is slow and becomes complete at about 4 hours of firing time, while the inert phase diminishes as the luminescent phase grows.

(2) At the 1200 ° C. firing temperature, the blue-green emitting phase has already passed its peak growth and has begun to change into a different inert phase. **This** inert phase increases with firing time. Thus, we have discovered that it is possible to **"overfire"** a phosphor.

What we have learned is that there are at least four(4) stages which occur in the formation and development of the luminescent phase, to wit:

1. Solid state reaction to form the host structure - inert
2. Sintering to form the crystalline lattice - inert
3. Formation of the luminescent phase, including rearrangement of host crystal components.
4. Continued reaction to form a different luminescent phase, including that of a separate post- luminescent inert phase.

In the case of our strontium fluoroapatite phosphors, we can explain the various changes rather easily. Initially, the host lattice forms and grows more crystalline. However, the host-activator combination does not become luminescent until the activator cation is incorporated at the correct lattice site to form the luminescent center. Originally, there is a fluorine-rich structure, but as firing time continues, oxygen diffuses into the structure, thereby changing the emission color. This may be due to a change in the site-structure of the activator, Sb^{3+}, or the formation of oxygen- vacancy sites, as in the example of CaF_2:Dy, given above. The chloroapatite phosphors are even less stable, and are subject to loss of chloride atoms forming the host lattice, thereby resulting in an inert material. The presence of oxygen in the lattice may also be a major contributor to obtaining the "over-fired" inert phase.

WHAT THIS ALL MEANS IS THAT THERE IS AN "OPTIMUM" FIRING CYCLE REQUIRED SO AS TO OBTAIN MAXIMUM ENERGY EFFICIENCY IN THE SO-PRODUCED PHOSPHOR.

d. Size and Mass Fired

Many cases will be encountered during the preparation of phosphors where the activator forms a volatile compound. If we use As^{3+} , Sb^{3+}, Bi^{3+} , Pb^{2+} , Tl^+ and Zn^0 or Cd^0 as activators, then the probability is high that volatile compound formation will ensue. This will lead to loss of activator, thereby causing a lower concentration than intended. However, there is also a beneficial side-effect in that the activator becomes better dispersed throughout the host structure when a volatile compound is formed.

We can control this loss by control of the mass being fired. Many times, this critical parameter leads to improved energy efficiency of the phosphor thereby produced. The significant parameter is **surface area per unit mass.** Consider the following:

6.5.17.-   Surface Area to Volume Ratios (S/V)

Diameter	Cubes	Spheres
1.0 cm	6 cm^2/cc.	6 cm^2 /cc.
10 cm	0.6 "	0.6 "
100 cm	0.06 "	0.06 "

Note that shape is not a factor. Therefore, the lower the value of S/V, the lower will be the loss of any volatile component during firing. Consequently, we may deliberately choose a chemical form of the activator which is volatile so as to take advantage of the dispersion mechanism. **Then** we must adjust the mass parameter and sometimes the firing cycle as well.

e. Firing Atmosphere

If we examine the oxidation states of the activators of 6.3.5., it is apparent that most require one of their lower valence states to function as an activator. Therefore, one needs to provide a neutral, or slightly reducing, atmosphere during firing, since the normal air atmosphere would promote the higher valence states. Suitable neutral atmospheres include N_2, A, Ne, etc. A reducing atmosphere would usually consist of 95-98% N_2 and 2-5% H_2.

However, there are cases where it is necessary to fire in air. A case in point is the tin-activated phosphors, involving the Sn^{2+} activator. These are fired in air first, and result in an optically inert material. They are then refired in a reducing atmosphere to develop the luminescence of the Sn^{2+} activator. The reason for this is that most tin compounds are easily reduced to the metallic form. By first firing in air, the tin is incorporated within the host lattice in an optically inert form. Refiring in a reducing atmosphere then converts the tin to the required lower valence state, and phosphors having high energy efficiencies and brightness result. We may conclude that the chosen firing procedure will depend upon the chemical properties of the reactants which are to form the phosphor composition.

f. Effect of Host Structure

Little has been said concerning this parameter because of its complexity. The subject is so complicated that a separate treatise could be written on this subject alone. Essentially, it involves two factors: 1) the site symmetry at the cation site, and 2) the strength of the crystal field at that site. Consider the activator, Sn^{2+}. As we change the host, the emission colors change from the ultraviolet to the deep red. That is, we obtain a series of phosphors having various emission colors, even though the activator remains the same. This is easily explained by considering the following diagram:

6.5.18.-

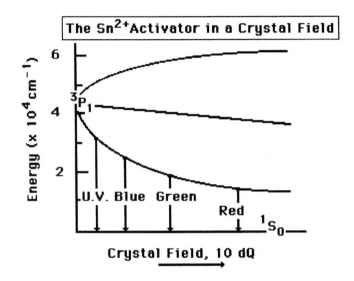

The emitting state for Sn^{2+} is the $^3P_1$ state. The free-ion energy level lies above 40,000 cm^{-1} . But in a crystal field, this level splits into three field components. For this reason, we can vary the emission band of Sn^{2+} by varying the nature of the host employed.

The alkaline earth pyrophosphates are polymorphic in nature. The low temperature form is called ß- $M_2P_2O_7$ while the high temperature form is α- $M_2P_2O_7$ for M= Ca and Sr. For Ba, α- is the low temperature form, while δ–$Ba_2P_2O_7$ is the high temperature form. The firing temperature required to change one form to the other (with differing crystal structure) varies from 1150 °C.(Ca) to 750 °C. (Ba). The requisite crystal structure factors for these crystal forms are shown in 6.5.18., given on the next page. Because of the similarity of these compounds, we can formulate mixtures of the cations to form a solid solution, and thereby "fine-tune" the strength of the crystal field at the activator site.

6.5.18.- Alkaline Earth Pyrophosphates

LOW TEMPERATURE FORMS

Crystal Structure	Lattice symmetry	Cation Site Symmetry	Cation Coordination	
ß- $Ca_2P_2O_7$	Tetragonal	C^4_4 (P4$_1$)	C_1	{7 oxygens or - 8 oxygens}
ß- $Sr_2P_2O_7$	"	"	"	{9 oxygens
α- $Ba_2P_2O_7$	Orthorhombic	C^9_{2v} (Pna2$_1$)	C_1	6 oxygens

HIGH TEMPERATURE FORMS

α- $Ca_2P_2O_7$	Orthorhombic	C^9_{2v} (Pna2$_1$)	C_1	6 oxygens
α- $Sr_2P_2O_7$	"	"	"	"
δ- $Ba_2P_2O_7$	NOT KNOWN			

Note that we have not included the magnesium pyrophosphates. The high temperature form is α–$Mg_2P_2O_7$ which has the thortvieite structure (monoclinic with space group: C^3_{2h} (C2/m) and cation site symmetry = C_1). It has the unusual feature that it transforms to the low temperature form at 68 °C. The low temperature form is β–$Mg_2P_2O_7$ (monoclinic with space group: C^5_{2h} (P2$_1$/c) with cation symmetry = C_1). Neither form produces a tin-activated phosphor as shown by Ropp (1960).

This brings up a **significant point**, namely that when we have a crystal host with a low transition temperature, we will **not** obtain a luminescent material. The reason for this is that once we have formed a given crystal structure during the firing cycle, the activator is situated in a cation site having a specific symmetry. We normally "quench" that phase by removing the fired mass from the furnace. The rate of cooling is generally fast enough to ensure that the desired crystal structure and site symmetry results. However, if a very low transition temperature exists, the mass cannot be cooled fast enough to prevent the crystal structure transformation from occurring. This results in a host lattice having considerable strain and distortion at the cation site. In our case, the problem is further compounded by the disparity in cationic size between Mg^{2+} and Sn^{2+}. Thus, another rule concerning phosphor preparation is:

6.5.19- IN THE SELECTION OF HOST COMPOUNDS TO PREPARE A PHOSPHOR, WE AVOID THOSE WHICH POSSESS A LOW TEMPERATURE FORM, ESPECIALLY IF THEY HAVE A LOW TRANSITION TEMPERATURE.

This explains the non-luminescent behavior of the $Mg_2P_2O_7$ crystal forms.

Note that as we change crystal structures in the other alkaline earth pyrophosphates, changes in lattice symmetry, and cation coordination occur, even though the local cation symmetry remains the same, i.e.- $C_1 =$ no symmetry. Nevertheless, there are major effects of these seemingly similar crystal structure parameters upon the phosphor so-produced, as shown in 6.5.20., given on the next page.

These data illustrate the several effects that the crystal lattice parameters have upon both the intensity (energy efficiency) and energy content (wavelength) of the emission bands. Thus, for the low temperature forms, $\beta-M_2P_2O_7$, both Ca and Sr have the same lattice symmetry and site symmetry but differ slightly in cation coordination. This results in essentially the **same** emission band for both.

The "cation-fit" parameter, however, plays a major role in determining relative emission intensity, i.e.- energy efficiency, in that the Sr-compound is by far the "brightest" phosphor.

For the $\alpha-Ba_2P_2O_7$ composition, the crystal structure and cation coordination differs from the others. This, plus the larger size of the Ba^{2+} cation, causes a high crystal field intensity to appear at the activator site, and results in a major shift in emission wavelengths.

6.5.20.- EMISSION BANDS OF Sn^{2+} IN ALKALINE EARTH
PYROPHOSPHATES

Emission		Cation Radius $(Sn^{2+} = 1.12 \text{ Å})$	Relative Intensity
Color	Peak		
$\alpha- M_2P_2O_7$ - High Temperature Forms			
Ca U.V.	3620 Å	0.99 Å	17
Blue	4290		15
Sr blue	4630	1.13	100
$\delta-$ Ba deep-red	6760	1.35	40
$\beta- M_2P_2O_7$ - Low Temperature Forms			
Ca U.V.	3690 Å	0.99 Å	14
Sr U.V.	3640	1.13	80
$\alpha-$ Ba green	5050	1.35	65

For the high temperature forms, $\alpha - M_2P_2O_7$, two emission bands appear in the Ca compound denoting the presence of **two cationic sites.** The reason for this is not known.

We cannot compare the effect of lattice symmetry or cation coordination on the emission bands, but can observe the effect of the cationic size on the energy position of the emission band (the strength of the crystal field at the activator site) as well as the "cation- fit" parameter on the emission intensity. A major shift in energy position of the bands is evident. Ba^{2+} has more electrons than Ca^{2+} so that the crystal field (10 dQ) at the activator site is stronger and the emission band is shifted to lower energies. However, part of the effect may also be attributable to a change in the

crystal structure. We cannot evaluate this effect because the structure of
δ– $Ba_2P_2O_7$ is not known.

For the general phosphor composition, $(Sr_x Ca_y Ba_z)_2P_2O_7$:Sn, where x ≠
y ≠ z, there is a linear relation between peak emission energy (the zero-
phonon line) and the cationic radius, namely-

6.5.21.- $E (cm^{-1}) = - 9610 R(Å) + 32,696$

What this means is that we ought to be able to "fine-tune" the emission
band between about 4300 Å and 5000 Å. by forming the requisite solid
state solution containing the various sized cations. In practice, this
approach has limited utility because as one shifts the ratio of : (Sr + Ca +
Ba) = 2.00, either the α– or ß– $M_2P_2O_7$ form becomes the dominant one.
However, if x>> y or z, that is- if the α– $Sr_2P_2O_7$ structure is maintained,
then one can shift the emission band to some degree.

The above example of the effect of the host crystal is generally applicable
to all cases we might encounter. Even our illustrative phosphor,
$BaSi_2O_5$:Pb, follows these rules. The structure is orthorhombic
(sanbornite) with a lattice symmetry of: D^7_{2h} (Pmna) and a site symmetry
of C_{2h}. Site coordination is six (oxygens). The cationic radii are: Ba^{2+}=
1.35 Å and Pb^{2+} = 1.20 Å. The emission band occurs at 3660 Å, with a QE
of > 75% (2537 Å excitation).

Our last example of the effect of host lattice on emission properties of
phosphors is that of Ce^{3+}. This activator has a ground state of $^2 F_{5/2}$ arising
from an electron configuration of $4f^15s^25p^6$. The lowest excited state is
$^2D_{3/2}$, arising from the $4f^05s^25p^6 5d^1$ configuration. This is an allowed
dipole transition with the excited-"d"-state lying outside of the shielding
electron shells. Thus, it is subject to the influence of the crystal field of
the host lattice. In YPO_4, the emission band is narrow and peaks at 3260 Å
(Ultraviolet). In Y_2SiO_5, the emission broadens somewhat and peaks at
4200 Å (Deep-blue). But in $Y_2Al_5O_{21}$, the emission peak is very broad and
peaks close to 5460 Å (Yellow). All of these phosphors have comparable

QE's close to 80%. Quite obviously, the strength of the crystal field at the activator site has a major effect on the position of the emitted band.

g. Effect of Preparation Method Employed

We now return to our phosphor example, $BaSi_2O_5$:Pb. There are at least three (3) methods we can employ in the general preparation of phosphors. these are:

6.5.22.-   POSSIBLE METHODS OF PREPARING PHOSPHORS

1. Reaction between Mixed Oxides - $BaO + SiO_2$
2. Precipitation of a Raw Material Containing the Activator
3. Direct Precipitation of the Phosphor

We have already discussed #1. To coprecipitate a raw material, we would use Na_2SiO_3 as a solution in water along with the proper combination of $BaCl_2$ and $PbCl_2$ in solution. We then precipitate at 100 °C. so as to obtain a crystalline product, adding the silicate solution to the cation solution. The precipitate is filtered, and the NaCl removed by washing. We then must add 1.00 mol of SiO_2 to the raw material and then fire the mixture. The solid state reaction is:

6.5.23.- $(Ba,Pb)SiO_3 + SiO_2 = (Ba,Pb)Si_2O_5$

Note that we are not addressing the cases where we use raw materials to prepare the phosphor, but of coprecipitated material containing both the cation and the activator. An example of a raw material used to prepare a phosphor would be $SrHPO_4$ which reacts:

6.4.24.- $2 SrHPO_4 = Sr_2P_2O_7 + H_2O$

But here we would have to add SnO to the $SrHPO_4$ before firing, to obtain the phosphor, $Sr_2P_2O_7$:Sn. In this case, we find that it is very difficult to attempt to coprecipitate the phosphates from solution because of the disparity in their chemical solution behavior.

There are a few phosphors which can be directly formed by precipitation from solution. One such is YVO_4: Eu. The chemistry is complex, since one must start with the ions, VO^{2+} , Y^{3+} , and Eu^{3+} , and convert these to the final product. The precipitate itself has a brightness of about 40% of the final fired phosphor.

This brings us to another rule for phosphor preparation, namely-

6.5.25.- FOR THE MOST PART, WE USE AN INTERMEDIATE PRECIPITATED RAW MATERIAL WHICH SERVES AS A BASE IN THE FORMULATION. IN THIS WAY, WE AVOID THE DIFFUSION-LIMITED SOLID STATE REACTIONS THAT MIGHT PRODUCE BY-PRODUCTS AND/OR LATTICE DEFECTS IN THE FINAL FIRED PHOSPHOR PRODUCT.

h. Ratios of components and Use of a Flux

We have already discussed the effects of varying the ratios of reacting components, both in this chapter and in Chapter 3. Reiterating, sometimes the solid state reaction does not proceed as predicted, particularly if we use mixtures of oxides, or if we use an exactly stoichiometric mixture of reactants, regardless whether the mixture includes a precipitated raw material which is to serve as a base for the overall solid state reaction. It is for this reason that we **always include** a small excess of the anion so as to drive the reaction to completion.

A flux is defined as an additive which does not affect the reactants during the firing cycle, but which promotes the crystal growth of the final product. It does so by providing a liquid phase for transport of material, including ions, so as to increase the crystallinity and size of the particles so-produced. The final product should also be non-reactive with the flux at the firing temperature. Because the flux must be removed after the reaction is complete, it also needs to be water-soluble. An example of the action of a flux is given as 6.5.26. on the next page.

6.5.26.-

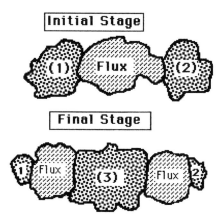

Particle (3) is the recrystallized product produced by liquid transport through the flux, from material transported from particles (1) and (2).

6-6 : COMMERCIAL PHOSPHORS

There are two classes of phosphors that we will discuss in the chapter. One is designed for photoluminescence and is used in fluorescent and high-pressure mercury lamps, while the other is designed for cathodoluminescence and is used in cathode-ray tubes, including Color Television. The former must have a high energy efficiency for 2537 Å photon excitation, the resonance radiation of a mercury-vapor discharge, while the latter must exhibit high energy efficiency when bombarded with a high-voltage electron beam. The design requirements for these two classes of phosphors are quite different, and we will discuss each in turn, including the special technology that has grown around each.

I. Cathode Ray Phosphors

The cathode-ray tube is now over 100 years old. In 1880, William Crookes had demonstrated that cathode-rays, i.e.- an electron beam, would cast sharp shadows on a fluorescent screen if a mask was present within the tube. It remained for Karl Braun to invent the device that changed the world, a working cathode-ray tube (CRT) having internal electrodes which could deflect the electron beam when a voltage was applied. To reach this phase of CRT technology, phosphors, vacuum technology and discovery of the electron and its properties had to precede the CRT. It was 1603 when Vincenzio Cascariolo described the first phosphor. Cascariolo's phosphor evidently was a barium sulfide, made by firing a barium sulfate ("Bologna Stone") with graphite. The first commercially available phosphor (1870) was "Balmain's paint," a calcium sulfide preparation. In 1866, the first stable zinc sulfide phosphor was described. Braun's CRT contained all of the components of the modern CRT including: an electron source, electron beam focusing and deflection, acceleration of the electron beam and a phosphor screen. These are condensed into two main components of the CRT, namely the electron gun and the phosphor screen assemblies. These two have received the most effort as the CRT has been improved over the years. We will address the CRT phosphors used up to 1990 in this chapter and the latest advancements in display technology in the next chapter.

a. How Cathode-Ray Phosphors Are Used

In a cathode-ray tube, an electron-beam is generated within an "electron-gun", and caused to sweep across a faceplate containing a phosphor "screen". This screen is actually a uniform layer of phosphor particles deposited by special techniques. The beam "paints" a series of lines on the phosphor-screen to form a rectangle called a "raster" which emits visible light, as shown in 6.6.1. on the next page.

The electron gun has a cathode which emits electrons. The electrons are controlled and bunched into a beam as it exits the electron gun. The deflection coil varies the motion of the electron beam, generated within the electron gun, to form the raster. The number of separate lines within the raster is about 525, but the separate rasters are interwoven so they overlap to form the complete picture.

The raster on the phosphor screen is excited in real time so that we see moving pictures. The human eye has a movement perception of about 1/20 of a second so that the raster, reformed about every 1/30 of a second

according to a JEDEC number (Joint Electron Device Engineering Council). This is the designation most often referred for any cathode-ray tube application. Table 6-2 gives a complete listing as follows:

TABLE 6-2

CHARACTERISTICS OF JEDEC PHOSPHORS

	COMPOSITION	COLOR	PERSISTENCE	APPLICATION
P-1	Zn_2SiO_4:Mn	green	medium-20 ms	oscilloscope
P-2	ZnS:Cu	blue-green	medium(70 ms)	oscilloscope
P-3	Zn_2BeSiO_4:Mn	green-yellow	medium(20ms)	oscilloscope
P-4	ZnS:Ag +ZnCdS$_2$:Ag	white	medium short (100 μs)	B&W television
P-5	CaWO$_4$:W	blue	short (25μs)	oscilloscope
P-7	ZnS:Ag + ZnCdS$_2$:Cu	blue + yellow	medium(57μsec) long (400 ms)	radar
P-10	KCl:V_{Cl}	dark trace	very long	electrochrome
P-11	ZnS:Ag	blue	med short (34μs)	oscilloscope
P-12	$(Zn,Mg)F_2$:Mn	orange	long (210 msec)	radar
P-13	MgSiO$_3$:Mn	red-orange	med long (52ms)	military radar
P-14	ZnS:Ag + ZnCdS$_2$:Cu	blue + orange	short(27μsec) + very long (500 ms)	radar
P-15	ZnO: Zn^0	blue-green	very short (2.8 μs)	flying spot scan
P-16	$(Ca,Mg)SiO_3$:Ce	purple	very short(0.12μ s)	flying spot scan
P-17	ZnS:Ag + ZnCdS$_2$:Cu	blue-white + yellow	short (5.2μsec) + very long (429 ms)	radar

TABLE 6-2(Continued)
CHARACTERISTICS OF JEDEC PHOSPHORS

	COMPOSITION	COLOR	PERSISTENCE	APPLICATION
P-18	$(Ca,Mg)SiO_3$:Ti + $CaSiO_3$:Pb	white	medium long(13msec)	monochrome television
P-19	$KMgF_3$:Mn	orange	long(220 msec)	radar
P-20	$ZnCdS_2$:Ag	yellow	med short(0.3 ms)	storage tubes
P-21	MgF_2:Mn	orange	long (220 msec)	radar
P-22	ZnS:Ag + $ZnCdS_2$:Ag + Y_2O_3:Eu	blue + green + red	m(200 μsec) medium short	color T.V.
P-23	ZnS:Ag + ZnCdS:Ag	blue+ yellow	medium short (200 μsec)	sepiatone TV
P-24	CaS:Ce	green	short (1.5 μs)	flying spot
P-25	$CaSiO_3$:Pb:Mn	orange	medium (60 μsec)	radar
P-27	$Zn_3(PO_4)_2$:Mn	red	medium (27 ms)	color TV
P-28	(Zn,Cd)S:Ag:Cu	green	very long (500 ms)	radar
P-31	$ZnCdS_2$:Cu:Ni	green	short (40 μsec)	high brightness monitor
P-32	$CaMgSiO_3$:Ti + (Zn,Cd)S:Cu	purple-blue	long (800 msec)	radar
P-33	$MgKF_2$:Mn	orange	very long (3.0 sec)	radar
P-34	ZnS:Pb:Cu	blue-green	very long(50 sec)	oscilloscope
P-35	Zn(S,Se):Ag	blue-white	short (0.85 μsec)	photography
P-36	(Zn,Cd)S:Ag:Ni	yellow-green	very short(10 μs)	flying spot
P-37	ZnS:Ag:Ni	green-blue	very short	flying spot
P-38	$MgZnF_2$:Mn	orange	long (1 - 4 sec)	radar
P-39	Zn_2SiO_4:Mn:As	green	med. long(400ms)	radar
P-41	$(Zn,Mg)F_2$:Mn+ $Ca_2MgSi_2O_7$:Ce	orange- yellow	long (100 msec)	light pens, etc
P-42	ZnCdS:Ag + $ZnCdS_2$:Cu	blue + green	short(27μsec) + very long (500 ms)	penetration screen
P-43	Gd_2O_2S:Tb	green	medium(20 msec)	displays

TABLE 6-2(Continued)

CHARACTERISTICS OF JEDEC PHOSPHORS

	COMPOSITION	COLOR	PERSISTENCE	APPLICATION
P-44	$La_2O_2S:Tb$	green	medium (20 msec)	displays
P-45	$Y_2O_2S:Tb$	green	short (1.8 msec)	displays
P-46	$Y_3Al_5O_{11}:Ce$	green white	very short(120n)	flying spot
P-47	$Y_2SiO_5:Ce$	purple blue	very short (82 nsec)	flying spot
P-48	$Y_3Al_5O_{11}:Ce$ + $Y_2SiO_5:Ce$	yellowish green	very short (120 nsec)	flying spot scanners
P-49	$YVO_4:Eu$	red	medium	graphic display
P-50	$Y_2O_3:Eu$	red	medium	graphic display
P-51	$Y_2O_3:Eu$ + $(Zn,Cd)S:Ag:Ni$	reddish orange	medium	Multicolor displays
P-52	$Zn_2SiO_4:Ti$	purplish	short (28 msec)	military
P-53	$Y_3Al_5O_{11}:Tb$	green	medium	Heads-up display
P-54	$Y_2O_3:Eu$	red	medium	displays
P-55	$ZnS:Ag$ + $Y_2O_2S:Tb$ + $(Zn,Cd)S:Cu:Al$	blue + green + red	medium	projection television
P-56	$Y_2O_3:Eu$	red	medium	projection tube
P-57	$Zn_2SiO_4:Mn$ + $MgF_2:Mn$	yellowish green	medium	radar

It should be noted that these JEDEC registrations, begun in 1945 by EIA in the U.S., i.e.- Electronic Industries Association, have been largely supplanted by WTDS designations. The various "P" registrations can be grouped as shown in 6.6.6., given on the next page.

Note that most of these phosphors are based on the zinc and cadmium sulfides. But there are some specialty phosphors based on silicates and

6.6.6.- Classification of JEDEC Phosphor Applications

Electrochrome display = 1   Oscilloscope display = 6
Radar display = 16   Television display= 6
Flying spot scanner = 8   Display monitor = 7
Storage monitor = 1   Photography display = 1
Light pen on display = 1   "Penetration" display = 1

P-6, 8, 9, 26, 29, 30 & P-40 are either obsolete or were withdrawn.

fluorides. The P-22 phosphors are those presently used for color television. Many of the phosphor screens used as display monitors for computers are the P-4 variety (white), or various silver-activated sulfides containing Zn/Cd ratios to produce green or yellow emission. Many of the JEDEC phosphors are not being used at this time because the newer rare earth activated phosphors have been proven superior to those based on zinc and cadmium sulfides. Table 6-3 shows a comparison of sulfide and rare earth activated phosphors as to their efficiency and performance in cathode-ray tubes.

Table 6-3

Comparison of Cathode-Ray & Rare Earth Phosphors in CRT's

Composition	Energy Eff.	Peak Wavelength	Color	Brightness Loss	Relative Strength
Zn_2SiO_4:Mn	8%	525 nm	green	~ 35 %	W
$CaWO_4$:Pb	4	425	blue	~ 30 %	VW
Y_2O_3:Eu	9 %	611 nm	red	3 %	S
Y_2O_2S:Eu	12	626	red	12%	S
Gd_2O_2S:Tb	15	545	Y-green	5 %	S
ZnS:Cu:Al	23	530	green	37 %	VW
ZnS:Ag:Cl	21	450	blue	31%	VW
Y_2SiO_5:Tb	1.5	545	Green	1.5 %	VW
$Y_3Al_5O_{11}$:Tb	2	545	green	2.1 %	VS
$Y_3Al_3Ga_2O_{11}$:Tb	1.5	545	green	1.7%	VS
CaS:Ce	22	520	green	30 %	VW
LaOBr:Tb	20	544 nm	green	----------	--------

Many of these measurements were made in conjunction with use in projection CRT systems. The projection television system uses three separate cathode-ray tubes, each containing a red or green or blue phosphor screen. Each CRT must be operated at its maximum output over a long period of time since the brightness output of the three are combined and projected upon a much larger viewing screen. Because the loss in brightness from the Schmidt-lens to the viewing screen is several orders of magnitude, any loss in screen brightness in the projection tube during operation is magnified in the viewing screen. The brightness loss given in Table 6-3 was accomplished using a standard electron beam current and voltage. Since many of these phosphors discolor, that is- they develop a dark cast or coating on the surface of the phosphor particles, this factor must be considered. It was arbitrarily defined to give a subjective "relative strength", applicable to each phosphor. As we can see, this factor shows how each phosphor screen maintains its relative brightness during operation of a CRT.

II. Fluorescent Lamp Phosphors

The fluorescent lamp is basically a low pressure mercury discharge lamp with a layer of phosphor particles on the inside surface of the glass tube, as shown in 6.6.6., presented on the next page.

The lamp has an internal gas-fill pressure of about 1 -2 mm pressure and contains a mixture of Ne, A, and Kr gases. The internal radiation generated is mostly (85%) 2537 Å , which is the resonance wavelength of the mercury vapor discharge. The other 15% is distributed between 1850 Å, 3150 Å, 3650 Å, 4300 Å, 5460 Å, and 5785 Å. The general phosphor system used has the apatite structure with the general composition:

$$Ca_5 F,Cl(PO_4)_3 :Sb:Mn$$

where Sb^{3+} is the sensitizer and Mn^{2+} is the activator. This system is used because the emission colors can be changed from bluish-white to reddish-white by variation in the Sb/Mn ratio and content. The phosphor coating is always a **blend** of phosphors with the major part composed of

6.6.6.-

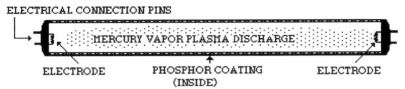

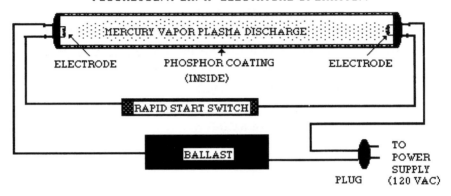

the apatite. In the operation of the lamp, the ballast controls the amount of operating current (it is essentially a "choke-coil") while the rapid-start switch starts the filaments. When the filaments glow so as to ionize the mercury at the electrodes, the voltage is switched, thereby initiating the discharge. Fluorescent lamps normally were manufactured in two sizes, a 48 inch lamp (40 watts) and a 96 inch lamp (80 watts). We will describe, in the next chapter, the latest changes taking place as a result of lamp and phosphor improvement.

Another type of lamp is the so-called high pressure mercury vapor (HPMV) lamp. If one increases the internal pressure of the lamp from a few mm. to several atmospheres, i.e.- >> 760 mm. pressure, the number of collisions per second between mercury atoms increases enormously in the discharge. This has the effect of shifting the resonance radiation to longer wavelengths, including the visible wavelengths: 4300 Å, 5460 Å, and 5785 Å., resulting in a greenish-white emitting lamp. There is also a considerable amount of 3650 Å (ultraviolet) radiation present. In order to achieve the required internal pressure needed for operation of this lamp, a quartz envelope is employed, as shown in the following:

6.6.5.-

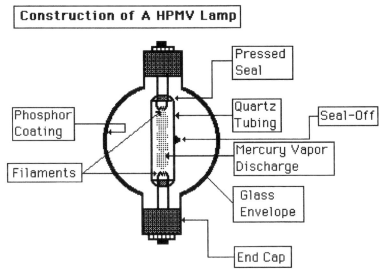

Construction of A HPMV Lamp

The HPMV lamp has been used extensively in street-lighting applications. Electrodes are sealed to each end of the quartz tubing to form a special pressed seal, thereby forming the lamp. Mercury plus a rare gas mixture is introduced into the evacuated tube at the "seal-off" point at a predetermined pressure. A phosphor coating is then applied to the inner side of the glass envelope, the quartz lamp is then mounted within and the whole is sealed with end-caps. This type of lamp runs very hot, typically between 700 °C. and 800 °C. at the outer surface of the quartz lamp. The reason for this is that a high temperature is required within the quartz tube to get the internal pressure up to the necessary operating conditions of several atmospheres of internal pressure. Therefore, this type of lamp dissipates about 400 watts of electrical power. However, the temperature at the surface of the glass envelope is never more than about 250-300 °C. Because the discharge emission is a greenish-white, a red phosphor is needed to correct for the lack of red emission in the discharge radiation. In addition to being red, the phosphor must also have good temperature dependence of emission and respond to 3650 Å.

Design of phosphors

Typical lamp phosphors are shown in the following table:

TABLE 6-4
COMMERCIAL LAMP PHOSPHORS

Composition	Emission			Excitation	Q.E.
	Color	Peak	Half-width	Peak	(2537 Å)
$Ba_2P_2O_7$:Ti	blue-green	4940 Å	1700 Å	2590 Å	85%
$Sr_2P_2O_7$:Sn	blue	4600	1050	2580	86
$(Ca,Zn)_3(PO_4)_2$:Sn	pink	6100	1360	2580	82
$(Sr,Mg)_3(PO_4)_2$:Sn	pink	6280	1250	2590	84
$MgGa_2O_4$:Mn	green	5040	300	2700	78
$(Sr,Mg)_2P_2O_7$:Eu	U.V.	3930	245	3250	---
$Sr_3(PO_4)_2$: Eu	deep blue	4060	340	3210	---
$Ca_5F(PO_4)_3$:Sb	blue	4760	1360	2550	71
$Sr_5F(PO_4)_3$:Sb	blue-green	5090	1530	2560	76
$CaSiO_3$:Pb:Mn	red	6150	940	2600	85
$BaSi_2O_5$:Pb	U.V.	3510	410	2600	75
$SrFBO_2$:Eu	U.V.	3710	190	3380	---
Zn_2SiO_4:Mn	green	5280	410	2580	70
$MgWO_4$:W	blue	4730	1360	2880	83
$Cd(BO_2)_2$:Mn	red	6180	750	2740	78
YVO_4:Eu	red	6200	40	3290	89
YVO_4:Dy	yellow	5750	60	3290	90
$CaWO_4$:W	blue	4330	1140	2540	75
Y_2O_3:Eu	red	6195	30	2600	92
$Ca_5(F,Cl)(PO4)_3$:Sb:Mn	various	varies		2600	85
$CaSiO_3$:Pb	UV	3300	560	2500	75
$CaWO_4$:Pb	deep blue	4330	1140	2600	76
Mg_4FGeO_6:Mn	deep red (5 lines)	6575	163	4200-3000-2200	82
$Mg_5As_2O_{11}$:Mn	deep red (5 lines)	6400	163	4400-3300-2300	80
$Sr_5Cl(PO_4)_3$: Eu	blue	4450	375	3000-2580-2300	91
$BaMg_2Al_{16}O_{27}$: Eu	blue	4540	600	4 peaks-2600	88
$SrMg_2Al_{18}O_{39}$: Eu	blue	4560	640	4 peaks-2600	100?
$Mg_3Al_{11}O_{19}$: Ce:Tb	green	5520	120	2800	92
$BaMg_2Al_{16}O_{27}$: Eu:Mn	blue-green	5150-4250	1300	4 peaks-2600	100?

Excitation bands are broad so that most of the incident 2537 Å photons generated within the lamp are absorbed. The electrical efficiency of the LPMV lamp for generating 2537 Å photons is only about 32%. Thus with a phosphor having a QE of 80%, a LPMV lamp would have a 22% energy conversion efficiency for generation of visible light. However, this is considerably better than its competitor, the incandescent lamp, whose energy efficiency is slightly less than 8%.

There are a number of specialty phosphors that have been used to make LPMV lamps for special purposes. These are listed in Table 6-5:

TABLE 6-5

COMMERCIAL SPECIALTY LAMP PHOSPHORS

Composition	Emission			Excitation	Q.E.
	Color	Peak	Half-width	Peak	(2537 Å)
$BaSi_2O_5$:Pb	UV	3510 Å	410 Å	2540 Å	76%
$(Ba,Sr)_2(Mg,Zn)Si_2O_7$:Pb	UV	3560	1000	2537	78
$Y_3Al_5O_{11}$: Ce	"white"	5000	very broad	3500-2600	82
$Sr_2P_2O_7$:Eu	deep blue	4200	290	2600	88
$YMg_2Al_{11}O_{19}$: Ce	UV	3400	465	2600	70
$MgGa_2O_4$: Mn	green	5040	120	2760	82

These phosphors are used primarily in "Black-Light" lamps and photocopy machines such as those marketed by Xerox Corp.

HPMV lamp phosphors include YVO_4:Eu , YVO_4:Dy and $(Sr,Mg)_3(PO_4)_2$:Sn. A better HPMV phosphor is the $Y(P_{0.20}V_{0.80})O_4$:Eu composition which has a superior temperature dependence of emission. Although Y_2O_3:Eu also has a good temperature dependence of emission, it does not respond to 3650 Å radiation. Before we leave this section, we should point out that new improvements have been made in both HPMV and LPMV lamps. For example, in HPMV lamps, the mercury vapor has been augmented with other metal vapors such as Tl, In and others. Advantage is taken of the fact that the iodides of these metals are volatile so that the metal-vapor emission spectra, containing a relatively high content of red radiation, results. This obviates the use of phosphors for most cases. The lamps are called metal-vapor lamps, even though they still rely on the mercury vapor emission for the major part of the visible light emitted.

For LPMV lamps, the newest lamps contain three (3) essentially "line-emitters, using Eu^{2+} for blue emission, Tb^{3+} for green emission, and Eu^{3+} for red emission. This is the so-called "high output lamp" which has line emission in contrast to the broad band emission of the phosphors given in Table 6-4. That is, the light output has large gaps in the wavelengths emitted. However, the human eye integrates the wavelengths so that the emission is perceived as "white" light. Note that the last four phosphors presented in Table 6-4 are based upon the magnetoplumbite structure, i.e.- aluminate compounds, which are related to the spinels. It is these phosphors which have caused a revolution in the manufacture of LPMV lamps. Such phosphors can be used at much higher current densities within the fluorescent lamp and are more efficient than the older lamp phosphors listed above. The current type of lamp is a 1.0" lamp, i.e.-33T5, where the first number is the watts of energy dissipation for a four foot lamp. The old lamp, which has nearly been replaced, is a 40T12 lamp. We will address the changes which have taken place in current lamp manufacture since 1991 (the date of the prior edition of this manuscript) in the next chapter.

It should be noted that all of the above phosphors were described in a companion book, "The Chemistry of Artificial Lighting Devices- Lamps, Phosphors and Cathode-Ray Tubes", published by Elsevier in 1993 (ISBN 0-444-81709-3). In this volume, the exact methods for manufacture of these phosphors were presented along with the exact formulas and materials required to do so.

6-7- MEASUREMENT OF OPTICAL PROPERTIES OF PHOSPHORS

Although methods of preparing several phosphor compositions have been explored, we have not yet examined methods of measuring phosphor properties. The need to measure the "brightness" or light output under a controlled excitation source should be apparent if one is to optimize any given phosphor composition. Other measurements needed are spectral energy distributions, color measurements and particle size. All of these are necessary before a lamp test is made to determine the stability of a test phosphor in an actual LPMV or HPMV lamp.

I. Measurement of Phosphor Brightness

Let us suppose that we have prepared a series of phosphor compositions wherein the activator concentration was the variable. We would need to measure these samples to determine which concentration was near to the optimum. Let us further suppose that we wish to use our phosphor in a LPMV lamp application. Although we **could** manufacture a test-lamp, usually this is not very feasible. What we do is to prepare a thick layer on a holder and compare it to a "standard" phosphor having similar emission characteristics. The layer is called a "plaque" and the apparatus is called a "plaque-tester", as shown in the following:

6.7.1.-

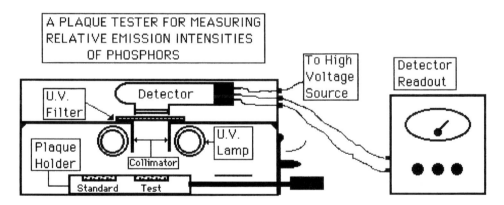

Essentially it is a light-tight box equipped with ultraviolet lamps, having a suitable wavelength for excitation of the phosphors being tested. For the most part, we use lamps with 2537Å output, although others can be used, as required. The plaque-tester also has a suitable detector to measure the intensity of the light emitted by the phosphor, a collimator to eliminate any "stray" radiation, and has a U.V. filter to prevent the excitation wavelengths from reaching the detector. Generally, an ordinary window pane will serve.

The detector can be one of several, including a photomultiplier tube (which is that illustrated above). The photomultiplier tube has the disadvantage of requiring a high voltage supply to be operative. The main requirement for the detector is that it must have the proper response to the wavelengths being measured. Other usable detectors include a GaAsAl diode or a barrier-layer cell. In addition to the apparatus shown above, we also need a regulated voltage supply so as to regulate the intensity of the 2537 Å illumination of the individual plaques. Note that this is just one of many possible configurations. However, it does show the required components needed to measure relative brightness of a series of phosphors. The next parameter we need to establish is the emission spectrum of any given phosphor.

II. Measurement of Spectral Energy Distributions

One of the more critical measurements required for phosphors is determining the spectral distribution of the emission and/or that of excitation, that is- the wavelengths which best excite the phosphor. Such an instrument is called a Spectrofluorimeter and has the general optical design, as shown in 6.7.2., presented on the next page.

Two (2) monochromators are required, an excitation monochromator and an emission monochromator. The source of excitation is generally a xenon lamp since it provides useful radiation from about 1750 Å to beyond 10,000 Å at intensities not obtainable from other types of sources. To obtain an emission spectrum, we set the excitation monochromator at a wavelength which provides good response from the phosphor sample. This can be done visually. We then operate the emission monochromator to obtain the spectrum of the emitted light, taking care that the switch is set for the correct side of the input from the operating monochromator. Having done this, we next set the emission monochromator at the wavelength of maximum emission of the phosphor, and operate the

6.7.2.-

Optical Design of a Spectrofluorimeter

excitation monochromator to obtain the excitation spectrum of the phosphor.

In this optical design, it is difficult to obtain an absorption spectrum. the best that we can do is to operate both monochromators simultaneously so as to obtain a reflectance spectrum. Even then, we need to compare the reflectance signal of the phosphor to that of optical-grade $BaSO_4$. Barium sulfate has a reflectance of 98.99% throughout the visible and ultraviolet portions of the spectrum and has the added advantage that it does not degrade under U.V. radiation. In our case, we must first obtain the reflectance of $BaSO_4$ and without changing instrumental setting, measure our phosphor. We will find that the instrument does have energy losses within the monochromators and detector (they are not linear in energy throughput). It is these losses, detected by $BaSO_4$, that we must apply to our phosphor reflectance spectrum to obtain the true spectrum.

There are commercial instruments which measure excitation and emission spectra directly in μwatts/Å bandwidth, thereby obviating the necessity of measuring absorption spectra. The quantity we wish to obtain is the total energy emitted (so as to be able to compare separate phosphors), namely:

a very low light intensity is present, i.e.- night-vision. When the luminance of the visual field is high, vision is said to "photopic" and the eye is "light-adapted". A low luminance condition results in "scotopic" vision, the eye being "dark-adapted". The range of luminance discernible by the human eye is about 50 :1. There are many other factors which contribute to human sight, such as binocular vision, but we are only interested in color perception at the moment.

The mechanism of color vision is fairly well understood. Basically, it involves three (3) types of chromophores (each having a unique molecular structure) which react with a photon to produce an electrical signal, detectable by the optic nerve. We will not delve into this chemistry. It has been demonstrated that three separate types of cones exist, giving three sets of color preceptors. Each set of preceptors is active over a different range of wavelengths, these ranges being centered in the **blue, green and red** regions of the visible spectrum. Thus, we speak of having three **primary** colors, red, green, and blue.

But the retina does not respond equally to all wavelengths. For equal energies, a yellow-green light produces a much stronger response in the human eye than a red or blue light. Thus, we say that the yellow-green light is "brighter" than the red or blue lights. This is called the luminosity response of the eye. By measuring a number of individual observers, we can obtain what we call a "Standard Luminosity Curve". Apparently, photopic vision relates to "sunlight", to which the human had adapted through evolution, while scotopic vision related to "moonlight", that is, sunlight modified by reflection from the Moon's surface.

Photopic vision peaks at 5500 Å whereas scotopic vision peaks at 5200 Å. The following, given as 6.7.7. on the next page, shows both the photopic and scotopic response curves for the human eye, as determined from a number of observers. In this case, the relative response of the observers are summed into a response called "THE STANDARD OBSERVER" and is normalized for easier usage. You will note that these eye-response curves are the result of an average of many human eye response curves.

6.7.7.-

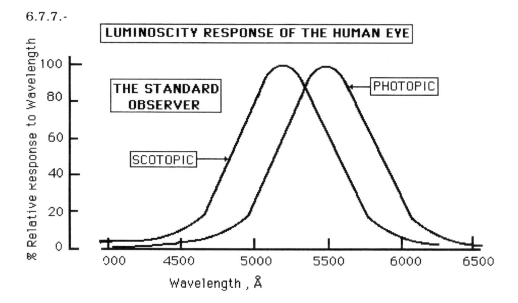

Now, let us examine the effects of colors as perceived by the human eye.

b. The Nature of Chroma

The visible spectrum extends from about 4000 Å to 7000 Å. We find that the eye acts as an integrating instrument. Thus, two colors may appear equal to the eye even though one is monochromatic light and the other has a band of wavelengths. This is shown in 6.7.8., given on the next page. In this case, we may see the same color, but the photon energies are much different.

It was Newton, using a glass prism plus slits, who first demonstrated that sunlight consisted of colors or chroma. However, the idea of monochromatic light compared to a band of color, i.e.- an assembly of lines, completely escaped the notice of investigators until the 20th century. Subsequent work then showed that colors could be duplicated by mixing the three primaries, red, green and blue to obtain the various chroma, including shades of "white". Actually, these shades involved the luminous intensity, that is- the amount of light falling upon a surface.

6.7.8.-

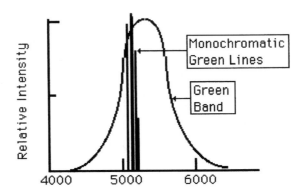

Let us first define some of the terms of measurement of luminosity and then proceed to determine how to measure the chroma, or chromaticity.

Consider a thin flat plate which has no absorption, only reflectance and transmittance. We will find that light falling upon its surface with a certain intensity has a luminance, L, while the light transmitted is defined as H, the exittance, or emittance. H will equal L if no absorption takes place or if there is no scattering at the surface of the thin plate. However, even at atomic distances, a certain amount of **scattering** does take place. We find that two types of scattering are possible. If the surface is perfectly smooth on an atomic level, then a light wave would be back- scattered along the same exact path, and we would have a **perfect** diffuser, which we call S_0 . However, there is always an angle associated with the scattering, which we call S_α (where α is defined as the scattering angle), and we have an imperfect diffuser, as shown in 6.7.9., presented on the next page.

If we view the thin plate from the left where it is illuminated with intensity, L, what we see is the scattered light, or light diffusion from the surface. If the plate is a perfect diffuser, then we will see the exact amount of L scattered back along the same plane as a diffuse component. Note that we are not speaking of reflection (which is an entirely different mechanism where the wavelength of the light is affected) but of scattering (where the light is absorbed, then reëmitted at the same wavelength). For scattering by a perfect diffuser, $L_0 = I_0 / S_0$. However, this is never the case.

6.7.9.-

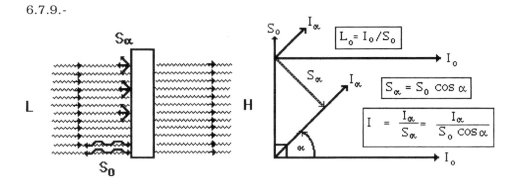

What we find is that there is an angle dependence of scattering, and that:

6.7.10.- $I = I_\alpha / S_\alpha = I_\alpha / (S_0 \cos \alpha)$

where α is the angle of scattering. Therefore, H does not equal L. If we define ϕ as the flux of light, i.e.- the number of photons incident per second, we find:

6.7.11.- $H = \phi / S_\alpha = \phi / S_0 \cos \alpha$ where: $\phi = 4\pi I$ (point source)
$$\phi = \pi I \text{ (flat surface)}$$

The intensity units for ϕ are related to a primary radiation source, the **candela**, Cd. The definition of a candela is:

6.7.12.-   CANDELA: *a unit of luminous intensity, defined as 1/60 of the luminous intensity per square centimeter of a black-body radiator operating at the temperature of freezing platinum (1772 °C). Formerly known as a candle.*

This gives us the following intensity units, as shown in 6.7.13., given on the next page. Thus, we have two units of measurement of intensity. One is related to scattering from a surface, L, i.e.- in foot-lamberts and the other is related to emittance, H, i.e.- in lumens per square foot. It is well to note the differences between these two units. Many times, they are confused with one another and not used correctly.

6.7.13.-   INTENSITY UNITS

$$1.0 \text{ Cd (at a one-foot distance)} \equiv L/\pi \text{ (foot-lamberts)}$$
$$\equiv H/\pi \text{ (lumens)}$$

Although we have assumed "white" light up to now, either of these two can be wavelength dependent. If either is wavelength dependent, then we have a pigment (reflective- but more properly scattering) with intensity in foot-lamberts, or a phosphor (emittance) with intensity in lumens.

In terms of color, we have additive processes (emittance) and subtractive processes (reflectance). If we wish to match colors, the primary colors are **quite different,** namely-

6.7.14.-   <u>ADDITIVE PRIMARIES</u>   <u>SUBTRACTIVE PRIMARIES</u>

red	magenta
green	yellow
blue	cyan

Let us now consider how to set up a color matching system.

c. The Standard Observer

Since color matching is meant for humans, it is natural to define color in terms of an average, or "Standard Observer". Our first step is to build an instrument which contains three colored lamp sources, a place for the observer, intensity detectors, and a monochromator. One design is shown in 6.7.15., given on the next page.

There are two (2) sources of light to be compared. One is from a set of three lamps whose emission is modified by means of suitable filters to give a red beam , a green beam and a blue beam. These are mixed at the screen to form a single spot (although we have not illustrated it in that way, so as to be more discernible). The other source comes from a monochromator so as to obtain a monochromatic beam of light.

6.7.15.-

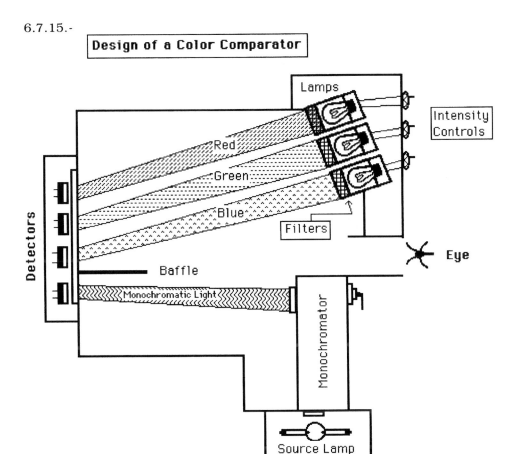

There are controls to adjust the individual beams of red, green and blue light, as well as that of the monochromatic beam. In this way, the mixed beams of light can be directly compared to the monochromatic spot. In the back of the screen are detectors to measure the energy intensity of the beams of light being compared.

We need about 5000 observers to obtain a satisfactory average, both for the dark-adapted and the light-adapted human eye. Note that we can compare any color in terms of red + green + blue to a monochromatic color.
There are three (3) things that we need to accomplish:

6.7.16.-   1. Define eye response in terms of color at **equal energy**
            2. Define shades of "white" in terms of % red, % green and %
               blue, at equal energies of those "whites".
            3. Define "color" in terms of % red, % green and % blue, as
               compared to monochromatic radiation.

The difficulty in setting up the initial system for color comparisons cannot
be underestimated. The problem **was** enormous. Questions as to the
suitability of various lamp sources, the nature of the filters to be used, and
the exact nature of the primary colors to be defined occupied many years
before the first attempts to specify color in terms of the Standard
Observer were started. But before we further delve into these problems,
let us first establish a better background into the factors controlling this
investigation. Some of this work we have already discussed. It is
presented again to create an integrated approach to the problem of color
specification.

It had long been known that an incandescent solid emits electromagnetic
radiation. Work by Planck (1889) on the concept of a "black-body" led to
Planck's Law: $E = h \gamma$, which was the beginning of the quantum theory.
The "black-body" is a theoretical concept of a radiating body which
absorbs all radiation incident upon it, but reëmits that radiation according
to certain laws, including its absolute temperature. Experimentally, we
approximate the black-body by a blackened sphere with a small hole in it
for the internal radiation to emerge. When the sphere is heated, the
energy emitted can be calculated from:

6.7.17.- $I (\lambda) = 3.703 \times 10^{-5} \lambda \, d\lambda / [\exp \{(1.432 / \lambda T) -1\}]$

which reduces to :   Total radiation = 5.73×10^{-5} T^4 ergs/sec/cm^2.

where T is in °K, and λ is the spectrum of the black-body emission. We
therefore speak of the "color-temperature" of an emitting source
(including lamps and phosphors) as related to a black-body at that same
temperature. The wavelength distribution of a black-body at several
temperatures is shown in 6.7..18., given on the next page.

6.7.18.-

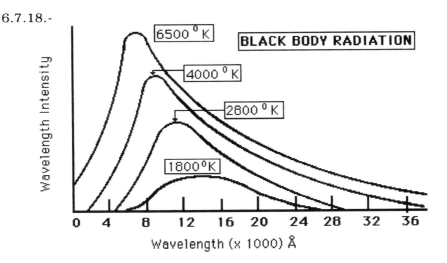

It is easily seen that temperatures around 6500 °K. are necessary to bring the peak maximum into the middle of the visible spectrum. Although the Sun is a black-body radiator of about 10,000 °K (as viewed directly from space), scattering and reflection within the Earth's atmosphere is sufficient to lower the effective black-body radiation perceived to 6500 °K. Thus, the Sun is a 6500 °K. source which we call "daylight". The direct viewed brightness of the Sun at the Earth's surface is about 165,000 candela/cm^2 , that of the Moon - 0.25 candela/cm^2 , and a clear sky is about 0.8 candela/cm^2. For these reasons, DAYLIGHT has been defined as: "The northern skylight at 11:30 am. at Greenwich, England on October 31, 1931". This is also the definition of ILLUMINANT- C. The other standard illuminant that we use is ILLUMINANT- A, which is the radiation emitted from an incandescent tungsten filament operating at 3250 °K. The spectra of these sources are shown in the following diagram, given as 6.7.19. on the next page.

Referring back to our Color Comparator of 6.7.15., we use these concepts to calibrate our lamps in terms of spectra and relative energy in terms of these standard sources. ILLUMINANT - B , by the way, was originally defined as "average sunlight" but it was soon determined that "average" is not the same at all parts of the Earth's globe.

6.7.19.-

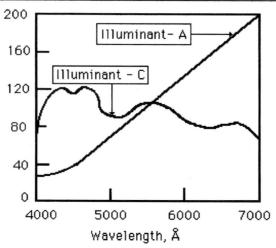

Our next step in using the Color Comparator is to set up proper filters so as to obtain and use "primary" color lamp sources. We find that by using the monochromator of the Color Comparator, we can approximate the wavelength response of our so-called "Standard Observer", in the red region, the green region and the blue region of the visible spectrum.

But we also find that we need a band of wavelengths for each color, since a monochromatic beam is not at all suitable. This is where the choice becomes subjective, since we are relying upon the spoken response of individuals. We find that by choosing a blue filter peaking at 4400 Å, a green filter peaking at 5200 Å, and a red filter peaking at 6200 Å (but having a lesser peak in the blue), we have a "Blue"-blue, a "Green"-green and a "Red"-red which will satisfy most observers. Note that the original single color was chosen by approximating the human eye response in the three-color regions, using monochromatic light to obtain a brightness response, i.e.- 6.7.7., and then adjusting the broad-band transmission properties of the three filters of the lamps to obtain the proper colors. A final check of these lamp filters would be to mix all three colors additively, and then to evaluate the "white" thereby produced. We can then substitute a Standard Lamp for the monochromator and see if we can reproduce its exact color temperature. If not, then we need to modify the transmission characteristics of our filters used on the source lamps.

Once we have done this, we now have our three primary colors in the form of standard lamps, and can proceed to determine Items 1,2 & 3, given in 6.7.16. To do this, we vary the wavelength of the monochromatic light, and determine relative amounts of red, green and blue light required to match the monochromatic color. This is done, as stated before, for about 5000 observers. The result is finalized response curves for the Standard Observer, also called "Tristimulus Response curves".

6.7.20.-

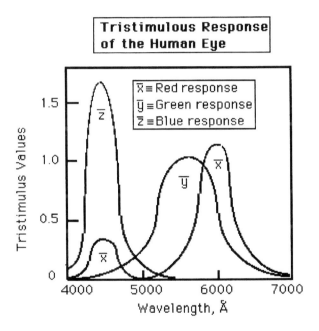

We finally arrive at the result we want, since we can now set up "Tristimulus Filters" to use in defining colors. We can now define $\bar{y}$ as our standard luminosity curve for the human eye (photopic vision). Note that $\bar{x}$, the red tristimulus value, has a certain amount of blue in it in order to duplicate the response of the red preceptor in the retina. Note that these colors are the result of the "Standard Observer" measurements that we started in the first place.

The above is known as the C.I.E. chromaticity diagram (Commission Internationale de l'Eclairage). First we note that the boundary of this x and y diagram is bounded, as we have already stated, by the values of monochromatic light. Thus, we can find any color, be it monochromatic or not, in terms of its x and y coordinates. It would be more dramatic to print the various hue areas in color but it is difficult to accurately print reflectance hues. It is better to name the colors directly. This was done by Kelly (1940). Note also that we do not use the term "color" anymore but use the term "hue".

Any hue can be specified by x and y. For example, we can specify the locus of black-body hues and even Illuminants A, B & C, namely-

6.7.25.-   x and y Chromaticity Coordinates

	x	y
6000 Å	0.640	0.372
5200	0.080	0.850
4800	0.140	0.150
ILLUMINANT A (3250 Å)	0.420	0.395
ILLUMINANT B (4500 Å)	0.360	0.360
ILLUMINANT C (6500 Å)	0.315	0.320

These are shown in the following diagram, given as 6.7.26. on the next page. Note that all of these are **emitters**, similar to that of a phosphor.

Let us now summarize the results we have achieved. We have measured the luminosity response of the human eye, in terms of photopic and scotopic behavior. We also defined a "black-body" and its wavelength emission, stipulating its absolute temperature. We then defined Standard Sources. We next designed a Color Comparator and then determined the transmission characteristics of three (3) filters required to duplicate the response of the three color preceptors of the human eye. These we called the tristimulus response of the Standard Observer.

6.7.26.-

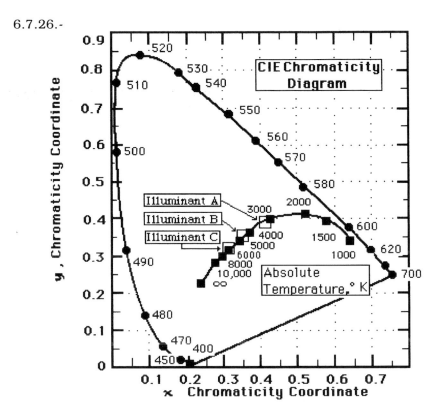

Finally, we obtained Chromaticity Coordinates which we could use to plot various hue values. If we have a Color Comparator, then we can measure any hue and compare it to any other. But, if we do not, we either build or buy one, or resort to an instrumental method. The Color Comparator uses a human observer to specify color. If we wish to use an instrumental method, then we must correct each component of the instrument to the chroma response of the Standard Observer. While the Color Comparator given above in 6.7.15. was perfectly satisfactory for setting up a system of chromaticity coordinates, it was difficult and awkward to use. What was really required was an instrumental method of color measurement.

The requirements for an instrumental method of specifying color include a light source, the colored object and a detector, arranged as follows in 6.7.27., shown on the next page.

6.7.27.- Components Required for an Instrument to Measure Color

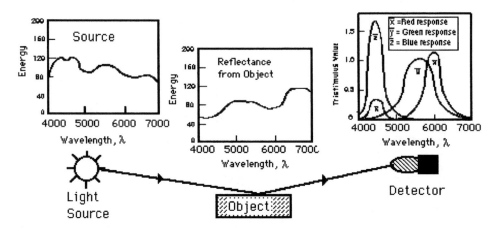

Note that we must be able to define the output of the source in terms of its wavelength. This includes any source, whether it is an incandescent lamp or a discharge lamp. The output must be stable. We also must define the reflectance of the object whose color we are trying to measure. This mandates a detector whose response to wavelength has been standardized. These are parameters of the instrument which are not always recognized until one begins to develop an instrument to measure color, particularly that of reflectance. While this measurement might seem simple enough to perform, it is not.

Since the response characteristics of these optical components are not linear, nor flat, we need an analogue system in order to be able to measure color. The analogue system simply corrects for the non-linearity of the source and detector, as shown in the following for emittance. This is shown in 6.7.28., on the next page.

This diagram illustrates the optical analogue process in which the properties of an instrument light source, a properly selected filter, and a photo-detector are combined to provide an optical analogue of the similar properties of a CIE standard illuminant (here ILLUMINANT C) and a CIE standard observer response function (here the $\overline{y}$ response).

6.7.28.-

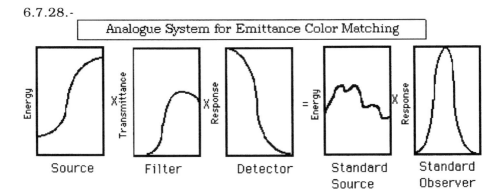

You might question why we need a light source to measure an emitter. The light source acts as a standard for comparison, using the reflectance from a standardized reflectance material so as to be able to use the STANDARD OBSERVER response we have already developed for color matching.

The corresponding optical analogue for reflectance is:

6.7.28.-

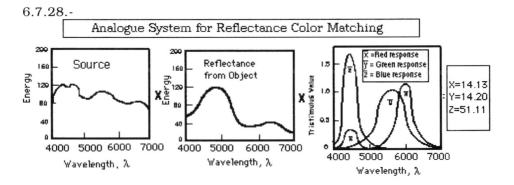

This diagram shows schematically how the spectral power distribution of a CIE source, the spectral reflectance, R, of an object, and the spectral color-matching functions, x, y and z, combine by multiplication (each wavelength by each wavelength), followed by summation across the spectrum, to give the CIE tristimulus values. However, these analogues are actually hypothetical. The actual values we need to measure reflectance are:

6.7.30.-

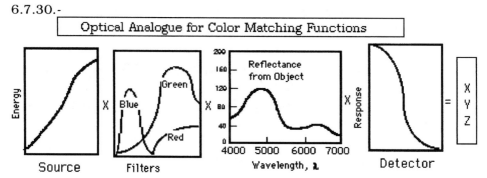

When we introduce the color response characteristics of the source, combined with the detector of our instrument, we find that we must drastically modify the transmission characteristics of our filters in order to duplicate the CIE color matching functions for the equal-energy spectrum. However, this is not an impossible task and we find that an excellent match can be obtained to the transmission functions of 6.7.20. This is typical for commercially available instruments. Now, we have an instrument, called a Colorimeter, capable of measuring reflective color.

The process for measuring emissive color is somewhat different. First, we obtain an emission spectrum by means of a spectrofluorimeter. We can now integrate I $d\lambda$ to obtain the energy and then specify this in terms of x and y (see 6.7.20. and 6.7.22.). However, it is nearly impossible to use each and every wavelength present. Referring to 6.7.20., we can see that if we draw **vertical** lines on each color function, the density of which are spaced according to the peak height of each function, we would obtain what we call "weighted functions". We would then have a set of lines (wavelengths) useful for calculating chromaticity coordinates. That is, We would have a set of lines whose spacing was inversely proportional to peak height (the higher the intensity, the closer would be the spacing), where the line spacing would have to be close between the wavelengths:

6.7.31.-   x = 5500 - 6500 Å   y = 5100 - 6000 Å
                z = 4200 - 4800 Å

A practical method for doing this is to **weight** the line heights by using tristimulus values times the height of our experimental curve, as determined by measurement at the specified wavelength, times the energy distribution of a Standard Source such as ILLUMINANT C.

In the following, we show sixteen (16) **weights,** spaced 200 Å apart, for all three (3) tristimulus values, X, Y, Z. By multiplying the line heights of a spectrum by these values and then summing them, we can obtain values of X , Y & Z.

6.7.32.-   STANDARD WEIGHTS FOR ILLUMINANT C

Wavelength	X-Weight	Y-Weight	Z-Weight
4000 Å	0.044	0.001	0.187
4200	2.926	0.085	14.064
4400	7.680	0.513	38.643
4600	6.633	1.383	38.087
4800	2,345	3.210	18.464
5000	0.069	6.884	5.725
5200	1.193	12.882	1.450
5400	5.588	18.268	0.365
5600	11.751	18.606	0.074
5800	16.801	15.989	0.026
6000	17.896	10.684	0.012
6200	14.031	6.264	0.003
6400	7.437	2.897	0.000
6600	2.728	1.003	0.000
7000	0.175	0.063	0.000
Sum	98.046	100.001	118.100

x = 0.3101   y = 0.3163

For example, the following, given as 6.7.33. on the next page, is a spectrum of a Cool-White fluorescent lamp, with the required wavelengths already marked off.

6.7.35.-

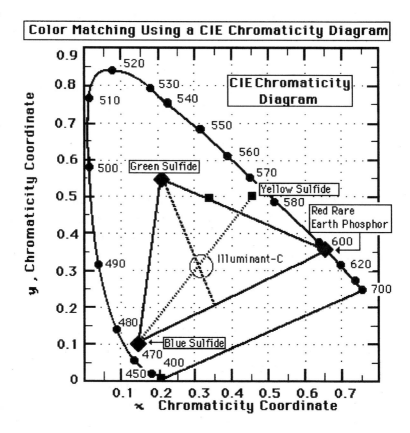

But such a phosphor is not currently available. It has been estimated that only about 85% of the hues found in Nature can be reproduced by a color picture tube.

We can also calculate and mix colors predictably and accurately. We use what has been called the "lever principle". We can mix phosphors or solid dyes or any other colored material, provided that we know, or can measure, the x and y values. Let us take the blue sulfide: x = 0.150 and y = 0.100, and mix it with a yellow emitting phosphor having x = 0.450 and y = 0.500. Note that we can reproduce any hue from yellow to blue, including bluish-white and yellowish-white. We draw a straight line between the two points and then measure the distance between the two points. If we start with 100% yellow, adding blue shifts the hue toward the blue. At 50%-50% blue to yellow, we are on the bluish white side. The exact figures required to reproduce the 6500 °K white are 45.7% blue and 54.3% yellow, providing the efficiencies of the phosphors are about equal. If we were using the three hues given in 6.7.35., we first use the baseline between the red and blue phosphor and draw a second line that intersects within the 6500°K. ellipse of ILLUMINANT - C.

f.- Color Spaces

If we have a certain color, a change in intensity has a major effect on what we see (in both reflectance and emittance). For example, if we have a blue, at low intensity we see a bluish-black, while at high intensity we see a bluish-white. Yet, the hue has not changed, only the intensity. This effect is particularly significant in reflectance since we can have a "light-blue" and a "dark-blue", without a change in chromaticity coordinates.

The concept of "lightness" involves the reflective power of materials. If the reflectance approaches 100%, we say the material is "white", whereas complete absorption (0.00% reflectance) produces a "black" material. Let us examine exactly what is meant by these terms, particularly those of **subtractive** color mixing. When colors are prepared by mixing dyes or pigments, the resultant reflective hue is controlled by a subtractive process of the three (3) primaries, whose reflectance spectrum is given above. When these are mixed, the resulting hue is that where the curves **overlap,** as is easily seen in the following diagram, given on the next page as 6.7.38.

When we mix these primary colors, their reflectances remove more of the incident light, and we see the part where the reflectances are reinforced. Thus, we can get red, green and blue, but they are not primary colors in the subtractive system. Intermediate hues can be obtained also in this process when the subtractive primaries are used in less than full concentration. That is, they are "lightened". Although we can explain this effect on a spectrophotometric basis, we do not have a way of specifying hues in terms of **saturation**, using the CIE system. This is one of the failings of this system

6.7.36.-

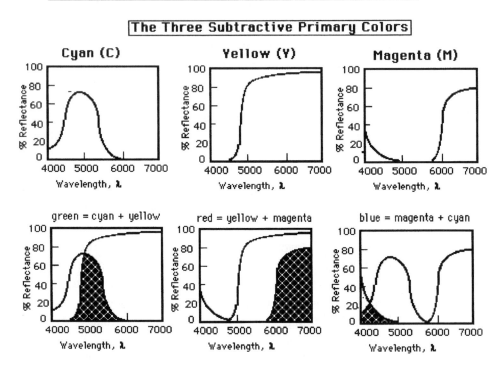

Nevertheless, we do not have a better system for specifying color at this time. There are other methods, which we will now discuss, in which this deficiency was addressed. However, as you will see, none have been entirely successful. It is the CIE system that is used mostly for color specification, particularly for emissive materials like phosphors.

g. The Munsell Color Tree

One of the first attempts to specify reflective colors, color mixing and saturation was accomplished by Munsell (1903). He devised a color system based on factors he called **hue, chroma and value.** Munsell set up a three-dimensional arrangement based upon *minimum perceptual color* *difference* steps. He based these upon direct observation because he did not have the instrumental means to do so. Therefore, his results are not the same as those we are now using for the CIE system.

Munsell set up a cylinder whose vertical axis began with "black" and ended with "white". This is the factor "Value" in Munsell notation, each layer of which is given a specific number. He also defined five (5) primary hues. These are: purple, red, yellow, green and blue. They are arranged around a circle as shown in the following, and a series of circles are "stacked" to form a "tree", namely:

6.7.37.-

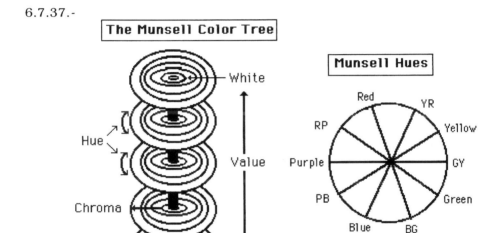

In Munsell's system, hues are specified in a circular fashion with the same set of hues on the same level. As the value, i.e.- "lightness", changes, one jumps to the next highest circle. The amount of color "deepness" is specified as chroma and becomes less as one approaches the edge of each color circle.

The advantage of this SYSTEM is that in addition to being able to specify hue, we can also specify Chroma, which is the degree of saturation of a specific hue, as we move from the center outwards. Value is the relative amount of "blackness" or "whiteness" (equal parts of black and white produce gray) in the particular reflective color being specified. It is this system that gives us access to the degree of "grayness" and/or saturation of any hue.

On a practical basis, if we wish to set up this system, we would assemble a set of "color-chips" at each point on the Color-Tree. Each color-chip would be specified by two factors, H = hue, and V/C, which is value (grayness) modified by chroma (saturation). The actual number of layers in the Munsell Color Tree was determined by "minimum perceptual difference". That is, the minimum change in color that produces a perceptible difference. This arrangement specifies all light colors as well as the dark ones. To use such a system, one would choose the color-chip closest to the hue and saturation of the test color and thus obtain values for H and V/C.

However, it was soon discovered that the system was not perfect. Reasons for this include the facts that the hues defined by Munsell are not those of the primaries of the human preceptor. Furthermore, Munsell was somewhat subjective in his definitions of hues. **And**, the system did not take into account the luminosity response of the human eye in regard to color. If one does so, then the Munsell System becomes "bulged" in the direction of the yellow-green regions and shrunken in the blue and red regions. We can illustrate this by taking the Munsell colors for Value-5 and Chroma-8 and plotting them on a CIE diagram, as shown in 6.7.38., given on the next page.

What we should get is a circular spacing of the ten Munsell colors, if the Munsell system is truly accurate in regard to the luminosity factor. **However, they are not**. On the CIE diagram, the spacing is considerably distorted. Since we already know the CIE method to be corrected for the luminosity factor, the conclusion is obvious. In 1920, Priest showed that if the Munsell-Chips were viewed on a white-background, the "brightness", i.e.- lightness as viewed by the human eye, could be related to the Munsell system by:

6.7.39.- $$V = 10 \, Y^{1/2}$$

6.7.38.-

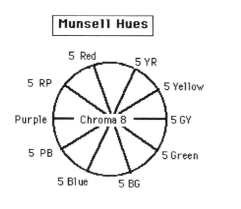

Munsell Hues

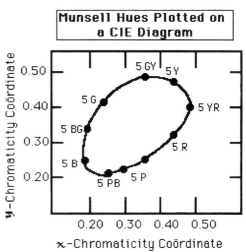

Munsell Hues Plotted on a CIE Diagram

where Y in 6.5.39. was defined as "brightness" or eye response. Over the years, the Munsell System has been modified in an attempt to agree with actual human perception, using a "middle-gray" background". Y was defined as the Tristimulus Value of the CIE color system and used in the following formulae, given as follows:

6.7.40.-   Munsell- Godlove (1927): $V = [1.47\ Y - 0.474\ Y^2\ ]^{1/2}$

N.B.S.- Judd (1943) : $Y/Y_{MgO} = 1.2219\ V - 0.2311\ V^2 - 0.2395$
$$V^3\ ^-0.0201\ V^4 + 0.008404\ V^5$$

Glasser (1958): $V = 25.29\ Y^{1/3} - 18.38$

The Judd formula given here is too difficult to use and it is the Glasser formula which came into general use to modify the Munsell Color System back towards the cylindrical form. Nevertheless, it is well to note that it was the subjective observation of the lack of correction for luminosity in the Munsell System that gave impetus to the original development of the CIE Color System. The major problem with the Munsell system was that

6.7.42.-   u   =   4x / (-2x + 12 y + 3)

            v   =   6y / (-2x + 12 y + 3)

Plotting these values gave the following chromaticity diagram:

6.7.43.-

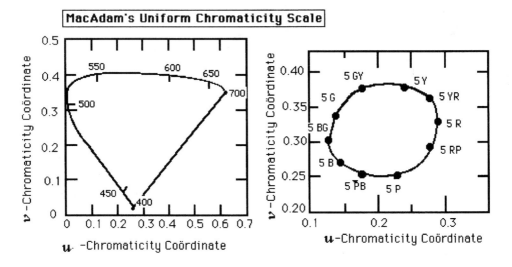

The true test is how well the Munsell hues plot out on the CIE diagram. As can be seen on the right hand side of the diagram, the Value 5 - Chroma 8 hues do construct a nearly perfect circle. Thus, the MacAdam transformation is a definite improvement over the 1931 CIE system. In 1960, the CIE adopted the MacAdam System, having defined the equations (with MacAdam's help):

6.7.44.-   u =   4 X / (X + 15Y + 3Z)

                 v =   6 Y / (X + 15 Y + 3Z)

The following diagram shows MacAdam's Uniform Chromaticity Scale, and has Minimum Perceptible Color Difference steps plotted as ellipses:

6.7.45.-

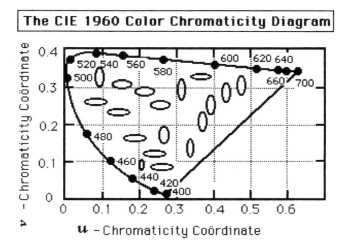

Note that the sizes of the ellipses are about equal in all parts of the diagram and in most regions of the color spectrum. This system was fully adopted by the CIE in 1964 and remains the recommended system for color matching.

As a final step, let us now return to phosphors and show the various types of fluorescent lamps manufactured in terms of emittance hues and tolerances. In the following diagram, given as 6.7.48 on the next page, we have plotted the Standard Specifications for fluorescent lamps, in terms of the chromaticity as defined by MacAdam ellipses.

It is these which are in use in the Industry for quality control of phosphors and fluorescent lamps. The ellipses show the color tolerance for emittance of each of the lamp hues. Thus, they are actually the Minimum Perceptible Color Difference, for each of these lamp colors.

Because the lanthanides ("rare earths") have become prominent in the manufacture of fluorescent lamps, we will address this technology next to show how they differ from the $^1S_0$ activators given above. We will find that these cations can be used in situations where no other activators can be applied.

Table 6-6

Name	Trivalent State	Investigator
Lanthanum	$La = 4f^0$	Mossander - 1839
Cerium	$Ce = 4f^1$	Berzelius - 1814
Praseodymium	$Pr = 4f^2$	Welsbach - 1885
Neodymium	$Nd = 4f^3$	Welsbach - 1885
(Promethium)	$(Pm) = 4f^4$	Fermi Labs - 1969
Samarium	$Sm = 4f^5$	de Boisbandran- 1879
Europium	$Eu = 4f^6$	Demarcy - 1869
Gadolinium	$Gd = 4f^7$	de Marignac - 1881
Terbium	$Tb = 4f^8$	Delafontaine - 1878
Dysprosium	$Dy = 4f^9$	de Boisbandran - 1886
Holmium	$Ho = 4f^{10}$	Cleve - 1879
Erbium	$Er = 4f^{11}$	Cleve - 1879
Thulium	$Tm = 4f^{12}$	Cleve - 1879
Ytterbium	$Yb = 4f^{13}$	de Marignac - 1878
Lutecium	$Lu = 4f^{14}$	Urbain, Welsbach, James - 1907
Yttrium	$Y = 4f^0$	Mosander - 1843
Scandium	$Sc = 4f^0$	Nilsson - 1879

II. Chemistry of the Lanthanides

Because each rare earth has essentially the same electronic configuration, it is very difficult to separate them from each other. Indeed, even a given rare earth said to be 99.99% pure will have several hundred parts per million of the others which comprise the ore from which they were refined.

The rare earths can be divided into two classes, the "lights" and "heavies", i.e.- $< 4f^7$ and $> 4f^7$, respectively. There are slight differences in chemical reactivity which follow this rule as well. Lanthanides form cations in solution and are not amphoteric. That is, they do not form anions. The oxides are the usual form, and most have the formula, Ln_2O_3. There are some mixed valence states which result in formulas like Pr_6O_{11} (which contains Pr^{3+} and $Pr^{4+)}$, and Tb_4O_7 (Tb^{3+} and Tb^{4+}, i.e. - $Tb_2O_3 + 2TbO_2$).

The rare earths follow the same pattern of solubility encountered in most other "non-rare earth" compounds, i.e.-

6.8.2.- SOLUBILITY OF THE RARE EARTH SALTS

SOLUBLE	INSOLUBLE
$Ln(NO_3)_3$	$Ln(OH)_3$
$LnCl_3$	LnF_3
$LnBr_3$	$LnBO_3 \cdot 2\ H_2O$
LnI_3	$LnPO_4 \cdot n\ H_2O$
$Ln_2(SO_4)_3$	$LnVO_4$
	$Ln_2(C_2O_4)_3 \cdot n\ H_2O$

This list is not all-inclusive but **is** illustrative. The oxalate is the most insoluble salt and is most often used for recovery of rare earths from solution. The commercial ores mined for extraction of rare earths include:

6.8.3.- LANTHANIDE ORES

NAME	GENERAL FORMULA
Monazite	$(Ce,La,Th,Ln)PO_4$
Bastnaesite	$(Ba,Sr,La,Ce,Ln)\ F_n\ CO_3$
Xenotime	$(Y,Gd,Ln)PO_4$
Euxenite	$(Ln,Y)NbTaTiO_6$

Only the major cations are shown in the above formulae, even though each ore contains **all** of the rare earths in various proportions. One commercial chemical process used for **initial** separation of the rare earths follows this sequence of steps, as shown in 6.8.4., presented on the next page.

Note that this involves ordinary laboratory procedures. The separation of the individual rare earths in a pure form is another matter. This problem was not solved until the late 1930's and early 1940's, when Spedding and co-workers (University of Iowa-Ames Laboratories) applied ion- exchange chromatography to the problem.

6.8.4. - CHEMICAL SEPARATION OF THE RARE EARTHS

1. ORE: Leach crushed rock with 96% H_2SO_4 = 98% dissolution.
2. Dissolve into water = dissolved rare earth cations.
3. Precipitate with oxalic acid = 99 % + Recovery
4. Fire oxalates to oxides (> 800 °C).
5. Product = Mixed rare earth oxides.

To do so, one uses a hydrogen-ion exchange resin which is loaded into a long glass column. For commercial separation, this consists of several 12" by 25 foot long columns in tandem, and the process takes several weeks (typically 4 to 6) to complete the process. After considerable work with several types of resins, Spedding was able to establish the following retention series on an ion-exchange column:

6.8.5.- $Th^{4+} > La^{3+} > |Ce^{3+}| > |Y^{3+}| > Lu^{3+}> Ba^{2+}>Sr^{2+}>Ca^{2+}>K^+>NH_4^+> H^+$

where $|Ce^{3+}|$ and $|Y^{3+}|$ represent the light and heavy fractions, respectively. The process includes loading a column with a lanthanide solution, and then eluting with a 5% ammonium citrate solution at a pH of 2.5 to 3.2 . Under these conditions, one gets a separation of each rare earth, which moves through the columns as overlapping bands, one following the other. We can characterize such a separation as:

6.8.6-   Ce (lights) Fraction
             Y (heavies) Fraction

The Ce-fraction is the familiar off-color beige powder, long used as a **glass-polishing agent.** Rare earth **metals** can be prepared by fused-salt electrolysis, starting with the fluorides. The product is a "mixed-metal" , or "mischmetal", used to produce cigarette lighter-flints.

However, this method was much too slow and costly. In the 1950's, a new method was developed by the Bureau of Mines in Colorado. This method involved solvent extraction with an aqueous nitrate solution of lanthanide cations, shaken together with an immiscible organic phase such as CCl_4

containing tri(n-butyl)-phosphate. Its extraction coefficient was $\alpha = 1.50$. The most useful, and expensive, rare earth is Eu^{3+} and it was soon determined that DEHPA extracted more europium from the ore. The latest process uses DEHPA, a di(2-ethyl hexyl) phosphoric acid, as a complexing and extraction agent. The extraction coefficient for DEHPA is $\alpha = 2.50$. By pH control, one can select and extract a single lanthanide as desired. This process has allowed the production of products containing 99.99% of a single lanthanide and even 99.999% purity (these levels of purity are in terms of the dominant rare earth present). All of the lanthanides are presently available in a purity up to, and including, 99.9999%. The commercial process used by Molycorp is shown in the following diagram given as 6.8.7. on the next page. The extraction process involves an organic insoluble phase in contact with an acidic aqueous phase of controlled pH.

Note that aqueous flotation (with appropriate detergent compounds) is used to separate the rare-earth-rich particles from the siliceous materials. After drying, the ore is calcinated (fired in air to form the oxides) and then treated with concentrated HCl. A cerium-rich concentrate is left after the other rare earths have dissolved. We have not shown how the cerium concentrate is processed since the main object is to obtain the more expensive oxides as a final product. The rare earth chloride solution is then adjusted to pH = 1.0, purified with activated carbon to remove some of the heavy metals and then extracted with DEHPA. The first extraction separates the "heavy" and "light" fractions of the rare earths. The resulting extracted solutions are then subjected to further extraction processes to produce the individual products, including La, Pr and Nd.

The organic solution contains the La concentrate which is purified, extracted further and then precipitated as single hydroxides. The other fraction is readjusted to pH = 3.5 and further extracted. After the first 2 extractions (which are returned to the pH = 1.0 step), concentrated HCl is added and the last 2 extractions are used to separate the Eu, Sm and Gd fractions.

Any heavy metals left over are precipitated and separated by means of H_2S.

6.8.7.-

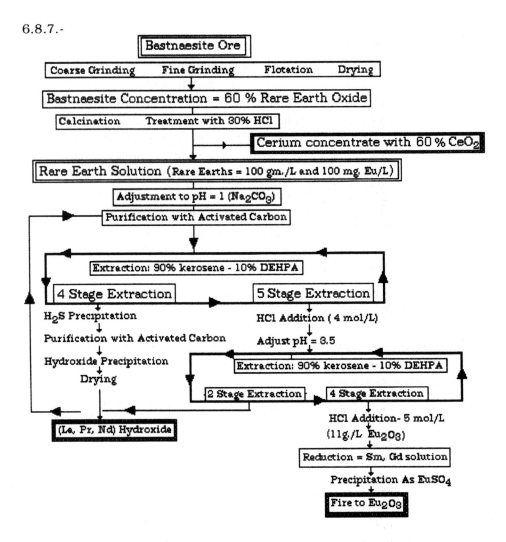

The details of this extraction process is shown in the following, given as 6.8.8. on the next page. A reduction step serves to form the Eu^{2+} state so that it can be further purified from the Sm and Gd remains that may contaminate it.

Precipitation as $EuSO_4$ is the final step of separation. Calcination forms the oxide, the usual commercial form of europium, i.e.- Eu_2O_3.

6.8.8.-

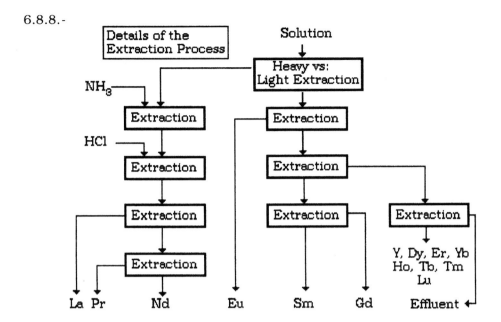

III. <u>Rare Earth Energy Levels and Electronic States</u>

Although the lanthanides can possess up to 14 identical f-electrons, each rare earth differs from the others in its optical and magnetic properties. For the general case, in terms of effective radius, <r>, we have the following relation of electronic energy levels, shown as follows:

6.8.9.-

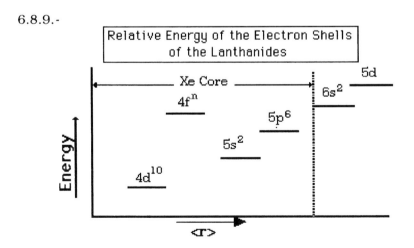

We can thus write the electronic configuration in terms of ascending energy as: Xe $4f^n$ ($5s^2 5p^6$) $6s^2$ 5d, where the 5d orbital is ionized first. then the $6s^2$ orbitals. Electronic transitions (ground to excited state) take place among the $4f^n$ orbitals. Because of the **shielding nature** of the $5s^2 5p^6$ shells, perturbation by surrounding neighbors does not take place and **sharp** line spectra result.

But, since l, the azimuthal quantum number, can take on values from 0 to 1,2,3,4, etc., with s = 3 1/2 (i.e.- 7 x 1/2), the observed spectra can become quite complex. The following table shows neutral atom electron configurations and observed valences:

TABLE 6- 7
Electron Configurations of the Rare Earth Elements

Number	Symbol	Electronic Configuration		Observed
		Neutral Atom	Trivalent Atom	Valence
57	La	$4f^0$ $6s^2$ 5d	$4f^0$	3
58	Ce	$4f^1$ $6s^2$ 5d	$4f^1$	3,4
59	Pr	$4f^3$ $6s^2$	$4f^2$	3
60	Nd	$4f^4$ $6s^2$	$4f^3$	3
62	Sm	$4f^6$ $6s^2$	$4f^4$	2,3
63	Eu	$4f^7$ $6s^2$	$4f^6$	2,3
64	Gd	$4f^8$ $6s^2$ 5d	$4f^7$	3
65	Tb	$4f^9$ $6s^2$	$4f^8$	3,4
66	Dy	$4f^{10} 6s^2$	$4f^9$	3,4
67	Ho	$4f^{11} 6s^2$	$4f^{10}$	3
68	Er	$4f^{12} 6s^2$	$4f^{11}$	3
69	Tm	$4f^{13} 6s^2$	$4f^{12}$	3
70	Yb	$4f^{14} 6s^2$	$4f^{13}$	2,3
71	Lu	$4f^{14}$ $6s^2$ 5d	$4f^{14}$	3

For any given atom, the number of **possible** states can be calculated from combinatorial theory, i.e.-

6.8.10.- $\qquad C^k_n = C^k_{14} = 14 ! / (f ! [14-f] !)$

where f is the number of electrons present in the particular electron state of the specific rare earth. This gives:

6.8.11.- NUMBER OF ELECTRONIC ENERGY LEVELS POSSIBLE

	Ln^{3+}		
f =	1	3	7
Spin and Orbit:	14	364	3432
J- Levels:	2	41	327

It turns out that the spin and orbit numbers are actually the degeneracy of the energy levels in question and that they represent the possible Stark State splitting pointed out in a previous chapter. The full set of values are shown in the following Table, namely-

TABLE 6-8

NUMBER OF STATES WHICH ARISE FROM THE 4f CONFIGURATION

f =	1	2	3	4	5	6	7
Total Number of J-Levels:	2	13	41	107	198	295	327
Multiplets:	1	7	17	47	73	119	119
Degeneracy : (spin and orbital)	14	91	364	1001	2002	3003	3432
GROUND STATE: MULTIPLET	^2F	^3H	^4I	^5J	^6H	^7F	^8S

Obviously, for f = 8, 9, the total number of J-levels will equal 295, 198, with the same degree of degeneracy, etc.

As we showed in Chapter 5, the Hamiltonian for the 4f - field will likewise consist of several parts:

6.8.12.- HAMILTONIAN FOR 4f ELECTRONS

Central Field   Coulomb   Spin-Orbit Crystal field

H_{4f} $= H_0$ $+$ H_c $+$ H_{so} $+$ H_{cf}

electrons   terms   J-levels   Stark components
                 (multiplets)   ($\sim$ 2000 cm^{-1})   ($\sim$ 200 cm^{-1})

The rare earths are unique in that their individual Stark States are separated by no more than 200 cm^{-1}. The following shows part of the diagram for the $4f^6$ electron configuration:

6.8.13.-

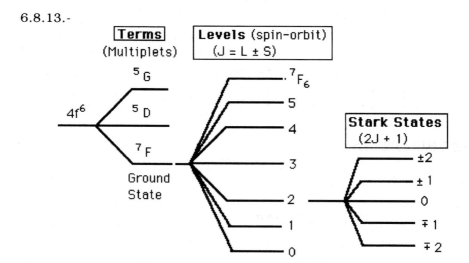

A Table is presented on the next page which shows how these states arise in L-S coupling, in terms of the number of levels and the number of multiplets. As we shall see, they are not valid because the only good quantum number for the lanthanides is J, the azimuthal quantum number.

The data in this Table lists the number of multiplets (Terms) and Levels (Spin-orbit: J = L ± S) for both odd and even electron configurations.

TABLE 6-11

Number of **J** Levels for Configurations Consisting of Equivalent 4f- electrons

ODD MULTIPLETS

Configuration	Multiplicity	Number of Multiplets	1/2	3/2	5/2	7/2	9/2	11/2	13/2	15/2	17/2	19/2	21/2	23/2	25/2	Total
f^1 and f^{13}	doublet	1	-	-	-	1	1	-	-	-	-	-	-	-	-	2
f^3 and f^{11}	doublet	12	1	3	4	4	4	3	2	2	1	-	-	-	-	24
	quartet	5	1	3	3	3	3	2	1	1	-	-	-	-	-	17
	TOTAL	17	2	6	7	7	7	2	3	3	1	-	-	-	-	41
f^5 and f^9	doublet	46	4	9	12	13	14	14	12	9	7	4	2	1	-	91
	quartet	24	5	10	13	14	14	12	9	7	4	2	1	-	-	92
	sextet	3	1	2	3	3	2	2	1	1	-	-	-	-	-	15
	TOTAL	73	10	21	28	30	29	26	20	16	9	5	3	1	-	198
f^7	doublet	72	7	12	17	20	19	18	16	12	6	3	2	1	-	142
	quartet	40	8	15	20	23	22	20	16	12	8	5	2	1	-	152
	sextet	6	2	4	5	6	5	4	3	2	1	-	-	-	-	32
	octet	1	-	-	-	1	-	-	-	-	-	-	-	-	-	1
	TOTAL	119	17	31	42	50	46	35	26	18	11	5	3	1	-	327

EVEN MULTIPLETS

Configuration	Multiplicity	Number of Multiplets	0	1	2	3	4	5	6	7	8	9	10	11	12	Total
f^2 and f^{12}	singlet	4	1	-	1	-	1	-	1	-	-	-	-	-	-	4
	triplet	3	1	1	2	1	2	1	1	-	-	-	-	-	-	9
	Total	7	2	1	3	1	3	1	2	-	-	-	-	-	-	13
f^4 and f^{10}	singlet	20	2	-	4	1	4	2	3	1	2	-	1	-	-	20
	triplet	22	3	5	9	9	11	9	8	5	4	2	1	-	-	66
	quintet	5	1	2	4	3	4	3	2	1	1	-	-	-	-	21
	Total	47	6	7	17	13	19	14	13	7	7	2	1	-	-	91
f^6 and f^8	singlet	46	4	1	6	4	8	4	7	3	4	2	2	-	1	46
	triplet	56	6	11	20	21	25	22	21	15	12	7	5	2	1	168
	quintet	16		3	6	10	11	12	10	9	6	4	2	1	-	74
	sextet	1	1	1	1	1	1	1	1	-	-	-	-	-	-	7
	Total	119	14	19	37	37	46	37	38	24	20	11	8	2	2	295

shell where L and S commute with H_{4f} , i.e.- the Hamiltonian in Schroedinger's equation of 6.8.12.:

6.8.15.- $H (L\text{-}S) \Psi_{4f} = E \Psi_{4f}$

Slater determinants for two equivalent f-electrons are written in terms of a ROOTHAN operator, $\hat{F}_{4f}$, where:

6.8.16.- $\hat{F}_{4f} U_i = \varepsilon_i U_i$

where U_i is a spin-orbital function relating to the Coulomb and Exchange operators, $\hat{J}_i$ and $\hat{K}_i$, respectively. ε_i is the energy of the state under calculation. Interaction energies for **pairs** of identical 4f-electrons are:

6.8.17.- $E_0 = F_0 - 4/195\ F_2 - 2/143\ F_4 - 1000/5577\ F_6$

The F- values are Slater radial integrals. Using these is easier than it appears, since Slater has tabulated all possible values for use in matrix diagonalization.

In our matrix of 6.8.14., **we set all non-diagonal terms to zero** and used only the diagonal terms to obtain J_z . Since this did not work, let us now use the **off-diagonal** elements to re-calculate adjusted- J_z terms and energies.

   This is equivalent to stating that we no longer have L - S coupling.

 For a given $4f^n$ configuration, we can get a number of J- values, many with the same labeling, but with **different energies** (remember what we said about intermediate coupling).

Our next step is to arrange all J's of the same term into a J-matrix, so as to calculate the energies. For $4f^2$ (Pr^{3+}) or $4f^{12}$ (Tm^{3+}), we are looking for 13 solutions of the j-matrix, as shown in 6.8.18, given on the next page.

6.8.18.-

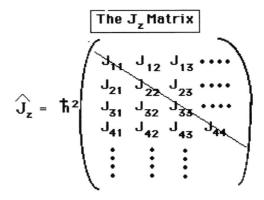

At this point, the Slater parameters can be described as:

6.8.19.- $E = F_0 - 2/45 \, F_2 - 1/33 \, F_4 - 50/1287 \, F_6$

It is convenient to use operators developed by Racah (1950) which considerably simplify the work. For 4f electrons, we have:

6.8.20.-   ENERGY PARAMETERS FOR CALCULATION

$$E_0 = F_0 - 2/45 \, F_2 - 1/33 \, F_4 - 50/1287 \, F_6$$
$$E_1 = F_0 - 14/405 \, F_2 + 7/297 \, F_4 + 350/11{,}583 \, F_6$$
$$E_2 = F_0 - 1/2025 \, F_2 + 1/3267 \, F_4 + 175/1656369 \, F_6$$
$$E_3 = F_0 - 1/135 \, F_2 + 2/1089 \, F_4 - 175/42471 \, F_6$$
$$\text{..etc.}$$

For $4f^2$, we can have up to 91 energy states, which illustrates the complexity to which we have subjected ourselves. The Racah parameters are given in the following:

$$A_k = F_0 - 49 \, F_4 \quad \{^q A_k\} \equiv \quad \text{crystal field parameter.}$$
$$B_k = F_2 - 5 \, F_4 \quad \{^q B_k\} \equiv \text{crystal field parameter} = {}^q A_k \, <r^k>$$
$$C_k = 35 \, F_4 \, \{^q C_k\} \equiv \text{spherical harmonic function} = (2k+1/4\pi)^{1/2} \, {}^q Y_k$$

where we have all the requisite values of k available for calculating these parameters. Nielson and Koster (1963) gave eigenfunctions and multiplet energies for all of the $4f^n$ configurations in terms of the parameters: ^qB_k and ^qC_k .

b. The Concept of Fractional Parentage

While we are carrying out these calculations, we find that we have introduced, through necessity, two new concepts, that of "intermediate coupling" and that of "fractional parentage". To explain what we are discussing, consider the following. While the original work on energy (spectroscopic) states of an atom was being formulated, it was realized that a free-atom's electronic states could be adequately described by either of these two states which differed by the method of coupling its azimuthal and spin states:

6.8.21.-   Russell-Saunders Coupling   j- j Coupling

$$L = \Sigma l \qquad S = \Sigma s \qquad\qquad j = l \pm s$$
$$J = L \pm S = (\Sigma l \pm \Sigma s) \qquad \Sigma j = j + j.... = \Sigma (l \pm s)$$

where l and s are individual quantum states of any given electron in the atom. In the first case, the individual electron vectors are first summed, and then coupled together to give the final J. In the second case (which has been found to apply to heavy atoms), the individual electron vectors are first coupled, and then summed to give j. Each j is then summed with the other j's of the separate electrons to give the final spectroscopic value. As we said, both of these methods represent the extremes in spectroscopic coupling notation. While Russell-Saunders coupling terms have been well documented, no such terms have been formulated for j-j coupling. Fractional parentage refers to the fact that the original system of specifying spectroscopic terms, i.e.- ^2G, ^4H, etc., was modified and adjusted using L - S coupling rules. In intermediate coupling, L and S are no longer **good** quantum numbers, and **only J is valid** (because of our method of calculation). However, even though we developed a new method of calculating energies, we did not, and have not, developed a new method of specifying **terms** in intermediate coupling. Therefore, we refer to the fractional parentage of L - S terms, namely:

6.8.22.- % SLJ (J = 15/2) = 93% ^{6}H, 7% ^{4}I

which happens to be the ground state of Dy^{3+}. The same fractional parentage occurs for the upper energy excited states as well. It is for this reason that the energy transitions in the lanthanides do not follow Hund's Rules for L-S coupling, and that we observe dd and dq transitions for rare earth transitions (with the correspondingly strong oscillator strengths) rather than the weak qq transitions expected. That is, because the excited states contain a fractional part of the ground state term, and vice-versa, the transitions actually involve transitions between **coupled** ground and excited states.

The complete intermediate coupling energy diagram for $4f^2$ electrons is presented in the following diagram, given as 6.8.23. on the next page.

Note that the energies are in terms of Slater parameters and the spin-orbit coupling parameter, ζ_{4f}, and are plotted against the crystal-field parameter, also in terms of ξ_{4f}. This parameter has been defined in terms of a set of constant values of the Racah parameters, as shown. On the right of the diagram ($\xi_{4f} = 1$), are shown the hypothetical levels of the infinitely strong cubic field, which are the j - j values. It has been shown that ζ_{4f} varies in a linear manner for $4f^n$ electrons. Using these data permits us to predict and use the proper values for establishing Intermediate Coupling diagrams for all of the rare earths, without having to begin the calculations from first principles.

We can then consolidate the effects of both the spin-orbit and crystal field parameters into a single diagram, as shown in 6.8.24., presented on a following page. In this diagram, we show the L - S states at the left hand side. The actual experimental energy values determined are plotted as points on the individual curves for the several Ln^{3+} ions. While the agreement is good, it is not perfect.

In comparing these values to those of 6.8.23., we see that overlapping states in 6.8.23. have been simplified to a single state in 6.8.24., thus illustrating the concept of fractional parentage.

6.8.23.-

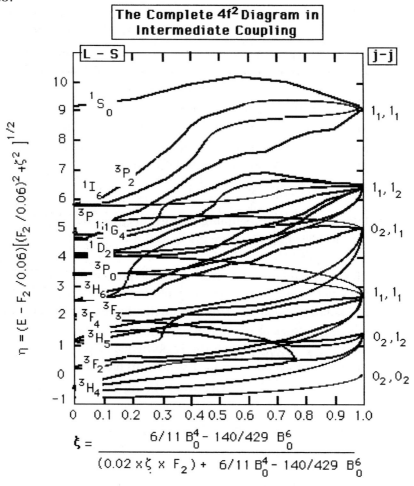

The most complicated spectra will be that of $4f^7$, which is the electron configuration for Eu^{2+}, Gd^{3+} and Tb^{4+}. By reference to 6.8.23. and the left hand part of 6.8.24., it is easy to see that if F_2 and ξ_{4f} are adjusted, one can get the energy levels for several ions. If we do this, we get the values shown in 6.8.25., presented on the next page. The formulas used to do the adjustment are:

6.8.26.- $\qquad \eta = (E - F_2/0.06)/[(F_2/0.06)^2 + \zeta^2]^{1/2}$

and: $\qquad \xi = X / 1+X$ where $X = 0.06 \, \zeta / F_2$

6.8.24.-

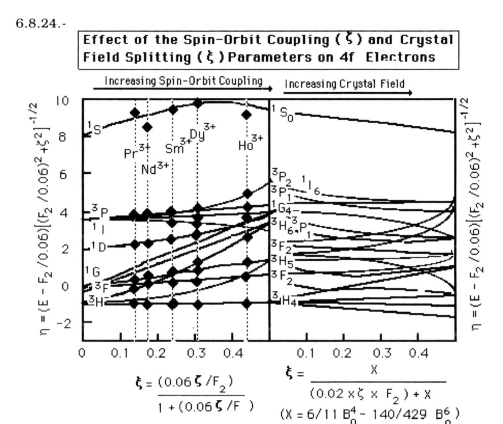

Effect of the Spin-Orbit Coupling (ζ) and Crystal Field Splitting (ξ) Parameters on 4f Electrons

$$\xi = \frac{(0.06\,\zeta\,/F_2)}{1 + (0.06\,\zeta\,/F\,)}$$

$$\xi = \frac{X}{(0.02 \times \zeta \times F_2) + X}$$

$$(X = 6/11\ B_0^4 - 140/429\ B_0^6)$$

6.8.25.- Spectroscopic Parameters for Several Related Ions

	F_2 in cm^{-1}	ξ_{4f} in cm^{-1}
Eu^{2+}	345	1220
Gd^{3+}	405	1581
Tb^{4+}	435	2080

These are the same as used in the above diagrams. We find that several overlapping states do exist so that we have difficulty in justifying any labeling. What has been done is to specify each of the energy levels in terms of their composite states. In the following Table, we list an abbreviated list of energy states for Dy^{3+} in terms of the fractional parentage of each state. J varies from J = 1/2 to J = 17/2 and we have only shown 3 of the 12 J-terms and 22 of the 56 energy levels.

Design of phosphors

TABLE 6-12

FRACTIONAL PARENTAGE OF Dy^{3+} ENERGY LEVELS BELOW 40,000cm^{-1}

After Wybourne (1963)

J	Calculated Energy	Observed Energy	% SLJ
1/2	13,723		91^{6}F , 9 ^{4}D
	31,991		54 ^{4}D, 28 ^{4}P, 11^{2}P, 7^{6}F
	37,764		70 ^{4}P, 28^{4}D, 2^{6}F
3/2	13,175	13,148	91 ^{6}F, 7 ^{4}D, 1^{4}F
	27,715		30 ^{4}D, 29 ^{4}P, 22 ^{2}P,9 ^{6}P, 5 ^{6}F,4^{2}D,1 ^{4}F
	32,129		55 ^{6}P,14 ^{4}D, 10 ^{2}D,10 ^{4}P,5 ^{4}F,3^{2}P,2 ^{6}F
	33,617		79 ^{4}F, 9 ^{2}D, 7 ^{6}P, 4 ^{4}D
	37,696		84 ^{4}F, 8 ^{2}D, 3 ^{6}P, 3 ^{4}P, 2 ^{2}P
	39,604		35 ^{4}P, 22 ^{6}P, 19 ^{4}D,11^{2}D,6 ^{4}F, 5 ^{2}P
5/2	10,155	10,155	92 ^{6}H, 7 ^{4}G
	12,389		93 ^{6}F, 4 ^{4}D, 3 ^{4}F
	28,756		27 ^{6}P, 26^{4}P, 20^{4}F,10^{4}D,10 ^{4}G,4^{2}D,1^{6}H
	30,013		38 ^{6}P, 28 ^{4}F, 16 ^{4}D,10 ^{4}G, 7 ^{4}P,1 ^{2}D
	31,874		38 ^{4}D, 33 ^{4}G,11 ^{2}F,9 ^{2}D,4 ^{4}P,3 ^{6}H,3 ^{6}F
	34,231		48 ^{4}F, 21 ^{4}G,14 ^{4}D,8 ^{2}F, 4 ^{6}F,2 ^{2}D,2 ^{4}P
	36,878		46 ^{4}G, 14^{4}F, 13 ^{2}F,10 ^{6}P, 8^{2}D,7^{4}P,1
15/2	0	0	93 ^{6}H,6 ^{4}I
	22,087	22,086	54 ^{4}I, 16^{2}K, 14 ^{4}K, 5^{6}H, 5^{2}L,4 ^{4}L,2 ^{4}M
	28,742		47 ^{4}M, 20^{4}I, 15 ^{4}L, 15^{2}L, 3^{4}K
	31,402		56^{4}K, 15 ^{4}M, 11^{4}L, 7^{2}K, 6 ^{4}I, 5 ^{2}L
	36,794		56^{4}L, 21^{4}K, 12 ^{4}M, 9^{2}K, 1^{2}L
	38,361		29^{2}L, 29^{2}K, 17 ^{4}M, 11^{4}I, 10^{4}L, 2 ^{4}K

We have already indicated the ground state and excited states to be coupled. i.e.- the spin-orbital states are mixed. The ground state level consists of: J = 15/2 (93 ^{6}H,6 ^{4}I). What we do is to label the energy states in terms of the spectroscopic label which makes the **major contribution** to its intermediate coupling makeup. Thus, for the J= 5/2 levels, we would call some of these ^4D and ^4G, even though the labels only contribute about half of their total makeup.

The data given in Table 6-12 for Dy^{3+} have been replicated for the other rare earths as well. Having this data allows us to form what is called a Groatrian Diagram of the indicated energy levels, according to the J-values (which are the only valid quantum numbers in intermediate coupling). Let us show this for Dy^{3+} so that we can examine the energy processes which take place for this trivalent rare earth ion.

In the following Groatrian Diagram, given as 6.8.27. on the next page, the energy levels have been plotted according to each J-value. The strongest emission for Dy^{3+} **in all hosts** consists of a set of lines around 5720 Å , a yellow emission. The energy of the emission transitions is centered at about 17,500 cm^{-1} and phosphors have been produced with quantum efficiencies in excess of 85%. The emission transition is always:

6.8.28.- $^4F_{9/2} \Rightarrow {}^6H_{13/2}$

Crystal field splitting accounts for the set of emission lines observed. In this diagram, **all** of the energy levels have been labeled with the term which contributes the greatest degree to that level.

C. Energy Transitions and Mixed States

Our next step is to further examine selection rules for "forced electric dipole" transitions. We need to account for the actual energy observed, in terms of the two terminal states possible, i.e.- yellow = [17560 cm^{-1} : $^6H_{13/2}$] vs: [grecn = 21,020 cm^{-1} : $^6H_{15/2}$]. We can see that if an excitation energy of 40,000 cm^{-1} were to be used, the $^2H_{11/2}$ level would be excited. This excited state would then relax, through phonon processes, to various states until the emitting $^4F_{9/2}$ state is reached. Then photon emission must result if the excited state is to return to its ground state.

You might ask why this state is the emitting state. Remember that a photon is a significant amount of energy and Nature would prefer to emit

6.8.27.-

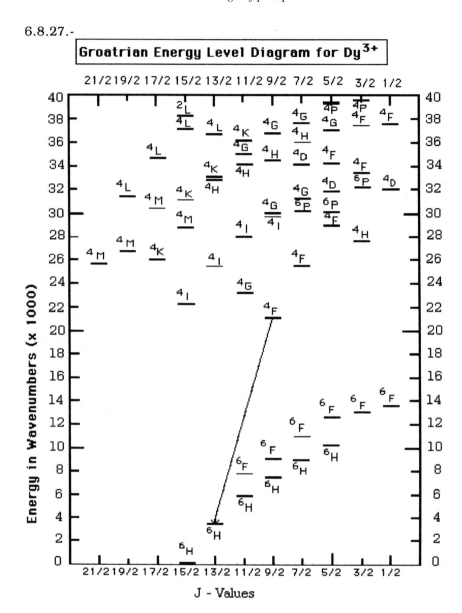

Groatrian Energy Level Diagram for Dy³⁺

phonons if possible. But the energy gap is too large for this. It turns out that photons are only emitted from large energy gaps. Note that other rare earth ions such as Nd³⁺, which emit in the infra-red, are used in a variety of lasers. Their ground states are generally at 0 energy. But, we still do not know why the $^6H_{13/2}$ state of Dy^{3+} is the terminal state, rather than $^6H_{15/2}$.

The composition of the three states, i.e.- emitting and terminal states, are shown in the following:

6.8.29.- MIXED STATES FOR Dy^{3+} IN INTERMEDIATE COUPLING

<u>EMITTING STATE</u> $^4F_{9/2}$ = 64 ^4F, 19 ^4G,9 ^6F, 4 ^2G,3 ^6H

<u>GROUND STATES</u> $^6H_{13/2}$:   96 ^6H, 3 ^4I
$^6H_{15/2}$:   93 ^6H, 6 ^4I

Note that the excited state has a ^6H component. It turns out that strong transition intensities in the lanthanides occur due to a transition termed: "forced-electric-dipole" transitions, where $\Delta J = 2$. This rule holds in all cases where strong intensities have been observed. Thus, for our case, the transition involves the $^6H_{13/2}$ state, because of the $\Delta J = 2$ restriction.

We have already shown in Chapter 5 that spectral intensities, I, are proportional to oscillator strength, f, which is inversely proportional to half-life, $\mathcal{T}_{4f}$, namely-

6.8.30.-   I $\approx$ f α 1/ $\mathcal{T}_{4f}$

Transition between 4f $\Rightarrow$ 4f energy levels are strictly forbidden, according to Hund's Rules, and are classified as quadrupole (qq) transitions. Their intensities ought to be: I $\approx$ 10^{-7}, but experimentally, I $\cong$ 1.0, i.e.- intensities rivaling dd- transitions.

D. <u>Spectroscopic Rules for the Rare Earths</u>

It was Van Vleck (1958) who showed that mixed states, i.e.- intermediate coupling, lead to "enforced dipole" transitions because of the above described coupling of states. He described this phenomenon as arising from lattice perturbation of the states wherein **the odd (u) crystal field**

terms become mixed in, thus making the transition "allowed". The Enforced Dipole Selection Rules are then somewhat similar to L - S coupling dipole transition rules. These rules are shown as follows.

6.8.31.- FORCED DIPOLE SELECTION RULES

Electric: $\Delta J = \pm 2$ (rarely ± 1)

                   $(0 \neq 0)$ $(u \leftrightarrows g)$

Magnetic: $\Delta J = \pm 1$ (rarely ± 2)

                   $(0 \neq 0)$ $(u \leftrightarrows u)$

Limits: $\Delta J \leq 6$

EXPERIMENTAL INTENSITIES OBSERVED

Electric: I ≈ 1.0

Magnetic: I ≈ 0.1 to 1.0

Using these rules, we can now reexamine the emission transition of Dy^{3+}. For example, a transition from the $^4 F_{9/2}$ emitting state to the $^6 F$ states is strictly forbidden $(0 \neq 0)$. The other possibility includes the $^6 H$ states, i.e.- $^6 H_{13/2}$, $^6 H_{11/2}$, $^6 H_{9/2}$, $^6 H_{7/2}$, and $^6 H_{5/2}$. However, the $^6 H_{7/2}$ and $^6 H_{5/2}$ states have no ^4I composition. This leaves us with the $^6 H_{13/2}$ and $^6 H_{11/2}$ states. While it is true that in absorption (excitation) ΔJ can equal up to six, the **change in J for emission** is generally limited to $\Delta J = 2$, with lesser intensities for $\Delta J = 1$. Because transition to the $^6H_{15/2}$ state involves $\Delta J = 3$, it is not possible. This accounts for the observed luminescent transition for Dy^{3+}.

The determining factors for **intense** rare earths spectra (dipole transitions) are:

6.8.32.- DETERMINING FACTORS FOR RARE EARTH SPECTRA

1. Mixed states of like parentage
2. Forced dipole selection rules
3. Symmetry at the lattice site

The last factor is rather easily explained. Suppose we put a Ln^{3+} ion into a site having inversion symmetry, i.e.- there are two or more mirror planes present at the site. Lattice perturbation is then uniform at the site, and no "odd" terms "become mixed in". Thus, the transition reverts to a (qq) transition, with resulting low transition intensity. Many examples exist in the literature that show that the emission intensity of Ln^{3+} ions are sensitive to the nature of the site symmetry of the host.

In general, low symmetry cation sites produce the most intensely emitting lanthanide phosphors.

While we have examined calculated levels, as compared to experimental ones, we have not determined exactly how the Stark States arise. Let us now examine this phenomenon.

e. Experimental Stark States

Up to now, we have adjusted the spin-orbit coupling factor, ζ_{4f}, to fit the experimental energy levels. One problem with this approach is that while most of the levels seem to fit, there are one or two which do not. If we examine 6.8.24. in terms of the "dots" representing a fit of experimental data, we see that correspondence is fairly good. However, the problem is more demanding than this. Let us further examine the fit of these energy levels, using Er^{3+} because its energy levels are far enough apart so that no discrepancy, or mistake, in identification of the individual levels can be made. To do so, we set up a spectrophotometer so that we can measure the reflectance of a powder containing the ion. We use the diffuse-reflectance mode so as to avoid spurious spectral effects. With a specially designed instrument, we find that we can obtain a resolution of about ± 2 Å of the spectral diffuse reflectance lines (Ropp -1970). This is sufficient to resolve the existing Stark States.

In order to minimize any contribution to the spectra from the host, we use a phosphate compound, $ErPO_4$, because the PO_4 group is transparent through out the visible and ultraviolet spectrum, up to about 1750 Å. The

following diagram, shows a high-resolution powder reflectance spectrum
of $ErPO_4$, reduced from a 30-foot chart:

6.8.33.-

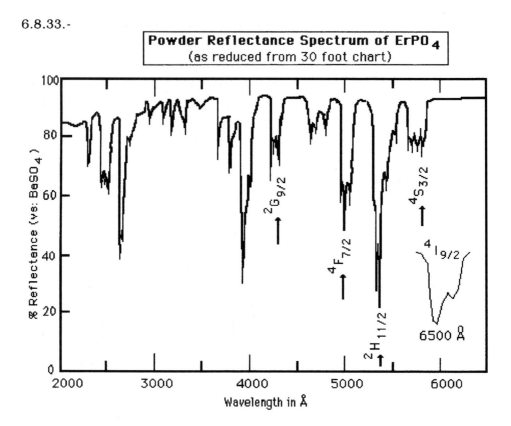

There are a number of bands which appear in absorption. Only a few of the
transitions are identified. However, it is clear that a number of individual
lines (Stark States) can be identified.

In this case, we have shown the calculated levels of Er^{3+}, according to
their J-value. Note that these levels are far apart, in terms of energy. This
corresponds to our experimental criterion given above.

The overall spectral transitions and their identification are shown in the
following Groatrian Diagram of Er^{3+}, given as 6.8.34. on the next page.

6.8.34.-

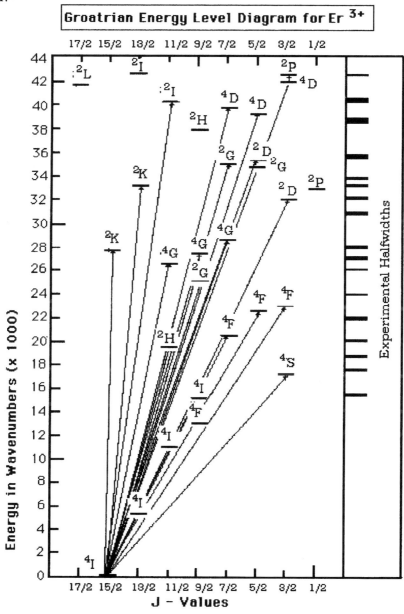

At the right is shown the experimental halfwidths of the bands shown in the main part of the spectra. Note that the correspondence between

6.8.38.- COMPARISON OF EXPERIMENTAL BARICENTERS TO
        CALCULATED ENERGY LEVELS AS A FUNCTION OF SEVERAL
        HOSTS AND SITE SYMMETRIES

Compound	Site Symmetry	Baricenter			
		$^4F_{7/2}$	ΔE	$^2G_{9/2}$	ΔE
$ErBO_3$	D_{3d}	4854 Å	27 Å	4070	53 Å
$Er_2Ti_2O_7$	O_h	4892	65	4077	60
$ErPO_4$	D_{2d}	4910	83	4083	66
Er_2O_3	C_{2v}	4920	93	4101	84

levels is nearly impossible to accomplish. Although the fit is close, it still
is not exact. This presents us with a dilemma since we want to be able to
identify and specify energy transitions in both absorption (excitation) and
emission. What we have done so far is to adjust the spin-orbit coupling
factor, ζ_{4f} , to fit the experimental energy levels so as to obtain a set of
energy levels.

Nonetheless, this approach has not been successful because of the factors
that we have presented above. It has proven impossible to **exactly** match
the energy levels of a given trivalent lanthanide in any given crystal host
because of the effect of the crystal field on the number of stark states that
can appear. Fortunately, another approach has arisen.

f. The "Free-Ion" Approach to Rare Earth Energy Levels

It turns out that another method exists where the approach has been to
calculate so-called "free-ion" levels, so that the spin-orbit coupling factor,
ζ_{4f} and the crystal field parameter, ξ_{4f}, **are not** adjusted to give a
"satisfactory" fit to the experimental data. Wybourne (1963) and others
have accomplished this task by using a Hartree-Fock approach (and a high
speed computer) to calculate the energy levels of all of the trivalent rare
earth ions.

These levels are shown in the following diagram, given on the next page
as 6.8.39. In general, these levels show many similarities and only a few

6.8.39.- Calculated Free-Ion Energy Levels of the Trivalent Rare Earths

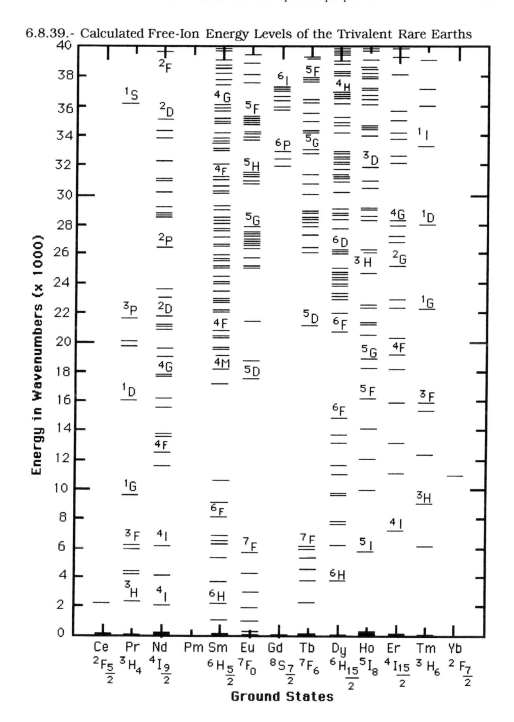

discrepancies to most experimental spectra. Still, they represent a better method of identifying energy transitions because **they are self-consistent,** and not dependent upon the crystal in which the rare earth finds itself.

Before we leave this subject, we should point out that the Ln^{3+} ions which produce the brightest phosphors are those which have the largest energy gap between excited and ground states. These are in general: Eu^{3+}, Gd^{3+} and Tb^{3+}, i.e.- $4f^6$, $4f^7$ and $4f^8$. These trivalent states also have the most complicated energy levels. This was first demonstrated by Thomaschek (1933), as shown in the following diagram:

6.8.40.-

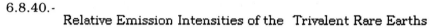
Relative Emission Intensities of the Trivalent Rare Earths

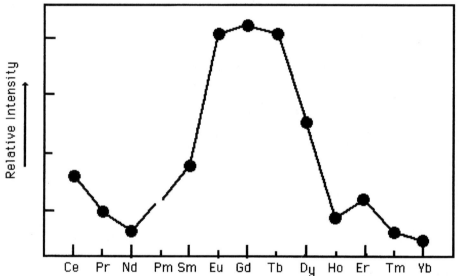

Excitation of these ions using 40,000 cm^{-1} energy results in various excited states for each ion. Each state relaxes by phonon processes until the terminal excited state is reached, from which emission of a visible photon results. But we need to remember that any time we put a trivalent rare earth ion into a host crystal to form a phosphor, we will modify its free-ion energy levels, depending upon the site symmetry and crystal field strength.

g. Charge Transfer States and 5d Multiplets

Although we have investigated the 4f electron energy states, we chose to ignore the possibility that other excited states might exist. It was Dieke (1964) who first pointed out that three (3) overlapping spectroscopic manifolds existed for the lanthanides. These are: the 4f -manifold, the 5d manifold and the charge-transfer manifold. All three exist within the same energy-space and it was Ropp and Carroll (1965) who first showed that these could be measured and sorted on the basis of the half-width of the measured band. But, it was Jørgenson (1962) who first showed that a specific trend exists for the 5d states which is exactly opposite to the trend shown for charge-transfer states in the trivalent lanthanides. The equations he derived were:

6.8.41.-   CHARGE - TRANSFER STATES

$$W - q(E - A) + K_1 \ D \ + K_2 \ \mathbf{E} \ + K_3 \ \zeta_{4f}$$

5d EXCITATION STATES

$$W_2 \ + (q - 1)(E - A_2) + K_4 \ \mathbf{E}^3 + K_6 \ \zeta_{4f}$$

where W and W_2 are reference standards for comparison of energies, q is the number of 4f electrons, $(E - A)$ is the difference between core attraction and interelectronic repulsion, D gives the change of q with spin-pairing energy, $\mathbf{E}^3$, and the various K's are adjustment parameters.

When the proper values are plotted vs the number of 4f electrons, the following curves result, as shown in 6.8.42., given on the next page.

It is easy to see that the lowest energy 5d-excited states are exhibited by Ce^{3+} and Tb^{3+}. The lowest energy charge-transfer excited states (which are also a function of the host crystal) are Eu^{3+} and Yb^{3+}. All the others are intermediate in energy to those extremes of the two curves. In addition, we find that charge- transfer states occur for those trivalent ions which

6.8.42.-

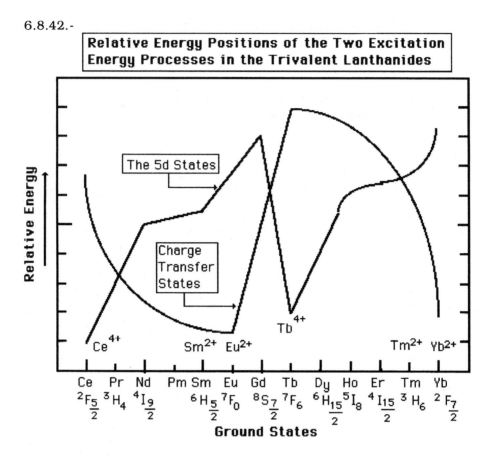

Relative Energy Positions of the Two Excitation Energy Processes in the Trivalent Lanthanides

also exhibit a **divalent** state while 5d states are prevalent for those ions which also have a **tetravalent** state.

In 1967, Ropp and Carroll demonstrated that the trivalent rare earths followed the trends given in 6.8.42. for both the 5d states and the charge-transfer bands. It is well to note that the following trivalent ions exhibit specific excitation bands in their spectra, and that these bands will be the dominant feature in the spectra being measured. That is, within the spectral range of 1800 Å to 4000 Å, , which is the usual range measured for excitation spectra of phosphors being developed for lamp purposes.

This is summarized as follows:

6.8.43.-   <u>5d Excitation States</u>   <u>Charge Transfer Excitation states</u>

Ce^{4+} , Tb^{3+} Sm^{3+} , Eu^{3+} , Yb^{3+}

As an example of a charge transfer band, the following shows the spectrum of the phosphor, Y_2O_3 : Eu, given as follows:

6.8.44.-

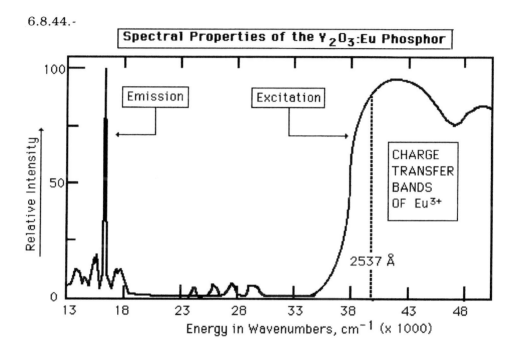

Since we cited Anti-Stokes phosphors as one area dominated by rare earths, we need to address this technological field. We will find that these phosphors are unique and only made possible by the unusual energy levels encountered in these 14 components of the Periodic Table.

h. Phonon Assisted Relaxation in Phosphors

Because the energy gaps for many of the trivalent rare earths are often no more than 2000 cm^{-1} and many of the Stark States are as close as 200 cm^{-1}, phonon-assisted relaxation (emission to, and absorption from, the host lattice) may be expected to predominate in certain cases. While this is certainly true in all phosphors, the close-spaced energy levels of the rare earths allow us to experimentally examine such electronic processes.

Examine the diagram, given as 6.8.40 above, showing calculated free-ion energy levels again. For the most part, the energy levels are rather closely spaced. When an upper energy level of a rare earth ion in a crystal is excited, it may decay to a lower state by:

1) Emission of a photon
2) Energy transfer to another site by a multipole process
3) Emission of several phonons to the host lattice.

When we say phonon-emission, we are actually stating that infra-red photons matching the phonon spectrum of the lattice are emitted and that lattice absorption then occurs- this is equivalent to **virtual** photon emission since the emitted photons never appear outside of the lattice). The last process has been studied extensively and we shall summarize the results herein. The rare earths are unique in that, for energy gaps of several thousands of wavenumbers, several **simultaneous** phonon emission processes may be involved.

Multiphonon relaxation processes are usually studied by determining both the transition rates for the process, and the phonon spectrum of the host crystal. Total transition rates (reciprocals of fluorescent lifetimes) are measured and multiphonon (MP) decay rates are extracted from these. In practice, one does not observe the MP transition rate between two energy levels, but that between two J- manifolds, i.e.- the sets of Stark States The decay rate, $\mathcal{W}$, obeys the relation:

6.8.45.- $\mathcal{W} = B \exp(\alpha\, \Delta E) = B \exp(\alpha\, \hbar\, \omega_n)$

where $\hbar = h/2\pi$ and n is the number of phonons of a given frequency, while ω_n, B and α are constants characteristic of the host lattice. The temperature dependence of the transition probability (decay rate) for processes involving the emission of n-phonons of a single frequency, ω, is given by:

6.8.46.- $\qquad W_n = W_0 [(1- \exp (-\hbar \omega / kT)]^{-n}$

Note that this equation is like a dispersion equation (which we have already discussed). Phonon relaxation rates follow these equations only at low temperatures, i.e.- 4,2 °K.

The phonon spectrum determined for YVO_4:Eu for the absorption transition, $^7F_0 \Rightarrow {}^5D_1$, is shown in the following:

6.8.47.-

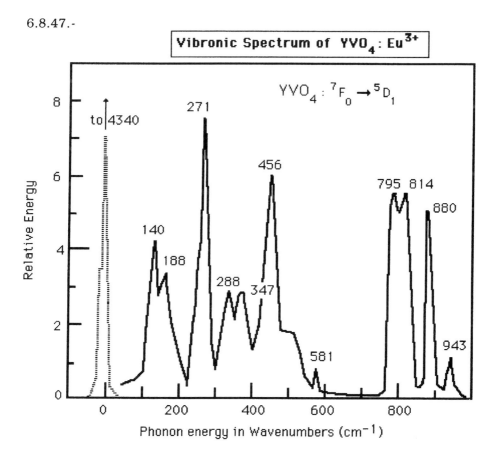

The zero-phonon line is also indicated, along with its relative intensity of 4340. This spectrum was obtained at 77 °K, by measuring the absorption of the $^7F_0 \Rightarrow {}^5D_1$ transition, while monitoring the $^5D_0 \Rightarrow {}^7F_2$ emission transition. The phonon lines shown are essentially vibronic side-bands. If we now assume that the rate of adjustment among the individual Stark States of a J-multiplet is much faster than the relaxation of the multiplet itself to the next lowest energy multiplet, i.e.- the lifetime of the emission, we can determine the actual number of phonons involved in the process (refer to 5.5.7. in Chapter 5). What we do is to arrange 6.8.46. into a Bose-Einstein occupation probability equation, namely:

6.8.48.- $n = [(\exp(-\hbar\omega/kT) - 1]^{-1}$

Combining this with 6.8.46. gives the form most useful for determining the multiphonon (MP) rate of the J-multiplet:

6.8.49.- $\mathcal{W}_{MP} = \sum W_0 \, g \, (\exp(-\hbar\omega/kT) / \sum g \, (\exp(-\hbar\omega/kT)$

We determine the experimental rate of phonon decay $\mathcal{W}_{MP}$, for several levels of various ions (activators) in YVO_4 and then plot them versus the energy gap. For the selected ions, we have the transitions:

6.8.50.- Eu^{3+} : $^5D_1 \Rightarrow {}^5D_0 + n \, \hbar\omega \Rightarrow {}^7F_2 + hv$
                            $^5D_3 \Rightarrow {}^5D_0 + n \, \hbar\omega \Rightarrow {}^7F_2 + hv$

            Er^{3+}: $^4S_{3/2} \Rightarrow {}^4F_{9/2} + n \, \hbar\omega \Rightarrow {}^4I_{15/2} + hv$
                            $^4I_{11/2} \Rightarrow {}^4I_{13/2} + n \, \hbar\omega \Rightarrow {}^4I_{15/2} + hv$

            Ho^{3+}: $^5S_2 \Rightarrow {}^5F_5 + n \, \hbar\omega \Rightarrow {}^5I_8 + hv$
                            $^5F_3 \Rightarrow {}^5S_2 + n \, \hbar\omega \Rightarrow {}^5I_8 + hv$

The phonon emission rates are seen to fit a straight line as shown on the following diagram, given on the next page as 6.8.51.

It is easily seen that the smaller the energy gap between levels, the larger is the phonon rate.

6.8.51.-

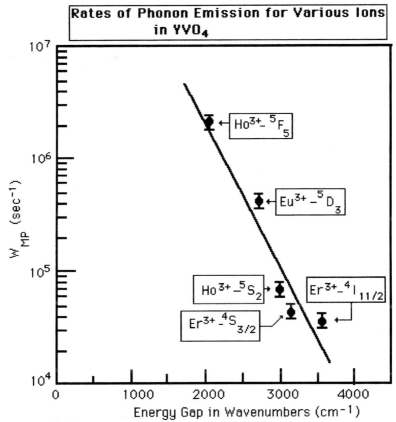

This is as we might expect since it should be easier for the phonon energy to span smaller energy gaps.

To complete the picture, we now need an estimate of the temperature dependence of the rate of phonon emission in YVO_4. The following measurements show experimental data obtained by Reed et al (1971) for the temperature dependence of the transition: $^4S_{3/2} \Rightarrow {}^4F_{9/2}$ of Er^{3+} in YVO_4. This is shown in 6.8.52. on the following page.

Note that when one increases the temperature in a lattice, the intensity of vibrations, i.e. density of phonons, increases correspondingly.

6.8.52.-

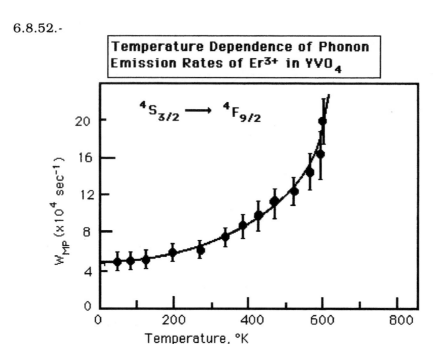

One can fit this curve above 50 °K by a function like 6.8.49. by using the data given in 6.8.50. We thus obtain:

6.8.53.-   Transition: $^4S_{3/2} \Rightarrow {}^4F_{9/2}$

$$W_{MP} = 5.00 \times 10^4\, [n(270) + 1]^2\, [n\,(880) + 1]^3/\, 1+3\, \exp\,(-700/kT)$$

where n(270) in 6.8.53. means the phonon-mode occupation probability. This result shows that **two** 270 cm^{-1} phonons and **three** 880 cm^{-1} phonons are involved in the relaxation process.

We can now summarize the numbers of phonons involved in the relaxation processes for all of the lanthanides given above in 6.8.50., as shown in the following, given as 6.8.54. on the next page.

It should be clear that the **sum** of the energies involved the controlling factor here. Note that a five-phonon process is indicated in two of the

6.8.54.- PHONON EMISSION TO THE LATTICE BY RELAXATION
PROCESSES FOR VARIOUS ACTIVATORS IN YVO_4

Ion	Phonon Emitting State	n	ω	Actual Energy Gap in Crystal
Eu^{3+}	$^5D_1$	2	880 cm^{-1}	1758 cm^{-1}
	$^5D_3$	3	880	2860
		1	271	
Er^{3+}	$^4S_{3/2}$	2	271	3091
		3	880	
	$^4I_{11/2}$	4	880	3464
Ho^{3+}	$^5S_2$	1	880	2914
		4	456	
	$^5F_5$	2	880	2108
		1	456	

transitions shown. These are phonons emitted during relaxation of an upper excited state to the **photon-emitting** state. It has been determined that a six (6)- phonon process represents the upper limit for any given relaxation process. This probably arises because of physical limitations. That is, **the lattice can only absorb only so many phonons during the time that the relaxation is taking place.**

It should be clear that we have now confirmed that phonon absorption and emission from the activator site **actually** occurs. We have alluded to this fact in prior chapters but no specific data had been presented which actually confirmed this mechanism. The above discussion has proven this.

Now that we have examined the case where the excited center **emits** phonons to the lattice, let us now study the opposite case, where phonon absorption **from** the lattice occurs, i.e.- a phonon assisted electronic transition. In this case, phonon **absorption** occurs so that the mismatched energy transition can take place.

As should be apparent by now, the rare earths are one of the few cases where such Stokes processes can be studied at room temperature. If we were to study phonon processes using other activators, we would find that we must study them at 4.2 °K. We would also find that the phonon spectra were rather complex and difficult to interpret, instead of being simple. Now we will address the so-called "Anti-Stokes" phosphors where the dominant mechanism is absorption of several phonons to produce a single photon of higher energy. We will find that only certain trivalent lanthanides are suitable when combined with a selected host lattice if we wish to have high efficiency, i.e.- "brightness".

g. Anti-Stokes Phosphors

The unique energy levels of the rare earths have been utilized to produce Anti-Stokes phosphors. These are essentially "infra-red to visible" light converters. Two, or more, infra-red photons are absorbed to produce one visible photon. Reexamine the energy levels of Pr^{3+}, Nd^{3+}, Ho^{3+} and Er^{3+} again. All of these ions absorb in the infra-red region of the spectrum. If we started with two ions, say Er^{3+}, each of which absorbed an infra-red photon of 6600 cm^{-1}, we would end up with two ions in the $^4I_{13/2}$ excited state. By an exchange process, one ion could then donate its energy to the other to produce one ion in the $^4I_{9/2}$ state, while the other reverted back to the ground state. This exchange process would look like the following, as shown in the diagram given as 6.8.55. on the next page.

The exchange process may be direct, or may occur by reabsorption of an emitted photon. A three-body process for Er^{3+} would bring us up to the $^4S_{3/2}$ state, from which we could expect a visible photon to be emitted. But the probability of photon emission from the intermediate states, $^4I_{13/2}$ or $^4I_{9/2}$, is just as high as that for the exchange process. Therefore, it is not surprising that we see little infra-red to visible light conversion in such a system. That is, photon emission in the infra-red is equally likely to that of the exchange excitation process. Remember, the exchange Hamiltonian depends upon the coupling coefficient, $\hat{J}_{eff}$, times the overlap operators, $S_{i,j}$. However, $\hat{J}_{eff}$ is only effective for nearest neighbors.

6.8.55.-

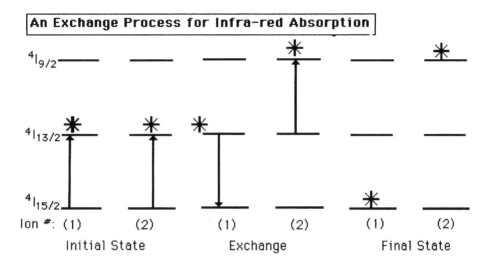

It drops off in relation to $1/r^3$, the radius of the ion in the crystal. A three-body exchange process has too low a probability to be of practical value in an Anti-Stokes phosphor. If we reexamine the calculated free-ion energy level diagram, we notice that Yb^{3+} has a single energy level with **no adjacent levels to which relaxation could occur.** Therefore, if we add this ion to our material, we should be able to obtain a phosphor which functions as an Anti-Stokes material. The energy level of Yb^{3+} lies at 10,300 cm^{-1} (9710 Å). It just happens that the output of a GaAs:Si laser diode occurs at 9500 ± 500 Å.

The "up-conversion" (Anti-Stokes) phosphors have been used in solid state indicator lamps incorporated within various electronic equipment for several years before being supplanted by the visible-emitting GaAlAs:Si solid state diode.

A large number of hosts were studied in which Yb^{3+} and Ln^{3+} were combined. Obviously, the phonon spectrum will be critical to the overall operation of up-conversion phosphors. Some of the hosts studied are given in the diagram shown on the next page as follows:

branches of the phonon spectrum) is very low. This arises because the atomic masses of the hosts are relatively low.

If we measure the phonon spectrum as deduced from vibronic coupling, we get the following:

6.8.61.-	Host	Color (for Er^{3+})	Phonon Cutoff Energy	
			Calculated	Measured

Transition = $^4S_{3/2} \Rightarrow {}^4I_{15/2}$

	Host	Color	Calculated	Measured
	$NaY(WO_4)_2$	green	360 cm^{-1}	350 cm^{-1}
	LaF_3	green	360	350
	YF_3	green	360	375
	$NaYF_4$	green	360	405

Transition = $^4F_{9/2} \Rightarrow {}^4I_{15/2}$

	Host	Color	Calculated	Measured
	Y_3OCl_7	red	580	600
	Y_2O_3	red	520	550
	$YOCl$	red	570	620

With these phonon energies, i.e.- less than 400 cm^{-1} for green emission and less than 600 cm^{-1} for red emission, the multiphonon **relaxation** process is nearly impossible because of the energy difference between multiplets would require more than a sextet of simultaneous phonon emissions. Therefore, these hosts exhibit a high efficiency for infra-red to visible light conversion.

The following diagram, given as 6.8.62. on the next page, shows the lower part of the Er^{3+} and Yb^{3+} energy level diagrams.

6.8.62.-

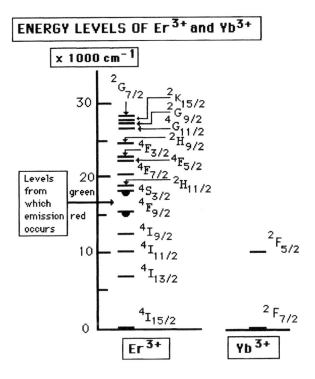

You will note that the two Er^{3+} levels, $^4I_{11/2}$ and $^2F_{5/2}$, are very closely matched in energy. Irradiation at 10,300 cm^{-1} will excite either of these centers. Deëxcitation by phonon emission to the lattice would take, for these hosts, a 7 to 9 phonon process for Er^{3+} ($^4F_{11/2} - {}^4I_{15/2}$), and a 16 to 25 phonon process for Yb^{3+} ($^2F_{5/2} - {}^2F_{7/2}$). Both are obviously physically impossible as a process.

Keep in mind that if both Er^{3+} and Yb^{3+} become excited and if Yb^{3+} transfers its energy, the final state of the Er^{3+} ion will be $^4F_{7/2}$. This state then relaxes to the emitting state, $^4S_{3/2}$, from which comes a green photon (about 18,400 cm^{-1} in energy).

Alternately, two Yb^{3+} ions can sequentially transfer energy to the Er^{3+} ion which then follows the same energy path.

The following diagram shows the excitation routes that have been cited for both green emission from Er^{3+} and for blue emission from Tm^{3+} in Anti-Stokes phosphors:

6.8.63.- Transitions Associated with Green and Blue Emitting Phosphors

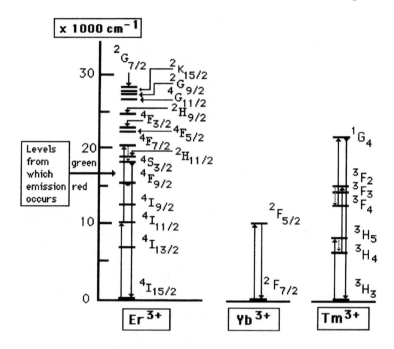

Note that the case for emission of a blue photon by Tm^{3+} is slightly different than that of Er^{3+}. A three-step process is indicated. This may consist of excitation of the Tm^{3+} ion and energy transfer from two excited Yb^{3+} ions, or excitation of the Tm^{3+} ion by exchange (sequential excitation) with three excited Yb^{3+} sites. Note that two relaxation steps (phonon emission to the host lattice) would be involved.

The excitation routes for red emission (energy = 15,200 cm^{-1}) are not as clear. Four (4) different paths have been postulated for excitation from Yb^{3+} to the $^4F_{9/2}$ level of Er^{3+}. None of these have been proven to dominate. The whole process for these 4 postulated mechanisms for red emission can be summarized as shown in 6.8.64., as follows:

6.8.64.- The Four Postulated Mechanisms for Red Emission in $YOCl_3$

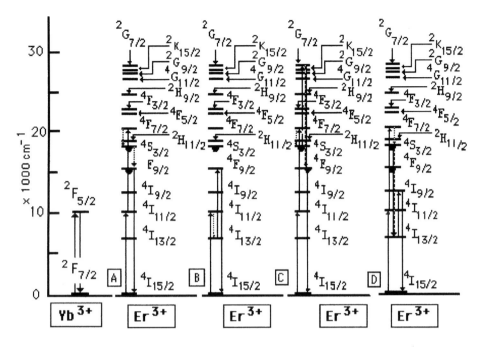

It shows the details of the four paths stated that have been postulated for excitation from Yb^{3+} to the $^4F_{9/2}$ emitting level of Er^{3+}.. Note that these mechanisms appear quite complex but actually involve only a few phonons. The exact mechanism remains today not fully defined.

Two relaxation steps are shown in (a), and one in (b). In (c), a three - **photon** (i.e.- infra-red photons) process is shown, with corresponding relaxation steps. Once the Er^{3+} center has reached the $^2G_{7/2}$ state, an exchange process (shown by the serrated line)with an unexcited Yb^{3+} ion may take place, leaving the Er^{3+} ion in the red-emitting state, $^4F_{9/2}$. In (d), still another possible Er^{3+} excited-state-exchange process is shown, wherein the terminal excited state is $^4I_{13/2}$, which then goes to the red-emitting state, $^4F_{9/2}$, by a final exchange process. Thus, (d) is a variation of (c). The whole process for (c) and (d) can be summarized as shown in 6.8.65., given on the next page, where the "stars" indicate the number of individual excitation-energy steps involved in the exchange process.

6.8.65.-

$$Er + 2\ Yb + 3\ h\ c\ v_1 \Rightarrow 2\ Yb^* + Er^*$$

$$Yb^* + Er^* \Rightarrow Yb + Er^{**}$$

$$Yb^* + Er^{**} \Rightarrow Yb^* + Er^{***}$$

$$Yb + Er^{***} \Rightarrow Yb^* + Er^{**}$$

$$Er^{**} \Rightarrow Er + h\ c\ v_2$$

$$\{h\ c\ v_1 = 10{,}300\ cm^{-1}\ ;\ h\ c\ v_2 = 15{,}200\ cm^{-1}\}$$

We can now compare the relative efficiencies of Er^{3+} for green emission, by up-conversion of infra-red radiation from a GaAs:Si laser diode, as a function of the host employed. This is shown in the following:

6.8.66.- Relative Efficiencies of Anti-Stokes Phosphors Using Er^{3+}

Host Lattice	Relative Green Emission Intensity
α- $NaYF_4$	100%
β- $NaYF_4$	40
Y_3OCl_7	105
YF_3	60
$BaYF_5$	50
$LaNaF_4$	40
LaF_3	30
La_2MoO_6	15
$LaNbO_4$	10
$NaGdO_2$	5
La_2O_3	5
$NaYW_2O_8$	5

Note that those hosts which have anions with **low** atomic numbers (and weights) perform best for Anti-Stokes conversion, i.e.- they have the lowest phonon-cutoff frequencies. There is a structure effect which may be ascertained by a comparison of α- $NaYF_4$ and β- $NaYF_4$. Actually, in β- $NaYF_4$, the red emission is 100% in intensity, not the green. One would suspect that differences already given for Y_3OCl_7 are due to symmetry changes at the Er^{3+} site, as affected by the surrounding Yb^{3+} neighbors.

One should also note that these phosphors were used for a time to enhance the output of LED light sources. With the advent of vastly improved LED components, they have been superseded and are not used anymore. We will address the LED's currently used and the applications to which they have found employment. The improvement has been such that LED's are likely to replace both incandescent lamps and fluorescent lamps in the near future, provided that a suitable phosphor can be found to use with them.

The final subject we will explore is that concerning solid state lasers, the spectral basis for establishing a solid state laser, and their properties.

h. The Solid State Laser

The solid state laser is actually a specialized phosphor. It consists of a host and an activator, but the host must be **single crystal**. What we have to say about the solid state laser applies equally to all lasers, be they gaseous or solid state.

Consider the properties of a photon in more detail. We know that it has both particle and wave-properties. We have determined that if the wave vibrates in one plane, i.e.- the direction of propagation, we have plane-polarized light. It may also vibrate in a circular motion, transverse to its direction of propagation, and then we have circularly polarized light. We also know that that if these photons are reflected from a surface, the polarization is destroyed, probably because each photon ends up on a different plane of vibration of the lattice. This effect arises because the atoms of the solid are vibrating, and the reflected waves will all have a slightly different orientation from each other, due to the reflection-interaction, i.e.- the Helmholtz scattering mentioned in Chapter 5. Furthermore, if they are not all of the same energy, we know that some photons will be scattered, and some absorbed and/or reflected (a non-resonant process).

6.8.68.-

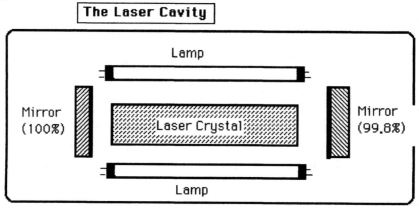

Under such conditions, any single photon emitted **must** have been reflected through the crystal many times before it finally escaped. We might now expect to obtain coherent emission, but we do not. This situation arises because we have an activator which emits a broad band of energies, and although we can restrict the emission so that coherent photons result, the intensity (and efficiency) is so low as to make the whole device almost useless. Actually, we get only about 1 coherent photon out of about 10,000 generated. The rest of the energy ends up as heat. Although we could cool our single crystal phosphor to minimize vibronic coupling, the extremely low quantum efficiency of the device precludes any practical usage.

This leaves us with two other possible choices. One, we can change to a rare earth where the energy transitions involve line emission, or two, we can choose an activator having a relatively long half-life in its excited state, so as to be able to build up a **sizable population** of activators in their excited state, before they are able to relax and emit a photon. If such a single-crystal phosphor is then placed in the resonant cavity, the resonance condition then precludes the emission of any photons other than those that conform to the resonance condition of the cavity. That is, the resonance wave builds in intensity and "stimulates" each excited activator site to revert to a specific vibronic state of coupling, dictated by the cavity, from which the stimulated emission emerges It is this mechanism. which is called "stimulated emission", that is the basis of the solid state laser. Our conditions for a solid state laser include:

6.8.69.- Conditions Requisite for Construction of a Solid State Laser

1. Construct a resonant cavity to match the chosen emission
2. Achieve a minimum spread of emission energies
3. Long half-life of the excited state
4. Achieve an inverted population, i.e.- of excited states over ground states, so as to obtain stimulated emission.

Of these conditions, the most important is #4 in which the activators centers must become excited and remain in the excited state until "stimulated" to emit under the resonant conditions to emit stimulated emission.

Rare earths as activators fit the first three criteria very nicely so as to produce condition (4). Remember that we are considering the solid state where the activator has essentially one degree of freedom, i.e.- vibration. Thus, if the activator has a relatively long half-life, has a narrow band of emission energies, and is incorporated within a host where the excited state electron wave states are **all** in the same site symmetry, then we will be able to achieve stimulated emission.

Gas lasers, on the other hand, have three (3) degrees of freedom so that each excited gas molecule can instantaneously orient itself to the resonance condition and thereby achieve population inversion, and coherent light emission. The last condition we find essential if we are to obtain any power in the beam.

It turns out that the central rare earths, which exhibit strong enforced-dipole emission, do not perform well as laser activators, where a long half-life is essential to obtain a population inversion. Population inversion requires an excitation source strong enough so that the number of exciting photons absorbed are sufficient to bring the population to the point where there are more activator centers excited than in the ground state. This means that there is a **threshold** of input energy, above which we obtain laser action, i.e.- stimulation of coherent light. Below this point, we obtain polarized emission, but not stimulated **coherent emission**. The best rare earth for laser application has been found to be Nd^{3+} which emits at 1.086 μ (10,860 Å). Other rare earths which have found some usage are: Er^{3+} and Tm^{3+}. These ions function best in fluoride-based materials. The hosts which function best as laser crystals are few indeed.

The criteria required are shown as follows. Of these four, the most important ones (for high powered lasers) is #1 and #4.

6.8.70.- REQUIRED CRYSTAL PROPERTIES FOR LASER CRYSTALS

1. High thermal conductivity
2. High site symmetry at the activator site, i.e.- a cubic crystal
3. An activator site having inversion symmetry so as to preclude enforced dipole transitions.
4. Be thermally stable under high power levels of excitation and emission.

The following table, presented on the next page, lists the single crystal hosts now being used for solid state lasers.

Table 6-13
Solid State Hosts Used to Construct Lasers

$Y_3Al_5O_{11}$ - Yttrium Aluminum Garnet (YAG)
Yttrium Lithium Fluoride - $YLiF_4$
La_2BeO_4 - Lanthanum Berylate (Alexandrite)
Yttrium Vanadate - YVO_4
$Al_2O_3 : Ti^{3+}$
TiO_2

with Nd^{3+} as the activator for YAG, YVO_4 and $YLiF_4$, and Cr^{3+} as the activator for La_2BeO_4. A YAG crystal is tough, has a high thermal conductivity and can withstand very high power levels. Although La_2BeO_4

is not so thermally stable, it has the advantage that the Cr^{3+} activator can be tuned among several wavelengths. YVO_4 shows the highest efficiency as a host. Also used was: $(Y, Sc)_3Ga_5O_{11}$, but it does not have the thermal stability of YAG.

The performance of a solid state laser is measured as "slope efficiency". Above the threshold, an increase in excitation energy density produces a like increase in laser power output. While phosphors may approach 90% in quantum efficiency, lasers are no more than 10%. The rest of the energy input is dissipated as heat. This means that we must cool the laser cavity. Slope efficiency is measured as an energy conversion figure. A good solid state laser will have a slope efficiency between about 0.5 to 5.0 %. This means that if we use an excitation source of 100 watts in energy, the best that we can expect is about 5.0 watts output The familiar gas laser based on neon (Ne) has a typical output of 0.5 watts (500 milliwatts) for a 100 watt input. A major limitation of the solid state laser is the power of the excitation source. Originally, Nd^{3+} was used because it is well-excited by the radiation emitted by the incandescent lamp. But, the inefficiency of the tungsten filament for conversion of energy to light is well known, and it is necessary to cool the laser cavity because the major portion of energy conversion occurs as heat.

The overall energy losses in a laser cavity are given s follows:

6.8.71.-   <u>Energy Conversion Efficiency</u>

Light source	10%
Crystal absorption	20%
Optical conversion losses	50%
Quantum efficiency	10%

By adding these numbers together, we get, for 100 watts input, about 100 milliwatts of output energy, or about 0.1% efficiency. However, the energy **is** in the form of coherent light.

The most recent "advances" in solid state laser technology has been the use of solid state laser diodes to "pump" the $^4G_{11/2}$ and $^2G_{11/2}$ states of neodymium at about 5860 Å. It is these states which exhibit the strongest absorption bands in the overall Nd^{3+} absorption spectrum. They are directly coupled to the emitting state, $^4F_{3/2}$, and the laser transition is $^4F_{3/2} \Rightarrow {}^4I_{13/2}$, having an energy of 10,580 Å . In a side-pumped design that uses a fiber-lens to couple light from diode-laser bars into a block of YAG:Nd (size - 5.9 x 0.8 x 0.3 cm), a power input of 11.3 watts to the block produced 3.5 watts of coherent emission, i.e.- 31% efficiency.

Let us now examine some of the mechanisms for achieving laser action in a crystal. Under conditions of thermal equilibrium, the number of atoms in any electronic state is proportional to a Boltzmann factor, so that for a total of N atoms, a particular state, N_i , would be defined by:

6.8.72. - $N_i/N = \exp(- E_i / kT) / \sum \exp(- E_i / kT)$

where E_i is the energy of that state. We can now define:

6.8.73.- ω_{ij} = spontaneous transition rate
                            W_{ij} = induced transition rate

where i is the initial state and j is the terminal state. Actually, we ought to multiply these rates by the appropriate Einstein coefficients, A_{ij} , the transition probability of spontaneous emission, and B_{ij} , the transition probability of induced absorption. The total number of states is:

6.8.74.- $N_T = N_1 + N_2 + N_3 + N_4 + \ldots\ldots\ldots N_i = 1.0$

If we have true equilibrium, then:

6.8.75.- $N_i/N_j = \exp(- h\nu/ kT)$ and $N_i \omega_{ij} = N_j \omega_{ji}$

        where $\omega_{ij} / \omega_{ji} = \exp(- h\nu/ kT)$

Let us now define the energy levels of a laser. From the ground state, energy absorbed transforms the activator to an excited state. In a laser, the excited state can relax **before or after** reaching the emitting state.

The relaxation step is needed so that our emitting state will have an appropriate **lifetime** before stimulated emission takes place. In other words, we cannot use a two-level laser, i.e.- excited state and ground state, because the excited state will not possess a lifetime consistent with that required to achieve an inverted population. Thus, a two-level system is fine for spontaneous emission (fluorescence) but not for a laser.

What we use is either a three-level or four-level energy system to obtain stimulated emission, as shown in the following:

6.8.76.-

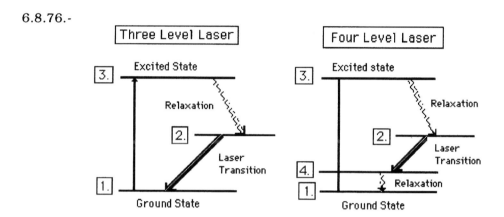

Note that the levels are numbered. $1 \Rightarrow 3$ is the excitation transition while $3 \Rightarrow 2$ and $4 \Rightarrow 1$ are relaxation transitions. Let us take the simpler case of the three level laser. Let N_1 be the number of activator sites in the ground state, N_3 in the excited state, etc. We can then set up a series of differential equations involving ω_{ij} and the number of sites changing per unit of time. This is shown as follows in 6.8.77., presented on the next page.

In this case, relaxation takes place from a level close to the ground state. And, we have specified rates in terms of Einstein probability coefficients. This allows us to determine under what conditions the threshold for stimulated emission can be reached. Note that in our three level diagram relaxation from Level 2. involves a phonon-assisted transition or a phonon emission to the lattice.

6.8.77.-

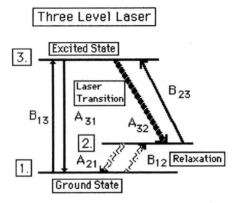

At a steady state of equilibrium:

6.8.78.- $dN_1/dt = dN_2/dt = dN_3/dt = 0$

But during the initial excitation stage, if we increase the excitation intensity, $h\nu_{13}$, i.e.- the density of photons, $\rho\nu_{13}$, then the number of excited activator centers will be a function of:

6.8.79.- $dN_3/dt = N_3 [B_{13} \, \rho\nu_{13}] - N_3 A_{31} - N_3 A_{32} + N_3 [B_{13} \, \rho\nu_{23}]$

Obviously, we want $B_{13} >> A_{31}$ and $A_{32} >> B_{23}$, so as to build up a population at State 3. This will occur if we pick our activator with energy levels far enough apart and with a sufficient half-life of State 3. to prohibit the transitions: $3 \Rightarrow 1$ and $2 \Rightarrow 3$. Likewise, for the other states, we can write:

6.8.80.- $dN_2/dt = N_3 A_{32} - N_2 [B_{23} \, \rho\nu_{23}] - N_2 A_{21} + N_{31} B_{12}$

$dN_1/dt = N_3 A_{31} + N_2 A_{21} - N_1 [B_{13} \, \rho\nu_{13}]$

In the last equation, we have included an exchange process, $\beta_{13} \, \rho\nu_{13}$.
This tells us that we must control the number of activator centers per cubic centimeter of crystal host or we will get a $2 \Rightarrow 3$ transition by exchange of resonance energy or by direct absorption of radiated energy

(and a lower efficiency laser crystal). Under steady state conditions, wherein **only** spontaneous emission is involved, we can solve for the ratios of the steady state populations for fluorescence:

6.8.81.- $N_3/N_1 = B_{13} \, \rho v_{13} \, / A_{31} + B_{12}$ $\qquad N_3/N_2 = A_{12}/A_{32}$

Now, for a population inversion to occur, the following conditions must hold:

6.8.82.- $B_{13} \gg A_{32}$ - strong absorption, long half-life of excited state

$\qquad A_{21} \gg A_{32}$ - rapid relaxation of terminal laser State 2.

This means that we shall have to choose our host-activator combination carefully. It is for these reasons that only a few combinations of host-activator such as YAG:Nd have been found to be satisfactory. To get an efficient laser system, an inverted excited state population must exist in order to get stimulated emission. If an oscillation is to build up, then:

6.8.83.- $N_3 - N_2 \, / N_1 = [B_{31} \, \rho v_{31} \,] \, / [A_{13} + B_{12}] \, \{1 - (A_{32}/A_{21})\}$

For this fraction to be high, the conditions of 6.8.78 must hold. This means that we need a strong and continuous excitation source to obtain a lasing action in our crystal.

Solid state lasers have become ubiquitous in our society. Pulsed YAG:Nd lasers are used for welding of metals and plastics. Among the applications are: medical apparatus catheters where the weld size must be very small; fiber optics communication systems; and hermetic sealing of integrated circuit packages.

Disc lasers based on YAG:Yb are used for welding and higher cutting speeds than other lasers. It uses an optical fiber to deliver the cutting power with improved beam quality and total power achievable. Fiber lasers, which are glass fibers doped with Yb^{3+}, are also being used in the same applications. Their advantage is that the fiber is flexible and can be placed in regions impossible to reach by other lasers.

In the next chapter, we will show how fluorescent lamp technology has matured and the area of studies that are being pursued in an effort to further improve or replace fluorescent lamps. We will also examine display technology and the many new types of displays now available that did not exist when the first edition of this volume was published in 1991.

SUGGESTED READING

1. "Some Aspects of the Luminescence of Solids"- F.A. Kröger, Elsevier Publ., Amsterdam, The Netherlands (1948).

CHAPTER 7

Current Phosphor Device Technology

In the last chapter, we emphasized how phosphors are designed, the parameters involved, and how phosphors can be measured. In this chapter, we will present the latest devices (some are 10 years old or more) and the phosphors currently employed in them. Included will be the many phosphors which have been developed for particular display purposes. We will also show how the light sources we now depend upon, i.e.- incandescent and fluorescent lamps, are going to become obsolete as the quality of light-emitting diodes (LEDs) improves. Nonetheless, it is clear that phosphors will continue to be used to make "white" emitting LED light sources. The main advantage of the LED lamp is the fact that it uses milliwatts of power, compared to watts of power for the incandescent and fluorescent lamp.

7.1.- CURRENT PROGRESS IN CRT DISPLAY AND DISCHARGE LIGHTING

We will begin our survey of present-day devices utilizing phosphors by first addressing recent improvements made in cathode-ray tubes followed by fluorescent lamps and then other lighting devices. We will then address the devices that depend upon thin film deposition for manufacture of the appliance. All of these utilize phosphors in some way for light output.

I. Cathode Ray Tubes (CRT)

In the last chapter, we presented a synopsis of the technological advances made in CRT's, particularly those for the color TV tube up to about the mid-1960's. Since that time, a number of innovations have been accomplished. This change has been in response to market demand for a less bulky display device as well as a larger display size, i.e.- the "flat" television tube. Major improvements have been created in the electron gun, deflection yoke and the glass faceplate. What we are addressing is the changes in type of display which has resulted. In the ordinary CRT of the past, the picture size was limited to about 27" diagonal because of the deflection limitation of the electron beam. The first color-TV tubes used a curved faceplate with a round tube construction of no more than 18". This was mandated by the shadow-mask which had to be carefully positioned so that the three color phosphors could be properly excited by the electron beam, i.e.- "color registration". The following diagram shows what we are discussing:

7.1.1.-

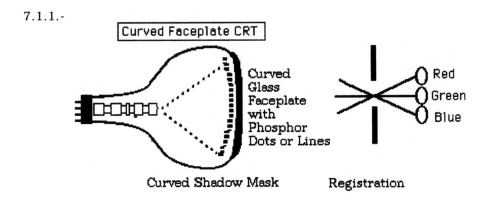

Curved Shadow Mask   Registration

Color registration became exacerbated as the shape of the faceplate changed from round to rectangular. This remained a significant problem. A rectangular tube was mandated by the semantics of the picture being viewed. That is, people were accustomed to a rectangular picture like a photograph and the round tube soon became obsolete. The innovation by Sony who introduced vertical stripes on the shadow mask, with accompanying phosphor lines, alleviated the problem somewhat. The Trinitron™ method of vertical phosphor stripes has become the norm today, along with the specialized electron gun needed to produce the TV picture (see 6.6.4. of the last chapter). Improvements such as a "black-surround" of the phosphor dots or stripes and darkening of the glass faceplate were made to improve "picture contrast" as well. Nonetheless, demand for larger TV displays continued.

In 6.6.1., we showed the typical CRT construction. What we did not state was that the maximum angle of deflection of the electron beam was about

30° vertically. This limited the size of the image that could be generated. The market demanded larger pictures. Work on the electron gun and the deflection yoke resulted in much larger faceplates and TV pictures. This is shown in the following diagram:

7.1.2.-

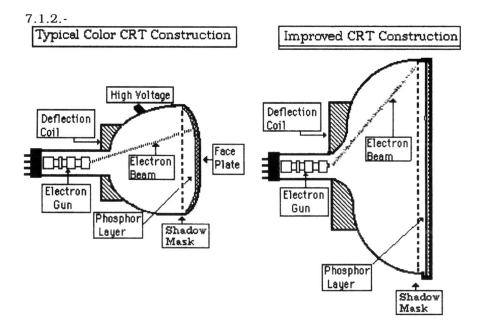

In this diagram, we have somewhat over-emphasized the curved glass faceplate on the left. You will note that the edges of the TV picture would be slightly distorted because of the curvature of the glass. This problem persisted until the advent of the flat glass faceplate of the CRT shown on the right of the diagram. This type of CRT was mandated for several reasons, one of which was color registration and minimization of achromatic aberration due to differences in refractive indices of air and glass. The difficulties in obtaining such a color CRT cannot be overemphasized.

Excerpted from United States Patents 5,536,995 (issued July 1996) and 5,964,364 (issued Oct. 2000- reissue 36,838- Aug. 2000), "Glass Panel for a Cathode Ray Bulb & Glass Bulb for a Cathode Ray and a Method of

Producing the Same", the following is a description of the inventions leading to this CRT innovation:

> "In the production of a flat face-plate panel, a differential stress (strain) pattern is imposed upon a glass panel with a glass-temper factor less in the skirt than in the face, both in the longitudinal mode and in cross-section (thickness) of the glass panel faceplate, by controlled cooling, immediately after press-shaping a mass of hot glass to a form, thereby obviating an annealing step, (which is usually) followed by a tempering step. Compressive stress layers are formed in the inner and outer surfaces of the skirt portion wherein the compressive stress value of the face portion is larger than the compressive stress value of the skirt portion and the compressive stress value of the outer surface of the face portion is larger than the inner surface. The so-formed glass faceplate is superior in impact strength as well as having a reduced thickness of glass at the center of the panel".

What this means is that the flat faceplate is a tempered-glass panel with reduced thickness at the center and increased tempering at the face of the tube. This configuration also reduced achromatic aberration at the edges of the glass faceplate. It was determined that this configuration was superior to the prior ones where the glass thickness was constant, as shown in the following diagram:

7.2.3.-

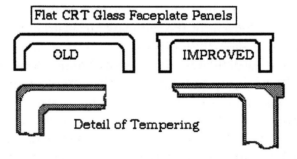

The flat faceplate panel on the left, with a constant thickness, was found to be subject to easy breakage during final fabrication of the color TV CRT whereas the new configuration was much stronger. It also allowed the construction of a color TV having superior color registration and less color aberration when viewed under normal light conditions. As shown in 7.2.2., the maximum angle of the electron beam of the tube having a flat-glass faceplate is near to 60°. Note that we are describing the vertical angle, not the horizontal beam angle which is much greater. It should be clear that the flat-faceplate CRT posed much greater problems in achieving color registration because the "holes" in the shadow mask could no longer be of uniform size. This mandated a major change in the manufacture of shadow masks, which depended upon both size and shape of the CRT viewing screen. In order to achieve such changes in electron beam generation within the CRT, a number of improvements were made in the electron gun, as well:

Table 7-1

Milestones in CRT Electron Gun Design

1. Change from "delta-shaped", i.e.- ⊙⊙⊙ to "in-line" configuration, i.e.- ⊙⊙⊙ substantially improved the relative alignment of the three individual guns, red, green and blue.

2. Realization that beam angle had significant effects upon spherical aberration of color registration, magnification of beam diameter and space charge increase at large angles, led to improvement in electron-grids within each gun.

3. In-line gun design with a common lens (electronic) critically reduced color aberration of the main lens and enabled most CRT's to achieve the limits of system performance in which the color registration remained even when picture contrast and brightness controls were set at their maximum.

4. A design of a shunt/enhanced module in the guns enabled the in-line color CRT to have perfect color convergence even at the corners and edges of the screen.

5. Dispenser cathodes were designed to have a current-loading density many times that of the conventional triple-oxide cathodes. This enabled the CRT to run at much higher beam currents with brightness and resolution much better than prior CRT's because the electron beam could be focused to a smaller spot.

The current "high-definition" color-TV's incorporate all of these improvements. However, major changes were made in how such improved CRT's were used. Most of these CRT's were manufactured in 21", 27", 31" and 36" sizes for use in homes.

For larger sizes, the CRT's were "stacked" so that each CRT showed only a part of the total picture. This is shown as follows:

7.1.4.-

Stacked CRT Display

Here, 16 CRT's are stacked to produce a large display picture. Many TV stations use such displays and they are used in outdoor displays as well. The flat-faceplates fit together seamlessly and no internal edges within the picture can be seen by the viewer.

With the advent of rare earth phosphors, an entirely different approach to a large screen TV became possible. Instead of generating all three color screens in the same CRT, three separate tubes were used and their images were combined on a projection screen so that a much larger picture could be viewed. This is shown as 7.1.5., presented as follows:

7.1.5-

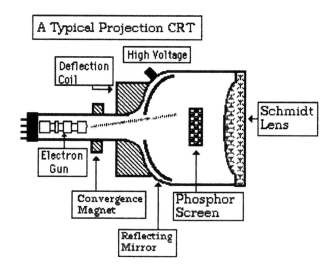

This made possible a viewing screen as large as 60 inches diagonally. Note that an internal reflecting mirror is used to enhance the emitted light coming from the CRT. Each color CRT is then combined with the other two to produce the picture (which must be aligned for proper color registration). The tube is operated under extremely high current density and, in most cases, the faceplate needs to be cooled by flowing air. The phosphors currently being used in cathode-ray tubes are:

7.1.6.-

Projection Tubes	Color Television
Y_2O_3:Eu - red	Y_2O_2S ; Eu - red
ZnS:Ag,Cl - blue	ZnS:Ag, Cl - blue

The green phosphors include Gd_2O_2S:Tb and Y_2SiO_5:Tb. Because stability, brightness output with current density and body color changes are a major problem in projection CRT's, phosphors for this application have been evaluated in terms of:

7.1.7.-   Relative brightness (Br) $= I_c \upsilon$

where I_c is the cathode current density and υ is a figure of merit.

The following compares several phosphors currently used for projection tubes, where the figure of merit has already been defined.

Table 7-2

Phosphor	Color	Fig. of Merit	Br. Loss	Rel. Stability
$Y_2O_3:Eu^{3+}$	red	0.92	3%	excellent
$Y_2O_2S:Eu^{3+}$	red	0.85	12%	excellent
$Gd_2O_2S:Tb^{3+}$	green	0.84	5%	excellent
$Y_2SiO_5:Tb^{3+}$	green	0.91	1.5%	poor
$Y_3Al_5O_{11}:Tb^{3+}$	green	0.91	2.0%	very strong
$Zn_2SiO_4:Mn^{2+}$	green	0.85	34%	very poor
$CaS:Ce^{3+}$	green	0.82	30%	very poor
ZnS:Cu:Al	green	0.67	37%	inferior
ZnS:Ag,Cl	blue	0.75	31%	inferior

Brightness loss is that incurred over a period of operating time of the CRT while Stability is a subjective measure based primarily upon brightness loss and development of "body color", i.e.- a color developed by decomposition of the surface layers of the phosphor particles (in many cases, the cation of the lattice is reduced to a dark metallic state). You will note that the best red phosphor is $Y_2O_3:Eu^{3+}$ while the best green phosphor is either $Gd_2O_2S:Tb^{3+}$ or $Y_3Al_5O_{11}:Tb^{3+}$. Only one blue-emitting phosphor has been found to be suitable for these CRT's. The phosphors currently being used in Color Television tubes are:

ZnS:Ag,Cl - blue

ZnS:Cu,Al - green

Y_2O_2S ; Eu - red

II. Fluorescent Lamps

There are two areas where discharge lamps are currently used in improved lighting applications: 1)   Home lighting, and 2) Commercial lighting (as in factories offices and medical facilities). The former uses fluorescent lamps while the latter uses both fluorescent lamps and "high-energy discharge" (HID) lamps. The latter are primarily high-pressure mercury vapor (HPMV) discharge lamps. One design was shown in 6.6.5. of the last chapter. We did not mention "metal halide" lamps which are HPMV lamps with selected metals added to the discharge. One such lamp was the Metalarc™ lamp which added: Indium, Thallium, Sodium and Scandium lines to the 3650 Å, 4060 Å, 4380 Å, 5470 Å, 5770 Å and 5790 Å lines of the HPMV discharge. The resulting spectrum is shown on the follows:

7.1.8.-

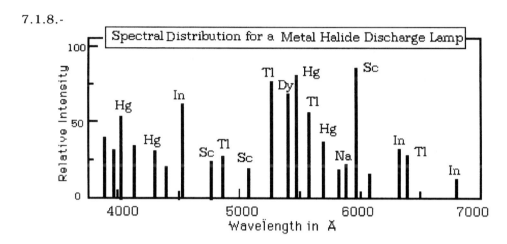

Note that these metals added the needed "redness" so that metal-halide lamps now produce a "white" radiation in contrast to the "greenish white" of the old HPMV lamp (you will note that, in the past, certain phosphors were coated on the inside of the glass envelope to improve the "redness", but this addition was never entirely satisfactory. The construction of the HID lamp is shown in the following diagram, given as 7.1.9. on the next page. Shown is the construction of a HPMV-HID lamp using a quartz arc-tube. A significant advance in HID lamps has been the replacement of the quartz arc-tube by a ceramic tube. Since the lamp runs hot, i.e.- 600 to 800 °C, quartz was never entirely satisfactory as a lamp envelope. The ceramic construction is the same except that end-caps are sealed upon the ceramic tubing whose composition is usually Al_2O_3. Zirconia is also an alternative.

7.1.9.-

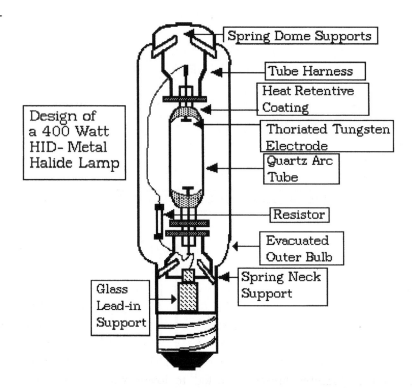

Design of a 400 Watt HID- Metal Halide Lamp	Spring Dome Supports
	Tube Harness
	Heat Retentive Coating
	Thoriated Tungsten Electrode
	Quartz Arc Tube
	Resistor
	Evacuated Outer Bulb
Glass Lead-in Support	Spring Neck Support

This type of lamp was used extensively for lighting in warehouses, offices and medical facilities until the advent of improved fluorescent lamps which used much less power.

Major advances in fluorescent lamps resulted with the advent of rare earth activated phosphors. In 6.6.6. of the last chapter, we showed the general construction and operation of a fluorescent lamp. The primary fluorescent lamp manufactured in the 1960's through the 1980's was the 40T12 lamp of various colors (see 6.7.46. of the last chapter). Cool White was the largest type of color purchased by consumers. 40 is the lamp wattage and "12" is the diameter in 1/8" inches, i.e.- 1.5 inches in diameter.

The following table, shown on the next page, gives a comparison of the various sizes of fluorescent lamps currently available.

Table 7-3

Lamp Type	Lamp	Watts	Length/ Diameter	Est. Life (hr)	Ambient Op. Temp	Rated Lumens	Lumens per Watt
T-12	F-40T-12	40	48"/1.5	20,000	77 °F	3200	80
T-8	F-32T8	32	48/1.0"	20,000	77 °F	2850	89
T-8	F-32T8 R	32	48/1.0"	24,000	77 °F	3100	97
T-5	F-39T5	28	46/.63"	20,000	90 °F	2900	103
T12HO	48T12HO	60	48/1.5"	12,000	77 °F	4250	71
T8HO	48T8HO	44	48/1.0"	18,000	77 °F	4000	91
T5HO	54T5HO	54	46/.63"	20,000	90 °F	5000	93

Whereas the first two lamps shown in the table used conventional phosphors, i.e.- halophosphate mixtures, the rest of these lamps utilized the tri-phosphor rare earth blends of three colors. These rare earth blends provide 17 to 28% improved brightness over the traditional phosphors, with high color rendition of 92% in most cases. The old "high-output" = HO lamps were much lower in lumens per watt light output and had much greater loss in brightness as a function of lamp life. The small diameter lamps have led to lower wattage at about the same brightness and lamp life. The tri-phosphors currently in use include:

7.1.10.-   Tri-band Phosphors Used in Fluorescent Lamps

$$BaMgAl_{10}O_{17}: Eu^{2+} - BLUE$$
$$MgAl_{11}O_{19}: Ce^{3+}:Tb^{3+} - GREEN$$
$$Y_2O_3: Eu^{3+} - RED$$

What has not been addressed is the operating characteristics of the two types of lamps, i.e.- conventional halophosphate lamps compared to tri-band rare earth lamps. Conventional lamps in 1950 lost up to 50% of their initial brightness at 4000 hours of operation and lasted no more than 8,000-10,000 hours. 50% of these 40T12 lamps failed by 8000 hours due to electrode problems. By 1990, these figures were improved to 75% at

4000 hours with about the same failure rate. In contrast, the F32T8 lamp, with improved ballasts, maintained its lumen output at 98% at 4000 hours operation and 91% at 30,000 hours. Lamp mortality was 1% at 4000 hours, 7% at 12,000 hours and 50% at 30,000 hours of operation. This, of course, depended upon the conditions of operation where the lamp was not turned off and on too many times. A 12 hour burn cycle per day was used to evaluate these lamps.

Commercial lighting has traditionally used so-called "high-intensity-discharge" (HID) lamps, i.e.- high-pressure mercury vapor discharge lamps. As we stated in the last chapter, such lamps were improved by the addition of selected metals to the discharge, resulting in "metal-halide" HID lamps. But even these 400 watt HID lamps are being replaced by rare earth fluorescent lamps. The following table gives a comparison of the two types of lamps.

Table 7-4 - Loss in Brightness of Metal HID and Rare Earth Fluorescents

			Hours of Operation			
Lamp	Watts	0	4000	12,000	20,000	30,000
Metal-halide	400	28,100	22,460	15,400	8,140	----
% Br. Loss		0	85%	72 %	68%	
Mortality			3%	17%	50%	----
6- F32T8 RE	192	20,580	19450	17,970	16,570	8,370
% Br. Loss		0	98%	95%	93%	91%
Mortality			1%	3%	7%	50%

You will note that the table compares six 32T8 lamps to one HID metal halide lamp. But, the wattage is about half for the 32T8 lamps. The total output as a function of time is a function of the loss in lumens with time and the remaining lamps. Thus, the 32T8 lamps lasted until over 20,000 hours when one of them failed. By 30,000 hours of operation (2500 days @ 12 hours/day = 6.8 years), three of these lamps had failed.

It should be clear that the 32T8 lamps with rare earth phosphors perform much better than the prior fluorescent lamps used for commercial lighting. Part of this performance is due to improved ballasts.

Four types of ballasts have been, or are being, used: electromagnetic, "rapid start" "instant start", and "electronic". Electromagnetic ballasts use a "choke-coil" to suppress current while the filament- electrodes are being heated to start the lamp. Once the electrodes are hot, the voltage is switched across the lamp to start the discharge. This method is relatively inefficient and was a major loss in fluorescent lamp operation. Rapid-start ballasts still preheat the lamp electrodes at a lower voltage but are very similar in control to electromagnetic ballasts. Instant start ballasts apply a voltage high enough to directly start the mercury arc. Many fluorescent lamp installations still use the "instant start" method of lamp control. The electronic ballast is essentially micro-processor controlled and is about 30% more efficient than any of the other types. The program allows time for filament preheat and then switches to a high frequency waveform of 25,000 to 40,000 cycles per second. Electronic ballasts can operate the 32T5 HO lamps more efficiently than any other ballast. The high frequency eliminates lamp flicker and extends lamp life. Internal losses are only 1-2 watts compared to 10-12 watts for the old electromagnetic ballasts. The newest fluorescent lamp ballast uses micro-wave energy to excite the lamp. Here, the mercury arc is maintained by electromagnetic radiation having a frequency within the range of 1 gigahertz to 1 terahertz (10^9 to 10^{12} cycles per second) with a wavelength between 1 mm. and 1 meter. The lamp has an increased lifetime of up to 100,000 hours.

The newest lamp technology, the T5 fluorescent lamp, is having a major impact on the lighting industry. The F32T5 lamp is brighter and more visible (color rendering index = 92 +) and produces 30% more light than the F32T8 lamp. They have become a viable alternative to HID lamps and particularly the HID metal halide (HID-MH) lamps in commercial installations since the cost of operating T5 lamps is about 50% that of the HID lamp annually. Whereas the HID-MH lamps lose up to 45% of their initial brightness over time, the T-5 lamps lose only 5%. The new fluorescent technology is now preferred for lighting warehouses and adjoining buildings, offices of all kinds and medical facilities. However, HID and HID-MH lamps are still preferred for outdoor lighting because of their ability to operate in and withstand a wide range of temperatures.

7.2.- CURRENT DISPLAY DEVICES BASED UPON ELECTRON CREATION

The next subject we will undertake will include those devices which depend upon thin film technology for their manufacture. Since the publication of this work in 1991, the use of thin film technology has burgeoned. In order to appreciate exactly what has happened since that time, we will address the many uses which have evolved for thin film devices in relation to phosphor usage. Thin films are formed by many methods and are used to make many devices that we now use in every day life. Many of these devices were not available in 1991 or were too expensive for many consumers. One good example is the portable cell phone which uses integrated circuitry made by thin film techniques. Another is the advent of flat display systems such as plasma displays. This has made possible the flat television tube which can be hung on a wall (although most people who have such a device usually set it up like an "ordinary" television display). We will emphasize display technology since this volume is meant to thoroughly describe the unique properties of phosphors as "energy converters" for use in lighting and display devices.

In the past 12 years, a number of illuminating and display devices have appeared in the marketplace. As we have already shown, only 2 types of excitation can serve to excite solid state phosphors. **These are the electron and the photon.** We have already presented the essential differences between these two types of energy sources in the last chapter. The following chart shows the current types of devices which depend upon electron energy for phosphor excitation:

Note that we have shown that phosphor excitation can occur via one of two paths, that of electron bombardment (cathode-rays) or by electric field generation of electrons. The former has been further divided into external (electron gun) and internal (electric field) generation of an electron beam. Since we have already described color TV CRT's, we will first address the other types of CRT's first and then internal electron generation in selected devices such as the vacuum fluorescent display.

7.2.1.-

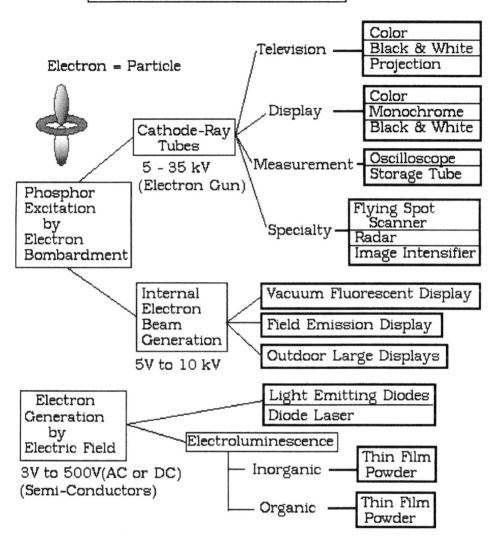

Finally, we will describe the properties of phosphors needed for devices in which the electron excitation is generated by electric fields. This includes LED's, Laser LEDs and electroluminescence (EL). We will first address the CRT's used for specialty applications.

I. CRT's for Display, Measurement and Specialty Types

We have already addressed CRT's for television, including color, black & white and color projection systems. Color display CRT's are used primarily as computer monitors. Here, the same rare earth phosphors are used as for color TV. Monochrome displays refers to CRT monitors where only one color screen is used for special purposes, as described in the last chapter. The following is a repeat of 6.6.6. given in the last chapter:

7.2.2.- Classification of JEDEC Phosphor Applications

Electrochrome display = 1	Oscilloscope display = 6
Radar display = 16	Television display = 6
Flying spot scanner = 8	Display monitor = 7
Storage monitor = 1	Photography display = 1
Light pen on display = 1	"Penetration" display = 1

Note that all of the classification of CRT applications of 7.2.1. are included here.

a. Oscilloscopes and Storage Tubes- Measurement

These CRT's are used to detect and analyze electrical signals and waveforms. The same basic structure is used similar to CRT's employed as color TV's except that the electron gun is highly sophisticated with special electron optics. The oscilloscope CRT is required to have superior beam deflection and sensitivity, linearity of response to voltage and frequency changes and linear response to very high frequencies. The latest oscilloscopes use a micro-channel plate (which is an electron multiplier- that is, the signal from just one electron is multiplied many times by secondary electron emission) to further improve deflection sensitivity. The microchannel plate is placed just between the electron beam and the phosphor screen to increase the beam current so that frequencies above several hundred megahertz can be detected easily. Most oscilloscopes are useful up to about one giga-hertz (10^9 cycles per sec.) frequency.

The storage tube is similar to the oscilloscope CRT except that it has a memory function fabricated within the CRT. There are two electron gun components, a writing gun and a reading electron gun. The writing gun uniformly scans the dielectric target which is placed between the gun and the phosphor screen. The target consists of a gold-plated mesh on which a thin layer (1-5 μ) of MgF_2 is deposited. A picture image of the signal is drawn on the target by the writing beam and stored as a pattern of electrical charges. That is, the target becomes charged only where the writing beam strikes the matrix, which retains the image of the signal being analyzed. Thus, quite fleeting signals can be detected. The electrostatic charge patterns remain on the target and are reinforced by the **non-focused** electron beam of the reading gun so that those areas remain charged. The image is transferred onto the phosphor screen as the reading gun maintains the charge. The excess is transferred to the phosphor screen. Since the electrostatic image is not erased but reinforced by the reading gun during operation, the stored image can be maintained for extended periods of time until it is intentionally erased. The image is updated every 1/30 of a second. Since the reading beam is supplied on a continuous basis, the screen brightness can be maintained at a high level. Thus, a storage tube has been used in aircraft cockpits for so-called "daylight display".

Phosphors used for these applications are shown as follows and include the following phosphors:

7.2.3.- Phosphors for Oscilloscopes and Storage Tubes

Oscilloscopes	Storage Tubes
P1- $ZnSiO_4$: Mn	P20- ZnCdS:Ag, Cl
P2- ZnS:Ag, Cl	
P31- ZnS:Cu, Cl	

Since high resolution is required for these screens, they must be extremely uniform in phosphor particle size. That is, the PSD must be narrow in size and no more than 6-8 μ in average size. You will note that these devices still employ the old JEDEC phosphors. More recently, the

or underground probing. The most common type of radar signal consists of a repetitive train of short-duration pulses. A ground-based radar system with a range of about 50 to 60 nautical miles (or 90 to 110 kilometers), such as the kind used for air traffic control at airports, would have a pulse width of 1.0 µsec. However, there would be 1,000 cycles within the pulse which corresponds to a pulse repetition frequency of 1,000 hertz. In this example, the average power is 1,000 watts (1 kilowatt). The average power, rather than the peak power, is the measure of the capability of a radar system. Radar systems have average powers from a few milliwatts to as much as one or more megawatts, depending on the application. A weak echo signal from a target might be as low as one trillionth of a watt (10^{-12} watt). In short, the power levels in a radar system can be very large (at the transmitter) and very small (at the receiver).

Timing of the extremes encountered in a radar system is critical. An air-surveillance radar (one that is used to search for aircraft) might scan its antenna 360 degrees in azimuth in a few seconds, but the pulse width might be about one microsecond in duration. (Some radar pulse widths are 1,000 times smaller--i.e., of nanosecond duration.) The range to a target is determined by measuring the time that a radar signal takes to travel out to the target and back. Radar waves travel at the same speed as light--roughly 186,000 miles per second. The range to the target is equal to c times $T/2$, where c = velocity of propagation of radar energy, and T = round-trip time as measured by the radar. From this expression, the round-trip travel of the radar signal is at a rate of 150 meters per microsecond. For example, if the time that it takes the signal to travel out to the target and back were measured by the radar to be 600 microseconds (0.0006 second), then the range of the target would be 90 kilometers or 56 miles.

To preserve the reflection signal, a long decay phosphor is required. The phosphors used in the past were generally composed of two phosphors, one with a long decay. Thus, the screen exhibits two colors, that of the target and the other forming the background which is refreshed in every 360 ° rotation of the radar antenna. Radar phosphors include:

7.2.6.- Phosphors for Radar CRT's

	Phosphors		Colors	Decay Times
P-7	ZnS:Ag + ZnCdS$_2$:Cu		blue + yellow	med.(57µs)+ long(400ms)
P-12	(Zn,Mg)F$_2$:Mn		orange	long (210 msec)
P-14	ZnS:Ag + ZnCdS$_2$:Cu		blue + orange	short(27µs)+v.long(500ms)
P-17	ZnS:Ag + ZnCdS$_2$:Cu		blue + yellow	short(5.2µs)+v.long(429ms
P-19	KMgF$_3$:Mn		orange	long(220 msec)
P-28	(Zn,Cd)S:Ag:Cu		green	very long (500 ms)
P-39	Zn$_2$SiO$_4$:Mn:As		green	med. long(400ms)
P-57	Zn$_2$SiO$_4$:Mn+MgF$_2$:Mn		yellow- green	medium

All of these were given in Table 6-2 of the last chapter. More recently, digital technology has replaced this type of CRT using long decay phosphors. A computer-controlled display utilizing a full color CRT has been developed for aircraft traffic control. Therein, the direction, size of target and altitude of aircraft are directly displayed upon the phosphor screen.

IV. Image Intensifier Screens

The image intensifier tube is one type of CRT but it does not utilize a scanning electron beam. Image intensifiers can be separated into two classes: 1) electronic tubes that incorporate an apparatus designed to operate in very low light levels, and 2) phosphor screens intended to detect and amplify radiation like x-rays. The former is used primarily as "night-vision" goggles and relies upon electron emission from a photocathode to amplify the available light as a picture. The latter is used for x-ray tomography in medical establishments wherein the phosphor screen is passive but responds to x-ray excitation.

The image intensifier tube, shown on the next page as 7.2.7., generally has the construction shown. All internal spaces in the image intensifier are evacuated to a high vacuum of at least 10^{-8} mm. to ensure that the operation of the device is not degraded by residual gasses.

7.2.7.-

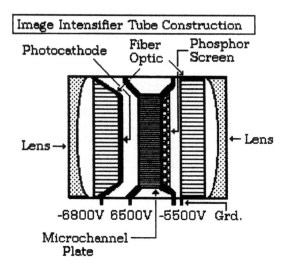

Note that an external lens focuses light upon a fiber optic having a photocathode on the internal side of the fiber optic. A 300 VDC potential channels emitted electrons directly onto the microchannel plate. What we have not shown is the transparent electrodes between the photocathode and microchannel plate and that on each side of the plate. This electrode is usually In_2O_3-SnO_2 and is formed by a sputtering process. This is the first time we have encountered a thin-film in a device. The next devices that we will describe utilize thin film processes for their manufacture and we will address this method in more detail below.

As we have said, the microchannel plate multiplies each photo-electron by causing secondary electron emission, thereby multiplying the cathode current. Once the photo-electrons reach the phosphor screen via the 1000 VDC drop, the resulting emission passes through the viewing fiber optic to form a picture. Since the incoming photons are kinetic, the resulting scene is viewed in real time. Note that fiber optics are used both in the front and rear of the tube. This ensures that the very low level of photon excitation on the photocathode which forms the image remains coherent so that the resolution of the picture viewed remains clear and not blurred. This type of passive cathode ray tube can be reasonably large or can be miniature as in night-vision goggles used world-wide.

The phosphors used for this type of CRT include:

7.2.8.- Phosphors for Image Intensifier Tubes

P-31	$ZnCdS_2$:Cu:Ni	green	short (40 μsec)
P-36	(Zn,Cd)S:Ag:Ni	yellow-green	very short(10 μs)

Because the image light intensity is often very low, these tubes use phosphors which have emission close to the peak of the eye sensitivity curve, i.e.- green or yellow-green. This maximizes the response of the device for its intended application, namely night-vision under lack of visible light. Note that some night-vision goggles also detect near infra-red radiation as well. This is a function of the composition used for the photocathode.

We will now summarize the techniques required to form thin films.

II. Thin Film Technology

The devices utilizing internal electron beam generation require the formation of thin films in most cases. Although we have described the growth of crystals in some detail in a previous chapter, we have not considered the fact that we can grow single or polycrystalline crystals in the form of a thin film. Such films can exhibit much different optical and electrical properties from the same bulk crystal. This is particularly true since we can form a series of stacks of thin films of varying composition whose overall behavior will be a combination of those films comprising the stack. It is the electronic properties of such materials that are useful in industrial applications. Therefore, it would behoove us to consider the factors that cause change in electronic properties of solids as a function of structure and bonding. It is evident that this aspect is very important, since the electrical nature of solids varies according to both composition and arrangement of atoms in the structure. Some materials are semi-conducting and others are dielectric. Still others are good conductors of charge (metals). Still others have unique optical properties, due to arrangement of electrons and atoms within the solid state. In addition,

but it was not until 1937 that an all metal oil-diffusion pump came into practical use in industry. All diffusion pumps up to that time were made from glass. One of the early major problems with oil diffusion pumps was that the oil was seriously carbonized and degraded during operation of the pump. W. Gaede invented the modern oil-sealed rotary-vane mechanical vacuum pump in 1904. By 1910, electric motor-driven rotary-vane pumps were in common use. Before that time, various types of pumps, including the mercury diffusion pump, were very inefficient in terms of "pumping power". The mechanical rotary-vane pump, operating in an oil-bath where the spinning vanes were sealed, increased the pumping power by many hundreds of times over the old vacuum-pumping systems. Rotary-vane pumps became the "fore-pump" in 1916 as used by Langmuir and Dushman of GE in their experiments on incandescent tungsten filaments. The forepump lowered the internal pressure to about 10^{-3} torr at which point the diffusion pump takes over to lower the internal pressure to at least 10^{-6} torr.

Mercury diffusion pumps were used widely from about 1920 to 1946 when they began to be replaced by oil diffusion pumps. Today, oil-diffusion pumps use special oils in which the vapor pressure at room temperature is 10^{-10} torr (polyphenol ether) in contrast to earlier oils whose vapor pressure about equaled that of mercury at 10^{-3} torr. Other types of vacuum pumps in use today include: "Turbopumps" (which utilize a high speed turbine for molecular drag from the pumping volume); "Cryopumps" (which use low temperatures generated by liquid helium (- 268 °C) to remove gases from the pumping volume) and "Getter" or "Ion" pumps (which trap gas molecules by molecular bonding). Extreme vacuums have been achieved up to 10^{-15} torr by using combinations of these pumps. At this point, diffusion of helium gas through glass from the outside atmosphere becomes the major deterrent to achieving lower pressures. The gas pressure regions where these pumps are used are shown in 7.2.12., given on the next page.

Note that this list is not exhaustive. In all cases, a forepump (oil-sealed rotary vane) is needed in addition to the specific high-vacuum pump.

7.2.12.- Types of Vacuum Pumps Needed to Produce a Specific Vacuum

Type of Vacuum	Rough= 1 to 1000	Med= 1 to 10^{-3}	High Vacuum = 10^{-3} to 10^{-7}	Ultrahigh= 10^{-7} to 10^{-10}
	Diaphragm	Vane rotary	Turbomolecular	Turbomolecular
	Vane rotary	Getter	Vane rotary	Sputter ion
	Piston	Oil diffusion	Oil diffusion	Cryopump
	Diffusion	Adsorption	Adsorption	Sublimation

We will now address the specific categories of methods used to produce thin films. As we said, thin films are formed by:

7.2.13.- Methods Used to Form Thin Films

Evaporation - This is a technique where a material is heated in an electrode to its boiling point in a vacuum. It works best with metals and uses a tungsten coil as the electrode. A few low-melting compounds can be used to form a thin film but decomposition during melting has remained a serious problem. During evaporation, all internal surfaces of the vacuum chamber are coated. Color TV's and the like use a thin film of Al as a charge- dissipation mechanism to prevent charge accumulation in the phosphor screen or the shadow mask within the CRT.

Chemical-Vapor Deposition - This technique involves the use of metal-organic compounds and the like which can be decomposed upon a hot substrate. They are generally delivered to the hot surface via a flowing inert gas. Molecular-beam epitaxy, commonly known as MBE, is a form of vapor growth. The field began when the American scientist John Read Arthur reported in 1968 that gallium arsenide could be grown by sending a beam of gallium atoms and arsenic molecules toward the flat surface of a crystal of the compound or a similar material. The amount of gas molecules could be controlled to grow just one layer, or just two, or any desired amount. This method is slow, since the gas stream has a low density of atoms. Chemical vapor deposition (CVD) was developed to replace MBE and is another form of growth that makes use of the vapor technique. Also known as vapor-phase epitaxy (VPE), it is much faster than MBE since the atoms are delivered in a flowing gas rather than in a molecular beam. Synthetic diamonds are grown by CVD. Rapid growth occurs when methane (CH_4) is mixed with atomic hydrogen gas, which serves as a catalyst. Methane dissociates on a heated surface of diamond. The carbon remains on the surface, and the hydrogen leaves as a molecule. Liquid-phase epitaxy (LPE) is similar and uses the solution method to grow a thin film on a substrate. The substrate is placed in a solution with a saturated concentration of solute. This technique is used to grow many materials employed in modern electronics and optoelectronic devices, such as gallium arsenide, gallium aluminum arsenide, and gallium phosphide.

Precursors like tri-ethyl gallium, arsenic hydride and phosphine are generally used in the gaseous phase. The composition is controlled by the rate of gas flow to the hot target substrate. CVD processes can take place over a wide range of temperatures and pressures, from 600 °C to 1100 °C and from 760 mm to a few milliTorr. It has found wide use because of its versatility.

3. Mechanical Film Deposition - In this technique, a doctor blade is used to mechanically deposit a thin film (usually particles in a liquid matrix) upon a surface as shown in the following diagram:

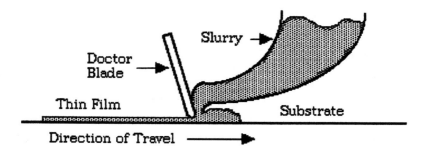

This diagram is a representation of the actual process. The actual machine is usually much more complex. The particle slurry (like an ink) flows upon the substrate but its depth is controlled by the doctor blade. The thin film thus formed is then dried and then fired to form a composite film. Obviously, any thin film thus formed must be stable to heating in air or atmosphere. Reducing atmospheres during firing are sometimes used.

4. Sputtering - This is a process in which atoms, ions, and molecular species in the surface of a target material are ejected under the action of an ion-beam irradiation. Energies typical of ion implantation are employed and, while any ion type may be used, rare gases such as argon and neon are most common because they avoid unwanted chemical interactions between the ions of the beam and the substrate. A simple apparatus is shown in the following:

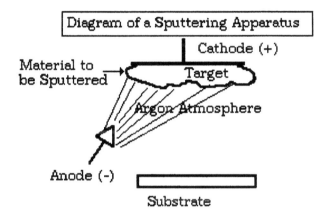

In this case, the argon atoms are ionized and attracted to the cathode having the material to be sputtered. Layers of material are stripped off the surface of the target by the action of the kinetic energy of the gas molecules. The released atoms and/or molecules are attracted to the substrate to form a thin adherent layer. Note that the material is not vaporized or ionized but is transferred to the substrate with the same composition of the original material. Sputtering results from several interaction mechanisms. Conceptually, the simplest is rebound sputtering, in which an incident ion strikes an atom on the surface of the target, causing it to recoil. The recoiling atom promptly collides with a neighboring atom in the target, rebounds elastically, and is ejected from the surface. By means of any of these various mechanisms, several atoms may be sputtered for each ion incident on the target. The number of atoms sputtered per incident ion is called the sputtering yield. Note that sputtering does not require a high temperature like evaporation techniques to form a thin film. Materials like alumina (with a melting point of 1840 °C) are easily sputtered. Even tungsten (MP > 3560 °C) is amenable to this method of forming a thin film.

The latest technique for sputtering is the use of a "magnetron" which is a high frequency AC device operating at 10 to 100 mega-hertz. The induced gaseous ions are controlled by a magnetic field instead of an electric field (shown in the above diagram). Large "Climate-Control" windows are coated by this technique. Sputtering to form thin films has become one of the most common processes used to do so.

5. Ion Beam Deposition Methods - This method involves formation of an ion beam of high energy. Compared with thermal energy processes, a 100 ev ion has a temperature equivalent to 1,000,000 °K. Ion beam systems have been used for: cleaning substrates, in-situ cleaning, etching of low volatility materials, reactive etching, sputter deposition, reactive sputter deposition, ion beam assisted coating of evaporated films, CVD coatings and modification of surface properties of materials before they are coated. All of these involve thin film formation. The following lists some of the areas where ion-beam thin-film technology has been employed:

> Magnetron Sputtering Systems
> Electron-Beam Ion-Beam Pre-clean Systems
> Pulsed Laser Deposition System
> Inert Ion Beam Etching
> Chemically Assisted Ion Beam Etching
> Reactive Ion Beam Etching
> Dual Ion Beam System Configuration
> Ion Beam Sputter Deposition
> Ion Beam Coating System

These ion-beam systems have been used to manufacture: superconducting films, thin film heads for electronic equipment, optical thin films, ring-laser gyro-mirrors, dielectric films of all sorts, encapsulation films of all sorts, optical anti-reflection coatings of all sorts, interconnect films on integrated circuits, sensors of all kinds, advanced magnetic heads for DVD players, x-ray optic devices, laser facets, high reflectance mirrors and field-divisional wave multiplexers.

6. Electron Beam Deposition- this technique uses an electron beam instead of an ion beam. In this respect, it is not as versatile as the ion-beam method or other methods used to manufacture thin films.

This completes our survey of the methods of achieving a vacuum, the types of vacuum pumps available and the techniques used to form thin films. We are now ready to continue our description of Devices which utilize phosphors in their operation. This was shown in 7.2.1. given above and includes displays that are not very prevalent at the present.

III. Devices Using Internal Electron Beam Generation

Whereas the devices described above used an external source or mechanism to generate electrons for phosphor excitation, the devices described in this section use an internal electric field to do so.

a. Vacuum Fluorescent Displays

The vacuum fluorescent display is a type of CRT except that the electron beam is diffuse instead of being focused. Its construction consists of: 1) an anode substrate, 2) an electrode assembly and 3) a front glass plate. One construction is shown in the following diagram, given as 7.2.14. on the next page.

In this diagram, we have shown one segment of the display which can indicate only one number. The anode segment is made by evaporating metal strips on the glass substrate (including the electrical lead-ins) and then coating them with a suitable phosphor. Conductive leads are attached, sometimes during the formation of the metal strips. The cathode electrode is formed from a tungsten wire coated with tri-carbonates. The whole is sealed together to form the display with the grid

7.2.14.- Construction of a Flat Vacuum Fluorescent Display

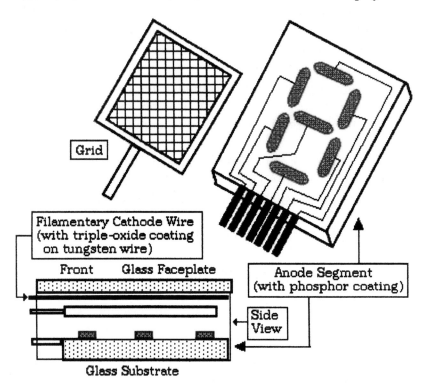

positioned between the two electrodes. The grid is present as an electrode for controlling and accelerating the diffuse electron beam. We have not shown the exhaust tube used to form the vacuum since several of these segments would be joined together to form the actual display used to indicate a set of numbers. Once the vacuum is achieved, the triple carbonate is then heated to form the triple oxide coating which is the active layer emitting electrons. When a voltage is applied between the anode and cathode assemblies, emitted electrons are accelerated to the anode segment connected at that moment, exciting the phosphor coating. The complete hermetically sealed assembly is essentially a triode vacuum tube with multiple anodes and directly heated cathodes. The usual number of segments is no more than six. However, since alphabetic characters can also be displayed, the total number of segments can be as high as 10-12. Vacuum fluorescent display panels have been used since

1972 in cash registers, "point of sale" displays and other similar applications. In operation, the filamentary cathode is heated to about 600 °C thereby releasing thermally-excited electrons which are directed toward the number segment to be displayed. Since low voltages are involved, i.e.- 2 to 50 VDC, the best phosphor for this tube is conductive phosphor, $ZnO:Zn^0$, which has a blue-green emission. Demand for other colors resulted in modification of other CRT phosphors. The phosphor characteristics needed for this device include:

1. Conductivity to prevent surface charging during operation (this prevents electron excitation of the phosphor by negating low voltage electron penetration of the phosphor surface and generation of secondary electrons).
2. Low threshold voltage for emission.
3. Efficient luminescence by low voltage electrons without emission saturation during operation.
4. Long lifetime stability without degradation during operation from harmful materials from the hot cathode.

The first requirement resulted in mixtures of a conducting medium as $InSnO_2$. The mixture proved more efficacious than either of the components. Phosphor mixtures such as $ZnS:Au,Al + InSnO_2$ proved to be much better than $Y_2O_2S:Eu^{3+} + InSnO_2$. The latter mixture degraded to almost zero after a few hundred hours of operation. The following table shows a comparison of some of the phosphors used:

7.2.15.- Phosphors Used in Vacuum Fluorescent Displays

Composition	Color	Effic.- L/watt	Peak,nm	Stability
$ZnO:Zn^0$	green	10.1	505	excellent
$ZnS:Zn^0$, $Cl + In_2O_3$	blue	1.2	465	fair
$ZnS:Cu,Al + In_2O_3$	yellow-green	2.8	548	good
$ZnS:Au,Al + In_2O_3$	green-yellow	3.2	554	very good
$ZnCdS:Au,Al + In_2O_3$	yell.-orange	2.8	592	very good
$ZnCdS:Ag,Cl + In_2O_3$	orange	2.2	605	very good
$ZnCdS:Ag,Al + In_2O_3$	red-orange	1.8	662	fair

Note the rare earth based compositions are not represented. None of these phosphors, except the ZnO:Zn° composition have been entirely adequate. It is, perhaps, for this reason that this display has become nearly obsolete. With the advent of flat panel displays such as the computer controlled plasma display or the back-lighted liquid-crystal displays (which we will describe later), the vacuum fluorescent display has been largely superseded. Such computer controlled displays can show the item purchased, the amount of the sale and keep a running total for the buyer.

b. Field Emission Displays (FED)

The field emission display is a relatively new type of display device. It relies upon "cold-cathode" emission from special electrode "tips" and is aimed at the color TV market. As such, it has not yet found complete acceptance and is still in the developmental stage. The following diagram shows the general structure that has evolved:

7.2.16.-

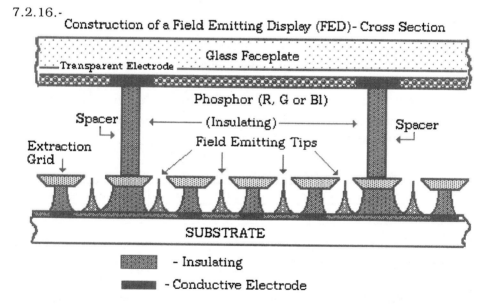

Construction of a Field Emitting Display (FED)- Cross Section

Note that this device depends upon cold-cathode emission of electrons to excite the phosphor dots. The field-emitting tips have been constructed from a variety of materials but diamond (tetrahedral carbon) seems to be the best material. It is hard, resistant to field effects and is conductive. The methods used to form diamond tips have been many. There are four types of carbon structures. These are based upon the bonding between s- and p- orbitals of the carbon atom: diamond (sp^3 - tetrahedral), graphite (sp^2- two dimensional layers), linear (sp^1 - linear and polymeric carbons) and amorphous (where a mixture of bonding types exist). The sp^3 bonding type is typical not only for diamond but for aromatic carbon compounds in general. The linear sp^1 type is typical for aliphatic and polymeric carbon compounds. When pure carbon is used to form FED tips, it has been found that the method used controls the type of field-emitting tip that is produced:

7.2.17.- Field Emitting Tips and Type of Tip Produced

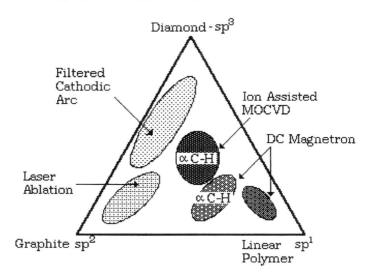

The tetrahedral diamond structure is preferable for several reasons. However, the actual carbon tip structure is always a mixture of the three s-p hybrid electron bond types. The closer that one can get to the pure sp^3 tetrahedral diamond structure, the better performing are the tips produced. This has remained a significant problem in the manufacture of FED displays. The C-H in this diagram indicates that methane was used as the precursor which was decomposed to a carbon layer, but that a certain amount of hydrogen remained in the formed composition. This does not affect the conductivity of the tips but the product is not as hard or conductive as is the diamond-like composition. Nevertheless, the pure diamond structural composition has not been achieved when thin-film techniques are used to form the tip structure.

Note that an array of several thousand tips is used to excite each individual phosphor dot or "pixel". Note that each tip in 7.2.16. is insulated from the voltage control electrodes to prevent "cross-arcing". The extraction-grid is voltage-biased to control the direction of electron acceleration. The advantages of the FED are that it can be made in large display sizes, even over 72 inches across, no scanning is involved, and that it is thin. The disadvantages are the complexity of manufacture of the tips and the extraction grid, as well as an upper limit to the accelerating voltage (< 6 KV DC) that can be applied between the two electrodes. At very high accelerating voltages, internal breakdown (discharges and shorting between components) can occur. Therefore, the FED is limited to fairly low voltages such as a few thousand DC volts at most.

At such voltages, the phosphors that can be used are limited and conductive phosphors seem to perform best. Although the type of phosphor mixtures given in 7.2.15. for vacuum fluorescent displays have been tried, none seem to be entirely satisfactory. The phosphors that have been tried or are being used in the FED display are:
7.2.18.-

$ZnO:Zn^0$ - blue-green = 505 nm.

$(Zn,Mg)O:Zn^0$ - blue = 476 nm.

$ZnGa_2O_4:Mn^{2+}$ - green = 510 nm.

$CaTiO_3:Pr^{3+}$ - red = 610 nm.

$Y_2NbO_4:Bi^{3+}$ - blue

$SrGa_2S_4:Eu^{2+}$ - green

$BaZnS_2:Mn^{2+}$ - red

Of these, the brightest and most efficient phosphor is $ZnO:Zn^0$, but not even this phosphor is completely satisfactory since it saturates at 200 VDC and beyond. The area where the FED has found to be advantageous is the small display and the very large display. For military purposes, the "heads-up" display, where the display is worn by the user, has been satisfactory because the screen brightness does not have to rival that of the CRT TV screen.

c. Large Outdoor Displays

Large outdoors displays such as a "billboard" constitute the last category using internal electron beam generation as a method of phosphor excitation. Actually, the type of display used for this purpose is one where small individual cathode-ray tube's are employed. Each small CRT employs a "Flood-Gun" approach to excite the phosphor screen. A standard screen is about 35 x 24 feet in size. Since the individual tubes are about 1.125 inches in diameter, 38,400 tubes are required to assemble the display. Note that these tubes are of a simple design and do not require to be turned off or on continuously. Nor do they require a scanning raster of the phosphor screen. Once the billboard picture is established by lighting selected tubes, the picture remains until deliberately changed. The advantage to this type of outdoor display is that it is easily viewed under full sunlight as well as night-time conditions. Two types of Outdoor Displays have been made available, one a mono-color and the other a tricolor display.

The other competitor for billboard size displays is that of a modified FED where the display utilizes millions of "tips" to form a pixel-display having sufficient brightness for daylight viewing. This display operates continuously. No scanning electronics or electrical connections are required in contrast to that which is intended for color-TV viewing. Thus, the large outdoor FED display has high brightness since the picture remains constant and the emission tips operate on a constant emission basis. Nevertheless, the CR tube outdoor display has seen more use because of monetary reasons. However, the lifetime of the FED outdoor display does not depend upon the tungsten cathode like the CR tube, but upon the degradation of the emission tips (which is minimal during operation unless they become "poisoned"). However, the ultimate usage of these two types of outdoor displays will depend the relative "lifetime" of each, a factor still has to be determined.

d. Light Emitting Diodes (LED)

LED's became commercially available in the late 1960's with the advent of the green-emitting GaP:N and the red-emitting GaP:Zn,O diodes. GaP is an indirect band-gap semi-conductor. That is, there is a "trap" that exists near the conduction band that acts as a recombination center for light emission. This mechanism is shown in the following diagram:

7.2.19.-

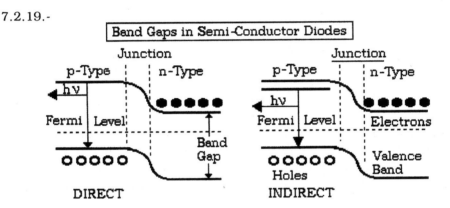

When a voltage is applied, electrons are forced through the p-n junction where they are available for recombination with the holes. The N and Zn,O centers in GaP are iso-electronic traps that provide very efficient recombination centers for electron-hole pairs and subsequent light emission. Nonetheless, the actual brightness for these LED's left something to be desired in relation to other light sources. The only use found for these devices was in signal lights on instruments or on kitchen appliances in homes.

More recently, the advent of improved LED devices, emitting red, orange, green, yellow and blue emission have become available. The progress made is shown in the following diagram, shown on the next page:

7.1.20.-

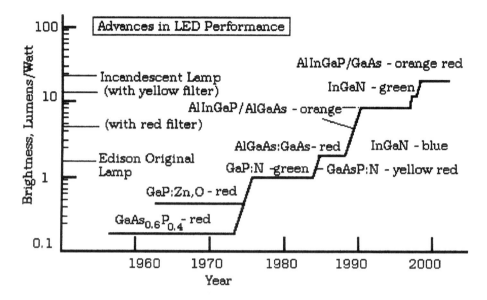

All of these LED's are based on direct- band gap materials. Thus, LED's covering the entire visible spectral range are now available. The AlGaInP double-heterostructure LED's have been fabricated using a technique where the GaAs absorbing substrate was removed and replaced by a n-type GaP substrate which was non-absorbing. This made possible luminous efficiencies that exceed unfiltered 60W tungsten bulb light in the yellow green to red regions (> 15 lumens/watt). This advance opened the way for use in traffic signal lights where the 60 watt bulb with yellow or red filter was replaced by a 20 watt LED bulb. In 1996, an actual installation showed a 38% energy savings and a 1.7 million watt reduction in electrical usage. Lifetimes of such LED's exceed 50,000 hours while incandescent lamps rarely exceed 1200 hours.

The LED's are grown by metal vapor phase epitaxy (MOCVD) on GaAs substrates for matched lattice growth. A p-type GaP window layer serves to form the semi-conductor. What we are addressing is shown in the following diagram, given on the next page as 7.2.21. Typical compounds used for MOCVD epitaxial growth include: tetraethyl silicate, trimethyl metal chlorides, hydrides like arsine, diborane or phosphine, and tributyl-gallium or triethyl-aluminum.

7.2.21.-

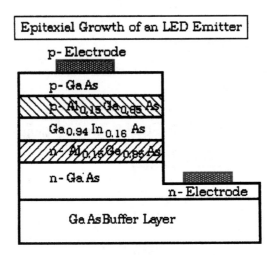

Originally, the LED shown above was grown by thin film techniques upon a sapphire substrate. Nowadays, it is usual to use a neutral substrate, i.e.- neither n- or p- doped, composed of the active lasing composition. We have already addressed the MOCVD method of forming thin films. The wavelength that any given LED will **emit** (in the case of the direct band-gap materials) is predicated upon the band-gap, that is, the energy gap between the valance band and the conduction band of the semi-conductor. Table 7-1, given on the next page shows these electronic properties.

As can be seen, a wide variety of energy gaps exists for these semi-conductors. However, it is the electron-mobility which is the deciding factor in the selection of materials for LED's. That and their relationship to the energy required for visible emission. These include: ZnSe, AlP, GaP, AlAs, AlSb, CdTe, GaAs, and InP. However, AsAs, GaP and AlSb are indirect band-gap semi-conductors and have not been found useful for high brightness LED's.

The family of III-V and II-VI semi-conductors has a wide-range of lattice constants and corresponding band-gaps. Those with smaller lattice

<u>Table 7-1</u>

Intrinsic Properties of III-V and II-VI Semi-Conductors

Semi-Conductor	Lattice Const Å	Energy Gap - electron volt	Electron Mobility *	Hole Mobility*
Si	5.4310	1.11	1,400	470
Ge	5.6461	0.67	3,900	1,900
GaP	5.4506	2.26	110	75
AlP	5.4625	2.45		
GaAs	5.6535	1.42	8,500	400
AlAs	5.6605	2.17	280	~ 20
InP	5.8688	1.35	5,000	150
InAs	6.0584	0.36	33,000	460
GaSb	6.0954	0.72	5,000	850
AlSb	6.1355	1.58	900	450
InSb	6.4788	0.17	80,000	1,250
ZnSe	5.6676	2.80	530	~ 70
ZnTe	6.0880	2.20	530	130
CdTe	6.4816	1.49	700	60

* Mobility @ 300 °K in cm^2/V sec.

constants tend to produce visible emission. Nonetheless, these "single" compositions have not proven to be satisfactory in other respects such as stability and change in refractive index with temperature.

It is for these reasons that the compositions producing the brightest LED's are always a mixed composite. The following diagram, given on the next page as 7.2.22., shows the wavelength range available as a tuned energy for several composite LED compositions. We have also indicated the visible range of radiation.

LED's are constructed from a p-type of semi-conductor and a layer of n-type semiconductor, deposited upon a substrate of a similar semi-conductor to form a p/n junction, as shown above. When voltage is applied in a forward bias (positive to the p-side and negative to the n-side) which

7.2.22.-

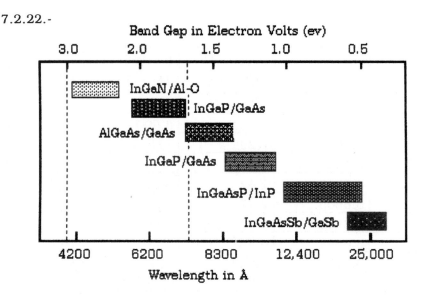

is slightly above the voltage corresponding roughly to the bandgap energy, current begins to flow and light is emitted. This recombination creates photons with energies corresponding to the composite semi-conductor's bandgap (Eg). This gives rise to the following relation:

7.2.23.- $\lambda \ = \ 1240 \ / \ Eg$

Devices comprise layers of specific composition like AlGaAs, an alloy of GaAs and AlAs in which the electronic bandgap and refractive index can be varied through changing the alloy composition. GaAs has an Eg of 1.4 ev. so it produces wavelengths of about 8500 Å. Changing the semi-conductor layers to alloys of AlGaAs or InGaP, which have higher bandgaps, created photons from the near-infrared (8500 Å) to the red (6300 Å). Changing the semi-conductor layers to a lower bandgap material creates 8500 to 16,000 Å radiation. Only "direct-bandgap" materials create radiation efficiently. The colors that can be produced by LEDs are shown in the chromaticity diagram given on the next page as 7.2.24. Note that the range of colors is larger than those covered by display (Television) phosphors. What is not stated is that the brightness leaves something to be desired. The current goal for LED's is to achieve an

7.1.24.-

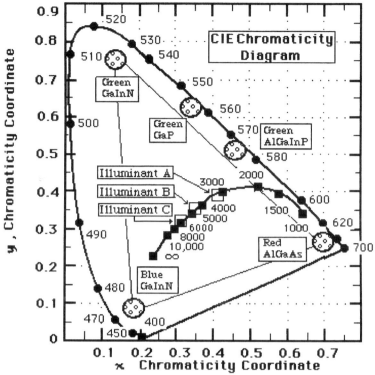

brightness equivalent to fluorescent lamps whose output is close to 80-90 lumens per Watt. The major impetus has been the use of a PHOSPHOR in conjunction with a blue LED which produces a "white" light. The phosphor currently being used is YAG:Ce^{3+} whose excitation band overlaps the output of the blue LED. Improvements in semi-conductor composition and packaging have produced a chip of 1.0 by 1.0 mm with a "white" brightness of 25 lumens using 0.350 amps, at 3.0 volts DC. A 5.0 watt lamp produces 120 lumens or 20 L/W. The LED of just a few years ago produced 1.5 L/W. The color- rendering index is just 70 but is expected to be improved with the use of improved phosphors. White LEDs are currently 15-20% more efficient in brightness than halogen or incandescent lamps. The main advantages of LEDs, in addition to their low power consumption compared to other lamps, is that they only lose 30% of their initial brightness after operating over 50,000 hours. Note

7.2.27.-

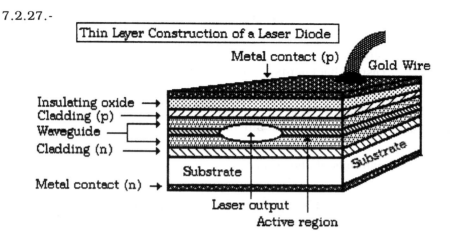

Thin Layer Construction of a Laser Diode

Note that waveguides and mirrors are provided to form a resonant chamber to generate coherent emission. When voltage is applied in a forward bias (positive to the p-side and negative to the n-side) above a voltage corresponding roughly to the active semi-conductor's bandgap, current begins to flow. Holes injected into the n-side layer recombine with electrons and electrons injected into the p-side recombine with the holes (see 7.2.1. given above). Resonant photons are thus emitted if the proper chamber is present.

First developed in 1962, laser diodes have achieved powers that allow them to be used in applications not thought possible a few years ago. The following table, presented on the next page, shows some of the laser compositions, wavelengths and applications now used for these versatile devices. Note that the bandgap of these materials could be calculated from the equation:

$$\lambda = 1240 \ / \ Eg$$

where λ is in nano-meters. This is the same equation already given in 7.2.21. You will note that diode lasers cover a wide range of coherent emission wavelengths, from the visible to the far infrared. Also, the same materials used as for LED's are employed here except that a resonant laser chamber equipped with totally reflecting mirror and an output mirror of

99.8% reflectivity is provided to achieve the resonant condition required for coherent emission of light.

Table 7-5

Semiconductor Laser Materials

Compound	Wavelength in nm.	Applications
GaN/AlGaN	448-490	Disc data storage
Zn(S,Se)	447-480	Disc data storage
ZnCdSe	490- 525	Disc data storage
AlGaInP/GaAs	620-680	Bar code scanners
InGaAsP/InP	1100-1650	Chemical sensors
$In_{0.73}Ga_{0.27}As_{0.58}P_{0.52}$	1330	Optical communication
$In_{0.73}Ga_{0.27}As_{0.90}P_{0.10}$	1550	Metrology
AlGaAsSb/InGaAsSb	2000-3000	Chemical sensors
GaAs/AlGaAs	6000-12,000	Quantum cascade laser
PbEuSeTe	330-5800	High res. spectroscopy
PbSnTe	6300-29,000	Cryogenic spectroscopy

Over recent years, diode lasers have achieved higher output powers, smaller sizes, higher efficiencies and increased reliability. They have made possible a broad range of electronic devices not possible in 1991.

These include: "compact discs" (CD) for read and write devices for computers, and playing music; laser printers; CVD (compact video discs) for use with TV sets to play movies and the like; hand held compact computers; fiber-optic switching chambers; and medical diagnostic devices.

One main advantage of diode lasers is their small size. They can be used singly or fabricated into a pattern for use as a display. One usage has been a circular arrangement for use as traffic lights. Another has been use as a single diode as a signal light. A typical construction is shown in the following diagram, given as 7.2.28. on the next page. Here, we show a "can" in which a single diode is installed. These can be quite small and are replacing the neon-discharge lamps of the past. Diode lifetimes are in excess of 20,000 hours, in contrast to incandescent lamps at 2,000 to 5,000 hours.

7.2.28.-

Typical Laser Can

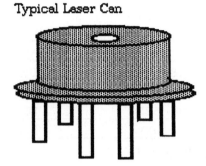

The can size can be quite small and no more than 1/8 inches across. This allows use in device applications not possible before.

Diode lasers also appear in other configurations. One area is their use in pumping solid state crystal lasers. The strongest absorption line of Nd^{3+} lies at 831 nm. That is, the YAG:Nd laser can be best excited at this wavelength. It is relatively easy to configure a diode having this output which would requires a bandgap of 1.5 ev. (See Table 7-1 and 7.2.20.). The construction used is shown in the following diagram:

7.2.29.-

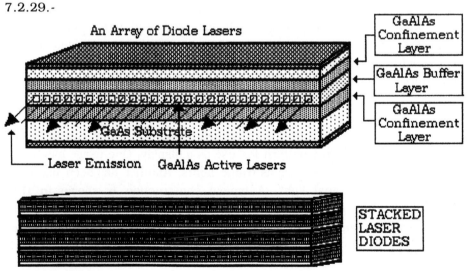

Here, we show how an array of diode lasers are arranged in a stack so that they can be used to generate very high optical power. What we have not shown is the details of the active layer comprising the individual diode lasers. This is shown in the following diagram:

7.2.30.-

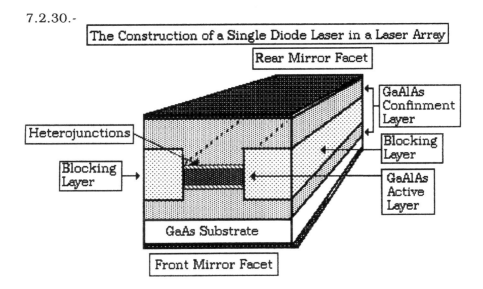

The Construction of a Single Diode Laser in a Laser Array

Rear Mirror Facet

Heterojunctions

Blocking Layer

GaAlAs Confinment Layer

Blocking Layer

GaAlAs Active Layer

GaAs Substrate

Front Mirror Facet

As you can see, a number of individual steps in the thin film process is required to form the individual diode lasers in the array. Such processes are used also to form traffic lights and the like. Note that a mask is needed for each individual step.

Diode lasers have virtually replaced flashlamps and the like as "laser pumps" for high power applications of solid state lasers including all solid state lasers based upon single crystals. What we have shown are "edge-emitting" diode lasers. There are also other configurations including "surface emitting" diodes such as would fit in the "Can" shown in 7.2.28. One such construction is shown in the following diagram, given on the next page as 7.2.31.

Note that the laser emission occurs from the front surface, in contrast to the edge-emission we have already illustrated. It is such configurations

7.2.31.-

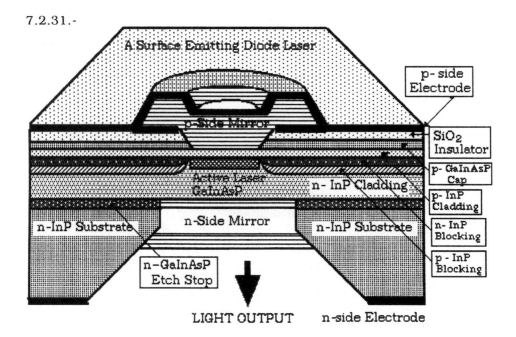

that will make possible the use of a blue diode laser, combined with a proper phosphor, to replace fluorescent lamps in home lighting. That, and the phenomenon known as the "quantum-well", discovered around 1988 in relation to singular diode lasers. The front surface diode laser also makes possible high brightness arrays in almost any configuration, including a round array suitable for traffic lights.

It was the "quantum well" structure that revolutionized the power output of the diode lasers. This is shown in the following diagram, given as 7.2.32. on the next page.

You will note that we have only shown the so-called "quantum wells" of the laser diode. These correspond to the "active layer" of 7.2.27. and 7.2.29. That is, it was found that the "active layer" of the laser could be made much thinner and could be stacked together in the same space as prior diode lasers. It has been determined that, in very thin layers of semi-conductors, i.e.- < 20 nm., quantum effects dominate as the electrons become confined.

7.2.32.- Semi-Conductor Laser Diode Progress

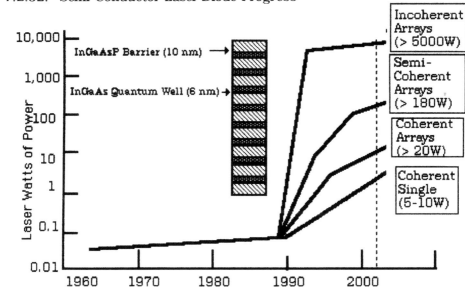

This results in the conduction and valence band becoming quantized into "sub-bands" or quantum wells. These strained layers result in a vastly improved electrical efficiency as the bias current is raised in the diode laser. Even a coherent single array has been improved markedly. The coherent array of 1988 had been improved by 100 times in 2003 while the semi- and incoherent arrays have been improved even more, up to 10^5 times. What this means is that instead of lining up the arrays as shown in 7.2.29., the arrays were arranged in a non-synchronous but cooperative fashion (it does not mean that the diode emission was non-coherent). This resulted in tremendous gains in laser power outputs.

However, a major part of this power improvement was also due to massive gains in diode laser efficiency. Diverse commercial lasers, based upon single crystals, are typically no more than 10% to 20% in efficiency, but usually much less. It is typical for a crystal laser to have a "slope" efficiency of 2-5% with outputs less than 10 milliwatts. InGaAsP diode lasers are efficient by these standards with efficiencies greater than 20%. Strained layer quantum-well structures have been measured with efficiencies in excess of 85% at 1500 nm. wavelength. Typical threshold currents for such diode lasers (the current needed to start the lasing action) is only 20-30 milliamps. Because of the thinness of the active lasing composition, it was found that the electrons and radiation emitted "tunneled" between the layers, thus multiplying the total output from a given current level operation. The optical output was multiplied by several orders of magnitude.

In the following table, we list the level of operating power that the several applications require:

Table 7-6

Wavelength	Application	Approx. Power (Watts)
600 -1000 nm	Illumination	0.04 to 0.20
	Commercial Processing	0.25 to 0.95
	Medical Devices	3.0 to 25.0
	Aviation Uses	15 - 100
1000 - 2000 nm.	Gas Sensing	0.35 - 2.5
	Eye Safe Illumination	0.90 - 7.5
	Lidar	15 - 90
2000 - 5000 nm.	Gas Sensing	0.30 - 2.5
	Eye Safe Illumination	0.90 - 7.5
	Missile Jamming	5.0 - 30.0
	Lidar	40.0 - 200

You will note that we have not listed all of the possible applications for high output diode lasers. Most of the applications listed as Illumination in the 600 to 1000 nm range are those already dominated by solid state or gaseous lasers like the familiar He-Ne red laser. Nevertheless, diode lasers are rapidly replacing those lasers in certain applications.

Diode lasers are revolutionizing medical diagnostics. For example, they have been used in brain-scanning where two wavelengths, 780 and 830 nm., are directed towards a patient's head. The light passes through the scalp and is reflected from the cerebellum. Differences in blood hemoglobin can be made which show the blood flow in tissues. Devices are being used in doctor's offices and hospitals to measure the amount of blood-borne oxygen, as a measure of the capability of a patient's respiratory system.

Two methods now under development include: 1) measurement of blood flow during epileptic seizures in the brain and 2) non-invasive measurement of blood glucose levels using three different wavelengths: 834, 1304 and 1554 nm. by diabetic patients. Compared to conventional methods of medical imaging, NIR (near infra-red) optical tomography is relatively safe (since it has no ionizing radiation) and cheap since it does not require large magnets or radiation shields.

f. Organic Light Emitting Diode Displays (OLED)

The OLED display is based upon certain organic compounds which emit photons in a manner already described for LED's. When a forward-biased current is applied, light is emitted according to the energy gap between the LUMO (lowest unoccupied molecular orbital) and the ground state, LOMO (lowest occupied molecular orbital). We have already discussed this mechanism in Chapter 5 (see 5.5.27.). As shown in the following, only certain types of organic compounds are useful in this application:

7.2.33.- Organic Electroluminescent Compounds

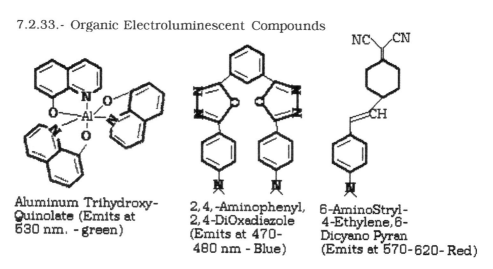

Aluminum Trihydroxy-
Quinolate (Emits at
530 nm. - green)

2,4,-Aminophenyl,
2,4-DiOxadiazole
(Emits at 470-
480 nm - Blue)

6-AminoStryl-
4-Ethylene,6-
Dicyano Pyran
(Emits at 570-620- Red)

These compounds are used to manufacture OLED displays that are used in numerous devices including cell phone displays. Note that all of these compounds are organic semi-conductors. The OLED display is formed by thin film techniques similar to those for LED's and Diode Lasers:

7.2.34.-

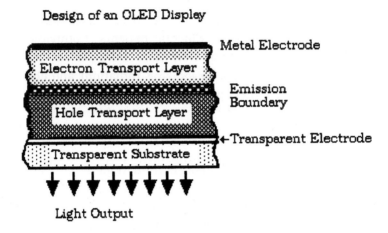

Design of an OLED Display

Light Output

As you can see, the fabrication is rather simple. Yet, the light output rivals many of the LED's and is certainly more than the output of phosphor screens in certain devices. A conductive layer is required for electron injection into the electron transport layer (ETL). The metal electrode (ME) is composed of Al, Ag or Ag-Mg, and is usually light reflective. The hole transport layer (HTL) injects holes from the transparent electrode (TE) into the HTL. The point where the excitons, i.e.- pairs of holes & electrons, recombine is the light emitting (LE) region. This can vary according to the type of compounds used to form the two organic layers. These devices can be classified as:

TE: HTL(LE): ETL: ME
TE: HTL: ETL(LE): ME

When bipolar materials are available, i.e.- have the ability to carry both holes and electrons, the device structure can have a three layer configuration wherein the LE layer is sandwiched between the HTL and ETL, i.e.-

TE: HTL: LE: ETL: ME.

An aromatic diamine: N,N'-diphenyl-N,N'-bis(3-methylphenyl), 1,1'-4,4'-diamine has been used as a HTL. Also used has been the compound shown to the right in 7.2.33. The organic compound, Aluminum tri-hydroxy-quinolate (also given in 7.2.33.) is probably the best ETL material. Laser dyes such as: 4-dicyanomethylene,2-methyl,6,9-dimethylstyrl,4H pyran, have been used as LE layers. Quinadone, i.e.- also is a good LE compound used as the emitting layer in OLED's.

The size of the display is limited only by the size of the thin-film apparatus used to make the film. A large number of organic materials have been examined for this application. Most of the successful ones are those based upon the modification of the basic backbone of those compounds given in 7.2.33. The requirements for a successful OLED are:

1. The compound must produce a stable and uniform thin film under vacuum sublimation.

2. The compound must have a large affinity for electrons, i.e.- be a "hole-conductor"- a p-semiconductor.

3. The compound must have a high electron mobility for transport of electrons to the emitting center.

4. The ionization potential of the compound must be high to prevent loss of hole-energy at the cathode before recombining with electrons, i.e.- the excitons formed during electron energy transport must remain stable until the energy is transmitted to the emitting center.

5. The compound should have large exciton energies to forestall energy transfer from the EM layer to the ETL, i.e.- the exciton energy should be large enough to prevent exciton recombination within the ETL and loss of excitation energy at the emitting centers.

A recent OLED design involved a three-color emission to produce a white-light emitting OLED, as shown in 7.2.35., given on the next page. Note that the construction is similar to that we have already presented. The only difference is that three different colored emitting strata are involved:

7.2.35.- Configuration of a White-Emitting OLED

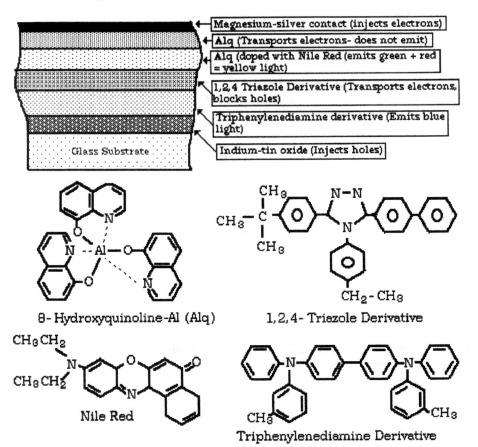

Magnesium-silver contact (injects electrons)

Alq (Transports electrons- does not emit)

Alq (doped with Nile Red (emits green + red = yellow light)

1,2,4 Triazole Derivative (Transports electrons, blocks holes)

Triphenylenediamine derivative (Emits blue light)

Indium-tin oxide (Injects holes)

Glass Substrate

8- Hydroxyquinoline-Al (Alq)

1,2,4- Triazole Derivative

Nile Red

Triphenylenediamine Derivative

This 3- layer device achieved a luminance of 4,200 candela/ square meter at a relatively low voltage of 16 VDC. In contrast, the surface brightness of a cathode-ray tube is only 100 candela/square meter, while a fluorescent lamp has a value of 8,000 candela/square meter. Whether such a device can reach a surface brightness of 10,000 candela/ square meter or not has not yet been determined.

An active research area is currently the use of conducting polymers as OLED devices. These polymeric-organic electronic materials are varied in compositions. In 1957, the first intrinsic electrically conduction polymer, polyacetylene was reported by Chiang et al. Since that time a number of such compounds have reported, as shown in the following:

7.2.36.- Electrically Conducting Polymers

trans-polyacetylene   Polythiophene   Polyaniline

Polypyrrole   Poly(para-phenylene vinylene)   Poly(para-pyridyl vinylene)

These are just a few of the many compounds that have been developed. However, they are **not** electroluminescent (EL) and cannot be used directly for OLED devices. What has been done is to modify the organic structure to produce light-emitting polymers. Polythiophene does not emit whereas derivatives exhibit strong EL emission. There is another advantage as shown in the following, given as 7.2.34. on the next page.

Here, we have shown four derivatives of polythiophene and their spectral properties. Note that the emission band can be shifted across the entire range visible to the human eye. The peak of the band can be shifted to optimize the light output in any desired wavelength, something that the OLED compounds described above cannot do.

The semi-conducting properties of conjugated polymers originate from the delocalized π orbitals formed in these carbon-containing compounds such as poly(phenylene vinylene), polythiophene and poly(paraphenylene). Like the OLED's described above, the polymer-OLED consists of a luminescent film, sandwiched between an anode and a cathode.

7.2.37.- Emission of Some Polythiophene Derivatives

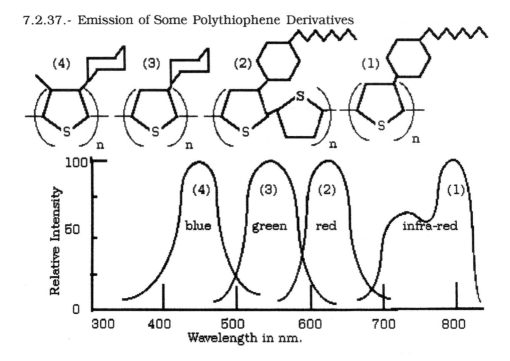

Glass coated with a thin layer of transparent metal and a transparent coating of indium-tin-oxide are most frequently used. The polymer can be applied by a number of techniques including use of a inkjet printer or sublimation methods as given above.. The films are stable and have high quantum efficiency. At this point in time, it is not clear which organic conducting films will be used. Lifetime studies, coupled with relative light output, will probably determine the final decisions made

OLED displays are replacing liquid-crystal displays in many areas. Among these are mobile phones and "personal assistant" displays. Conventional LCD's suffer from limited viewing angle clarity and are difficult to see in ambient light (unless they are "back-lit"). The possibility of using OLED's for full size TV's and computer displays has not yet become realized. However, there is a good likelihood that this display area will be realized. Since the OLED screen can also be printed by inkjet technology, the main problem of forming lines or dots on a TV screen has been circumvented. Because the OLED display is easier to manufacture than either the backlit color-LCD or the LED (mono-chrome or color) display, the impetus to replace these applications is strong. OLED displays also require much less power than either of these technologies. The major problem will be the range of colors that can be displayed and the color-reproduction possible in comparison to the already well-developed phosphor display now used in color-TV, and the filters used in the color-LCD displays. Note that all of these OLED compounds are organic in nature and are fabricated using organic reactions well known since the early 1800"s. Thus, the largest chemical companies like DOW, BASF, Eastman Kodak and DuPont are actively working on OLED displays.

It is well to point out that the use of phosphors has been circumvented in the OLED devices. It may be that, in future years, phosphors will become obsolete as the use of conductive organic compounds and polymers mature.

g. Electroluminescent Displays (EL)

The electroluminescence phenomenon was originally discovered in 1936 by Destriau. Light emission was observed from a ZnS phosphor powder. The device consisted of the powder layer dispersed in castor oil, with the layer between two high voltage electrodes. When the transparent conductive films, SnO_{2-x} or $InSnO_{2-x}$ became available, Sylvania Electric Corp. manufactured an AC device called "PanelLight" that was used as a night light. We have already described the technology of OLED materials (which are actually electroluminescent (EL) substances).

The first EL devices were based upon powdered semi-conductors like $ZnS:Mn^{2+}$ which emits green light. Next came LED materials as described above and finally, OLED EL-devices spanned the gap between the two types of materials. $ZnS:Mn^{2+}$ based EL devices require a rather high voltage to perform, in contrast to that needed for LED's and OLED's. The EL-device consists of an EL-phosphor, sandwiched between two insulating layers. When a DC-voltage is applied, light is emitted. This construction is very similar to 7.2.25. or 7.2.27., as shown in the following diagram, given as 7.2.38. on the next page:

7.3.- CURRENT DISPLAY DEVICES BASED UPON PHOTON CREATION

The types of illuminating and display devices that we will now address are shown in the following:

7.3.1.

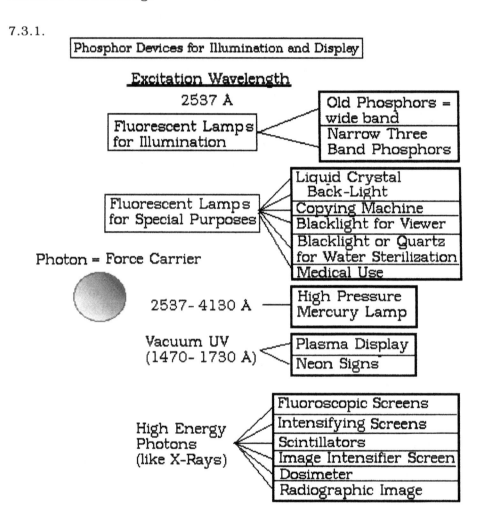

In this case, phosphors as screens or thin-films are excited by photons having various energies, from near UV to x-rays. We will address all of these devices in connection with the phosphors employed in their manufacture and the thin-film structure used to form the screen. We have

already addressed "Fluorescent Lamps for Illumination" earlier. This included substitution of 3-narrow-band phosphors for conventional broadband phosphors which resulted in smaller, but brighter, fluorescent lamps. Such lamps use much less power than their predecessors, as compared to the 40T12 lamp previously used for general illumination. In addition, the latest fluorescent lamps show much better maintenance of output over their entire life. Improved ballasts have also contributed to this remarkable improvement. The F32T5 lamp is now replacing the HID metal halide (HID-MH) lamps in commercial installations since the cost of operating T5 lamps is about 50% that of the HID lamp annually. Whereas the HID-MH lamps lose up to 45% of their initial brightness over time, the T-5 lamps lose only 5%. The new fluorescent technology is now preferred for lighting warehouses and adjoining buildings, offices of all kinds and medical facilities.

I. Fluorescent Lamps for Special Purposes

These lamps include those used for "back-lights" in computer terminals and those used in copying machines.

a. Liquid Crystal Back-Light Displays

In order to show how these lamps are used, it is necessary to describe liquid crystal displays (LCD). We will not describe this technology thoroughly since it does not directly relate to the use of phosphors in the display, except in the fluorescent lamps used for "backlighting" the LCD. A "liquid crystal"- (LC) is an organic compound whose molecules align themselves with an applied electric field. An LC substance flows as a liquid but maintains some of the ordered structure characteristic of a crystal.

Because the molecular forces producing liquid crystalline states are very weak, the structures are easily affected by changes in mechanical stress, electromagnetic fields, temperature, and chemical environment. Three main categories have been recognized: smectic, nematic, and cholesteric:

Smectic liquid crystals consist of flat layers of cigar-shaped molecules with their long axes oriented perpendicularly to the plane of the layer.

Nematic liquid crystals are also oriented with their long axes parallel, but they are not separated into layers, and they behave like toothpicks in a box, maintaining their orientation but free to move in any direction. Nematic substances can be aligned by electric and magnetic fields, resulting in a number of characteristics such as the ability in some cases to be electrically switched from clear to opaque. This property gives rise to many technical applications such as in display systems.

Cholesteric liquid crystals form in thin layers, each one molecule thick; and within each layer the molecules are arranged with their long axes in the plane of the layer and parallel to each other, as a two-dimensional nematic structure.

LCD displays are formed by thin-film techniques in which the liquid crystal layer is surrounded by electrodes which control the alignment of the LC's. If the LCD is a color display, color-filters are used to form a red, green or blue pixel, with a thin-film-transistor at each pixel to control the transmission of light at that point. This construction is shown in the following:

7.3.2.-
CONSTRUCTION OF A COLOR TFT-LCD

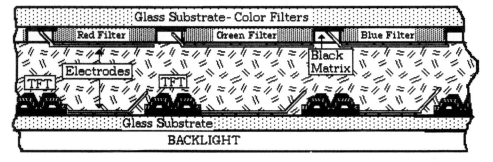

In this case, a "backlight" is used for the light source. A "twisted nematic" liquid crystal is used as a shutter for light transmittal through the glass and the Si TFT transistor controls the shutter. If colored filters are used, then we have a color television or computer monitor display. Note that we have presented a simple structure in order to promote understanding of how such configurations are formed by thin film technology. The actual structure is usually much more intricate. The TFT's control the electrodes which align the liquid crystals. We have not shown the polarizers which are sometimes used in conjunction with the color filters. These are placed just before the backlight and after the filters.

Let us now examine the fabrication of a color TFT-LCD display. A color television uses a cathode-ray tube (CRT) as a display as does a computer monitor. However, the cathode-ray tube is being slowly superseded by liquid-crystal displays (LCD) which are considerably thinner than the glass CRT. LCD devices can be transmissive or reflective. They are made with a demanding set of processes similar to those used for making IC's. But, while IC makers save money by making their chips smaller, display users insist on having their displays larger. This leads to a struggle between cost and utility for display makers. Most color displays rely upon a thin-film transistor (TFT) to control a liquid-crystal (LCD) bit or pixel. The cathode ray television tube uses a pattern of phosphor dots (pixels) with selective excitation to form an image. The TFT-LCD forms an image by controlling the transmission of an external light source by means of the TFT. For producing a red image, all of the red-filter TFT pixels are turned "on" so that the image seen is red. Obviously, color reproduction is dependent upon the quality of the color-filter. Nonetheless, the steps to achieve the final TFT array are many and include steps similar to IC fabrication except that glass is used as the substrate instead of silicon. The following diagram, shown as 7.3.3. on the next page, presents the diagram of a single TFT-pixel on the overall glass substrate. Obviously, the actual number of TFT pixels needed for any given display amounts to many thousands. The overall process involved the use of many "masks" similar to the many steps involved in Integrated Circuit manufacture.

Storage condensers are needed to operate the pixel (active-matrix). Some storage capacitors can be constructed by using part of the gate electrode as a storage capacitor electrode (which is called a Cs-on gate). Note that

7.3.3.-

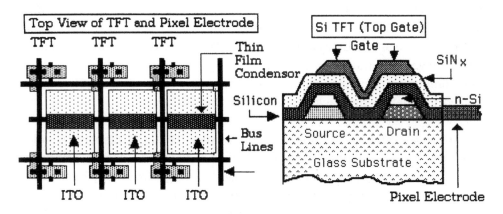

the ITO areas on the left are transparent and are the source of the transmitted light, i.e.- pixels. The three layers are "etched-backed" to define the α- Si islands which form the TFT part of the structure. This process thus defines the island, or "active matrix", which controls each pixel in the display. The transparent electrodes are also not shown in this diagram. Note that this process uses photolithography and plasma dry etching to define the islands by use of masks, the latter called an etch-protect mask. We have not described the steps used to produce a mask. A better view of a single pixel construction is shown in the following:

7.3.4.-

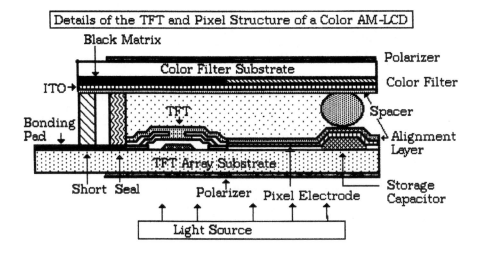

Note that we have now shown all of the details of a single pixel associated with the TFT. The overall construction was given in 7.3.3. and the single transistor and pixel in 7..3.4. Note that the entire space surrounding the TFT is filled with liquid crystal. The same TFT compositions are shown in both diagrams.

We have not yet described the Backlight. This is simply a serpentine fluorescent lamp made in a flat configuration:

A Serpentine Fluoresecent Lamp

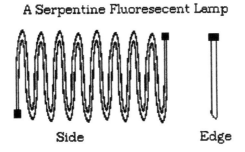

Side   Edge

This lamp is usually formed from a lead-glass which can be easily shaped at relatively low softening temperatures. The lead-glass tube is coated with a phosphor mixture, then evacuated and sealed. It is then bent into shape. The phosphors most used are the 3-line emitting rare earth compositions given above for the 32T5 lamps, i.e.-

$$BaMgAl_{10}O_{17}: Eu^{2+} - \text{BLUE}$$
$$MgAl_{11}O_{19}: Ce^{3+} :Tb^{3+} - \text{GREEN}$$
$$Y_2O_3: Eu^{3+} - \text{RED}$$

The electrodes are mostly of the "cold-cathode" type. That is, they consist of unheated metal electrodes, which emit electrons under a high voltage. More recently, electronic ballasts, described above, have been used as well as micro-wave high-frequency ballasts. The problems of getting these lamps to operate properly upon a DC voltage (using a battery) have been enormous.

Because of the complexity of manufacturing AMLCD displays, it is unlikely that they will ever be used for applications other than computer monitors or "point-of-sale" displays. Because of such complexity, these displays are limited in how big they can be made. This factor involves one of equipment size required more than the actual procedures needed to complete the display. One estimate of the ultimate size that any of the feasible display functions is shown in the following:

7.3.5.- Estimate of Ultimate Market Penetration of Several Display Types

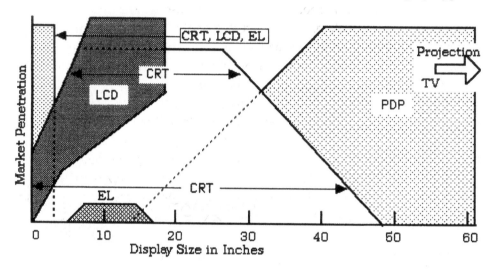

Note that the CRT covers a sizable portion of the current, and estimated future market for displays. This ranges from the very small (one inch or less) to the very large, up to 48 inches across (I have seen such a CRT tube (48") and it is enormous. I estimate that the vertical deflection is close to 80 °- see above discussion on CRT's). For computer monitor displays, the LCD display has taken over a major share of the market while the very small displays are a combination of these two, including EL displays. The larger "Heads-Up" displays are dominated by the EL type of display. Nevertheless, the PDP ("Plasma Display Panel) has become the major type for larger displays and large color-TV's. The color is superior and high resolution displays are easier to manufacture. For the very large displays, the projection TV seems to dominate at this moment. Whether this dominance remains will be determined by how much the PDP and/or projection-TV can be improved. Note that we have not yet discussed the PDP. This will be done below. It should be clear that each type of display is more suitable for certain applications than any of the others.

b. Lamps for Copying Machines

The xerographic process utilizes a toning drum coated with a photosensitive layer. This layer is first charged, usually positive, before scanning begins. Green radiation from a fluorescent lamp is most often used. The lamp is mechanically scanned over the object to be reproduced (most often a sheet of paper containing alphabetical characters and/or images) while the drum is simultaneously turned. When the light from the lamp is reflected from the sheet of paper being scanned, certain areas on the layer on the drum becomes discharged except at the points or lines where the black ink does not reflect the light. Toner particles, being negative, are attracted to these still charged areas and are then transferred to a blank sheet of paper to make the copy. Because the lamp is used in a scanning mode, it is usual to produce an aperture lamp in which a "slot" is induced by wiping a narrow line across the phosphor coating. The lamp looks like this;

Radiation from an aperture lamp is emitted in tight "fan" configuration. For scanning a sheet of printed paper, it is ideal since the emission is concentrated in a relatively narrow band. The phosphor most used for this application is: $Mg_2Ga_2O_4:Mn^{2+}$. More recently, some of the terbium activated phosphors like: $(Mg,Ce)Al_{11}O_{19}:Tb^{3+}$, $LaPO_4:Ce^{3+}:Tb^{3+}$, and $Y_2SiO_5:Ce^{3+}:Tb^{3+}$ have been employed because the major part of their emission is concentrated in the 543 nm line.

c. Blacklight Lamps for Water Treatment and Viewers

The "Blacklight" lamp is essentially a fluorescent lamp that emits in the near Ultraviolet. The $BaSi_2O_5:Pb^{2+}$ phosphor has no peer in this application. This UV-emitting lamp (emission peak at 355 nm.) is used at sewage treatments plants to destroy bacteria and fungi spores in the incoming sewage stream. A combination of lamps surrounds the stream and the ultraviolet emission penetrates the aqueous suspension of sludge to destroy all pathogens present. Although other lamp emissions have been tried, even the quartz low pressure emission at 254 nm., the rate of success has not equaled that of the 355 nm lamp. The phosphor emitting at 370 nm., $(Ba, Sr, Mg)_2Si_2O_7:Pb^{2+}$, is almost as good but the phosphor, $(Ba, Sr, Zn)_2Si_2O_7:Pb^{2+}$, emitting at 295 nm. is much inferior.

d. Medical Uses for Blacklight Lamps

"Blacklight" is also used to view luminescent paints and dyes. There are numerous medical uses where "fluorescent tags" consisting of selected organic complexes of Eu^{3+} and Tb^{3+} are used to identify certain compounds of medical interest, including tagged DNA fragments.

Note that we have been discussing lamps utilizing 254 nm. photon as an excitation energy. Since we have already discussed HPMV lamps which utilize 254-413 nm radiation to excite phosphors (see above), we are now ready to explore the vacuum-UV region which uses photons generated at 147 nm to 173 nm wavelengths to excite phosphor display devices. These include 2 types, plasma display panels (PDP) and Neon Signs.

II. Display Devices Based on Vacuum UV Photon Generation

These devices are based upon a so-called "plasma-discharge" which is a misnomer for the same type of discharge used to create 254 nm radiation. A stable plasma, in physics, consists of an electrically conducting medium in which there are roughly equal numbers of positively and negatively charged particles, produced when the atoms in a gas become ionized. Thus, the low pressure mercury discharge used in fluorescent lamps is a plasma. The modern concept of the plasma state is of recent origin, dating back only to the early 1950s. Its history is interwoven with many disciplines. Three basic fields of study made unique early contributions to the development of plasma physics as a discipline: electric discharges, magnetohydrodynamics (in which a conducting fluid such as mercury is studied), and kinetic theory. Of the two types of light generating devices utilizing plasma-generated vacuum-UV radiation in use today, the Neon Sign is the oldest.

Since both operate on the same principles, we will address the Plasma Display Panel (PDP) first.

a. Plasma Display Panels- Vacuum UV Photon Generation

Whereas most of the display types that we have described are based upon either electron- generation by electric fields of some sort or by mercury-discharge (254 nm photon generation), the newest display is based upon generation of vacuum ultraviolet photons to excite the phosphors. The gaseous discharge uses one or more of the noble gases, i.e.- Ne, Ar, Kr, and/or Xe. Of these, Xe (xenon) seems to be the best, but its strongest emission line lies at 147 nm., in the so-called "vacuum UV" region. Most materials strongly absorb this radiation but some phosphors are not excited at all by this radiation line. The resonant line of atomic Xe is 147 nm but the resonant line of Xe-Xe molecules lies at 173 nm.

The design of a PDP display is complex since the plasma itself is hard to start. Once it is started, it is necessary to maintain it by a voltage less than that needed for the full discharge but enough to keep some gaseous atoms ionized, i.e.- $Xe \Rightarrow e^- + Xe^+$. The following diagram, given on the next page as 7.3.6., shows one construction that is being used.

Note that this display is essentially two plates of glass in which separators are used to form a series of channels along the width and breadth of the device. Within the channels, a series of conductive metal "bus" bars are located. It is these that are covered with sequential layers of phosphors. Note that the entire inside surface is covered. The top glass faceplate consists of **transparent** bus-electrodes, along with "sustain" electrodes, at right angles to the phosphor-covered bus-bars. The dielectric layer is essentially a layer of low-softening glass particles. The final step to form the PDP is to sinter the dielectric layer to form the integral PDP product.

Note that the plasma is only generated at the point where the top and

7.3.6.- Construction of a PDP Display

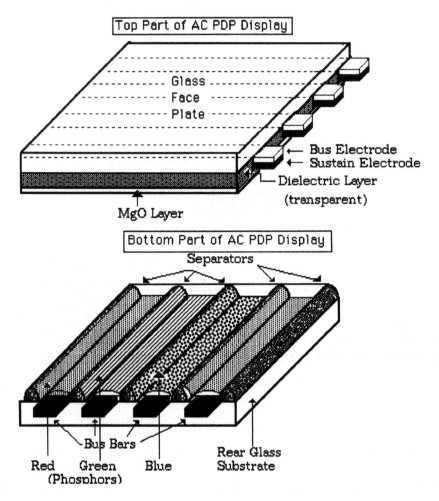

Top Part of AC PDP Display

Glass Face Plate

Bus Electrode
Sustain Electrode
Dielectric Layer
(transparent)

MgO Layer

Bottom Part of AC PDP Display

Separators

Bus Bars

Red   Green   Blue
(Phosphors)

Rear Glass
Substrate

bottom bus-bar electrodes cross each other. Actually, this is a very simplified diagram, since scan electrodes, priming slits, and other features are required for proper operation as a TV monitor and/or display. However, we have not shown these in the interest of simplicity.

The following, given as 7.3.7. on the next page, shows a simple version of the construction of one pixel in the PDP display.

7.3.6.-

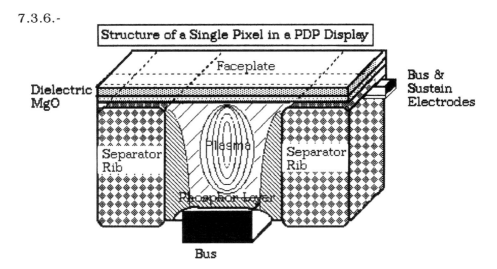

Here, we have only shown the basic parts of the pixel. Note that, in a 48" PDP display, there will be up to several million of these pixels present.

Originally, a PDP based upon the "eximer" discharge was proposed. This involves "Excimer" lasers which had been developed based upon the molecules: ArF, KrF, XeF, and XeCl, which exhibit laser action at: 193 nm., 248, 308 and 350 nm., respectively. The lasing action was obtained by passing an electrical discharge through a suitable mixture of a noble gas and fluorine gas. However, it was soon determined that silicate glass could not be used to form a PDP because of the well known chemical action of F_2 on silica. Although alumina could have been used as the containment for the plasma discharge, it proved to be too expensive, especially in the larger sizes. Thus, Xe gas remains the major one used, with small amounts of the other noble gases sometimes added. Although some of the original PDP displays were based upon a DC voltage to excite each pixel of the display, the most recent ones use AC voltage and current to originate and maintain the plasma at the pixels being excited. You will note that we have not attempted to describe the electrical characteristics and the electronic parts of the PDP design. For example, a 48 " display requires at least a 800 line-scan to address the display at > 200 volts, with access times of 3-4 μsec. That is, the repetition rate must occur in that time to sustain a moving color-picture without apparent blurring to the human eye. Large area high-resolution PDP panels are driven at about 1.0 µsec., and at a repetition rate of 3 µsec.

Although a great body of work had been directed towards the excitation characteristics of lamp and cathode-ray phosphors, little had been published concerning the vacuum-UV (VUV) excitation attributes, particularly at 147 and 173 nm. The major reason for this is that both nitrogen and particularly oxygen strongly absorb at those wavelengths. The obvious solution to this dilemma is to construct a VUV "plaque-tester" (see 6.7.1. of the last chapter. However, the design will be simpler since the plasma discharge is generated internally. One design is shown in the following:

7.3.8.- Design of a VUV Plaque Tester

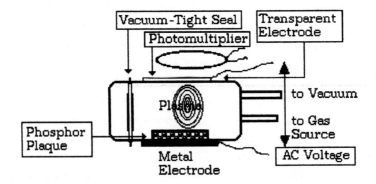

We have not shown the auxiliary equipment needed, just the essentials. One of the most important features is maintaining a vacuum-tight seal. The other is being able to evacuate to > 10^{-8} torr while eliminating **all** traces of water vapor. This means operating in a controlled atmosphere room where the humidity is kept low (or in a "glove-box" where the humidity is very low). The results of such measurements started the development of a spectroradiometer where the excitation-emission monochromators were evacuated as well. The final results are shown in the following table, given on the next page, compares several phosphors under VUV excitation.

This is not a comprehensive listing of all of the phosphors investigated. The most important part for a color-television display is the brightness produced at the "Daylight" color, i.e.- the Illuminant-C color

Table 7-7

Phosphor Composition	Color	Rel. Lum. Output
$Zn_2SiO_4:Mn^{2+}$	green	1.00
$Y_2O_3:Eu^{3+}$	RED	0.71
$YBO_3:Eu^{3+}$		1.00
$GdBO_3:Eu^{3+}$		0.92
$YGdBO_3:Eu^{3+}$		1.20
$Zn_3(PO_4)_2:Mn^{2+}$		0.34
$Zn_2Ga_2O_4:Mn^{2+}$	GREEN	0.23
$BaMgAl_{14}O_{23}: Mn^{2+}$		0.92
$BaMgAl_{12}O_{19}: Mn^{2+}$		1.15
$YBO_3:Tb^{3+}$		1.12
$CaWO_4$	BLUE	0.76
$BaMgAl_{10}O_{17}: Eu^{2+}$		1.65
$Y_2SiO_5:Ce^{3+}$		1.14

.Some of the combinations used are shown in the following:

7.3.9.- Tri-Color Phosphors Used in PDP Display

Phosphors			Rel. Brightness
Blue	Green	Red	
$CaWO_4$	$Zn_2SiO_4:Mn^{2+}$	$Y_2O_3:Eu^{3+}$	65
$Y_2SiO_5:Ce^{3+}$	$Zn_2SiO_4:Mn^{2+}$	$Y_2O_3:Eu^{3+}$	105
$BaMgAl_{10}O_{17}: Eu^{2+}$	$Zn_2SiO_4:Mn^{2+}$	$YBO_3:Eu^{3+}$	150
$BaMgAl_{14}O_{23}: Eu^{2+}$	$Zn_2SiO_4:Mn^{2+}$	$YGdBO_3:Eu^{3+}$	182

As you can see, the luminance of the PDP display is predicated upon the output of the phosphors when excited by the plasma discharge. What we have not stated is that some phosphors are not as stable under the processing conditions used to make the PDP. At the present time, the phosphors being used for the PDP display are:

$$BaMgAl_{10}O_{17}: Eu^{2+} \text{ - blue}$$
$$Zn_2SiO_4:Mn^{2+} \quad \text{ - green}$$
$$YGdBO_3:Eu^{3+} \quad \text{ - red}$$

Of the displays that can be used for color-television, the PDP panel outperforms all of its competitors. The screen is brighter, resulting in superior contrast for high definition display.

b. Neon Signs

Signs are ubiquitous in our society. The word "sign" originally meant a device used for advertising purposes. The ancient Egyptians, Greeks and Romans created signboards by whitewashing convenient sections of walls for suitable inscriptions. Early shop signs were developed when tradesmen, dealing with a largely illiterate public, devised certain easily recognizable emblems to represent their trades. Some signs like the three golden balls of the pawnbroker, and the red and white stripes of the barber early became identified with particular trades.

When practical electric generators were invented in the late 19th century, illumination became possible for shop signs and billboards, and by 1910 the French scientist Georges Claude was experimenting with the neon tube and other gas-filled illuminating devices. In less than a decade, signs were being fashioned of glass tubes bent to form words and designs that glowed red when the gases inside them were subjected to an electric current. A typical neon sign tubing consists of a 10-15 mm (0.59") glass tubing at least 3 meters (118.5") long. The tubing has end-caps attached and then is evacuated. The glass is usually a lead-glass with a low temperature softening point. The evacuated tubing is heated to remove all traces of water and air and then is "backfilled" with 0.8-2.3 torr of neon gas. The tubing was then bent into a desired shape and operated to produce a reddish orange plasma discharge. The plasma discharge originally used a cold cathode operating mode at high AC voltage (but low

AC current). Nowadays, the high frequency ballasts are used, similar to those used for fluorescent lamps. Because colors other than the red plasma discharge of neon were needed to make the sign more effective for advertising, the tubing was internally coated with phosphors. After bakeout, mercury and rare gases was added and the tubing was then sealed with endcaps similar to a miniature fluorescent lamp. The only difference was in the glass used since the soda lime glass used in fluorescent lamps could not be easily bent into shapes. As a matter of fact, the flat "backlights" described above were derived from neon-sign technology. The electrodes are mostly of the "cold-cathode" type. That is, they consist of unheated metal electrodes, which emit electrons under a high voltage. More recently, electronic ballasts, described above, have been used as well as micro-wave high-frequency ballasts. A variety of phosphors are used similar to those cited above for the PDP display.

The nighttime skyline of every city in the world has changed as electric neon signs came to dominate the main commercial streets. The city with the most "neon-lights" is, of course, Las Vegas, Nevada.

III. Devices Utilizing High Energy Photons

We will first describe devices which use phosphors which respond to high energy photons like x-rays and gamma rays (high energy x-rays).

a. X-Ray Intensifying Screens

There are two types of x-ray intensifying screens. One is passive and the other active. The passive type of X-ray phosphor intensifying screen is used to increase the speed of radiographic films using light emission from the phosphor. The phosphor must have high response to x-ray excitation since the x-ray absorption of the radiographic film is very low, at best 1%. Its construction is shown in the following diagram, given as 7.3.10 on the next page.

The construction of the cassette uses two reflective light layers, one a highly reflecting layer for the phosphor emission and the other highly absorbing. The film is sandwiched between these optical layers. This arrangement reduces blurring of the image due to light scattering and/or movement by the patient. It also makes it possible to use a small focus x-ray tube. Note that the film has two protective layers coated with silver halide emulsion layers on both sides. The image quality of x-ray intensifying screens is appraised by sharpness of the image, graininess of the image and contrast obtained. Contrast can be altered by changing the 7.3.10.-

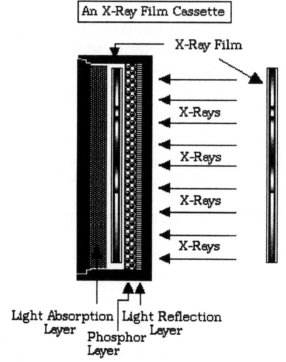

An X-Ray Film Cassette

X-Ray Film

X-Rays

X-Rays

X-Rays

X-Rays

Light Absorption | Light Reflection
Layer | Layer
Phosphor
Layer

x-ray absorption attributes of the phosphor. The characteristics of phosphors required for this application are:

1. High emission efficiency
2. Strong x-ray absorption
3. Emission spectrum matched to the spectral sensitivity of the x-ray film
4. Short emission decay time

5. Stability to x-ray stimulation- no degradation over time
6. Good physical properties including narrow PSD and low particle size, and good coating dispersion to form the screen.

The following lists some of the phosphors used for this application.

7.2.11.- Phosphors used in X-Ray Image Intensifier Applications

Composition	Color	Efficiency - X-ray Excit.	Peak,nm	X-ray Response
$BaFCl:Eu^{2+}$	violet	13%	380	49.3
$BaSO_4: Eu^{2+}$	violet	6	390	45.5
$CaWO_4$	blue	5	420	61.8
$Gd_2O_2S:Tb^{3+}$	green	13	545	59.5
$LaOBr:Tm^{3+}$	blue	14	360,460	49.3
$La_2O_2S:Tb^{3+}$	green	12.5	545	52.6
$YTaO_4:Nb^{3+}$	blue	11	410	67.4
$ZnS:Ag^0, Cl$	blue	17	450	9.7
$(Zn,Cd)S:Ag^0,Cl$	green	19	530	38.4

The best phosphors for this application have been found to be: $LaOBr:Tm^{3+}$, $Gd_2O_2S:Tb^{3+}$ and $YTaO_4:Nb^{3+}$. The active type of X-Ray intensifier screen, utilizing phosphors, uses a combination of excitation sources, x-rays and electrons. It is used to detect x-rays under a number of conditions such as direct medical viewing at low x-ray intensity and in airports to view baggage contents. The active x-ray device consists of a quasi-CRT, except that it is constructed so that electrons are emitted somewhat like the image intensifier of 7.2.7. However, instead of detecting photons, like the night-vision goggles, it detects x-rays in a coherent pattern to form a dynamic picture. Its construction is shown on the next page as follows.

The qualifications for this type of display include:
1. High image brightness;
2. Sufficient image resolution.

7.3.12.-

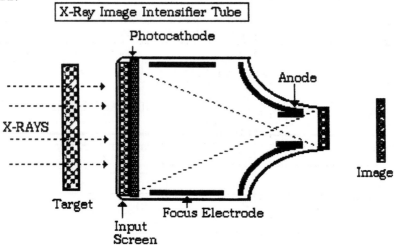

The phosphors used for the input screen must have:

    1. A large x-ray absorption coefficient;
    2. High luminescent efficiency when exposed to x-rays;
    3. An emission spectrum matched to that of the photocathode.

Since the focused beam current in this type of modified CRT display is much lower than in the ordinary CRT, the output phosphor must have:

    1. High luminous efficiency at low current inputs;
    2. High resolution within the image, especially if this type of display is used for medical imaging (Note that the primary usage is predicated upon maintaining a low x-ray dosage to the patient). In general the output screen is prepared by thin film methods in order to achieve the high resolution required.

The following, given as 7.3.13. on the next page, lists some of the phosphors used for this application. You will note that some of the phosphors used for the image intensifier tubes of 7.2.7. like P-31 and P-36 have been used for this application.

7.3.13.- Phosphors Used for Image X-Ray Intensification Screens

Input Phosphors	Formula	Emission color
	(Zn,Cd)S:Ag	green
	$ZnCdS_2$:Cu:Ni	green
	Gd_2O_2S:Tb^{3+}	green
Output Phosphors	ZnS:Cu	green
	$ZnCdS_2$:Cu	green-yellow
	$YTaO_4$:Nb	blue

It is well to note here that the differences between x-ray:

Fluoroscopic screens
Intensifying screens
Image intensifier screens
Radiographic image screens

is not great. All of these utilize phosphors that respond to x-ray excitation. The following lists some of the phosphors that have been developed for detecting x-rays in various applications:

7.3.14.- Phosphors for X-Ray Detection

Formula	Emission color-peak	X-RAY Application
(Zn,Cd)S:Ag	green	Fluoroscopic screens
$ZnCdS_2$:Cu:Ni	green	Fluoroscopic screens
Gd_2O_2S:Tb^{3+}	green-545 nm	Intensifying screen
BaFCL:Eu^{2+}	purple-380 nm	Intensifying screen
LaOBr:Tm^{3+}	blue-	Fluoroscopic screens
$YTaO_4$:Nb	blue-410 nm	Radiographic image
$(Gd,Y)_2O_3$:Eu^{3+}	red-610 nm	Fluoroscopic for CT

X-ray intensifying screens are used in radiological diagnosis and in industrial non-destructive testing. X-ray fluoroscopic screens are used mainly in health examinations and in radiological examination of luggage

They are emitted in all directions. The total energy represented by this light (given as the number of photons multiplied by the average photon energy) is a small fraction of the original particle energy deposited in the scintillator. This fraction, called scintillation efficiency ranges from about 3 to 15 percent for common scintillation materials. The photon energy (or the wavelength of the light) is distributed over an emission spectrum that is characteristic of the particular scintillation material, i.e.- modified by vibronic coupling. The excited species have a characteristic mean lifetime, and their population decays exponentially. The decay time determines the rate at which the light is emitted following the excitation and is also characteristic of the particular scintillation material. Decay times range from less than one nanosecond to several microseconds and generally represent the slowest process in the several steps involved in generating a pulse from the detector. There is often a preference for collecting the light quickly to form a fast-rising output signal pulse, and short decay times are therefore highly desirable in some applications.

3. Some fraction of the light leaves the scintillator through an exit window provided on one of its surfaces. The remaining surfaces of the scintillator are provided with an optically reflecting coating so that the light that is originally directed away from the exit window has a high probability of being reflected from the surfaces and collected. As much as 90 percent of the light can be collected under favorable conditions.

4. Only a fraction of the emerging light photons are converted to charge in a light sensor normally mounted in optical contact with the exit window. This fraction is known as the quantum efficiency of the light sensor. In a silicon photodiode, as many as 80 to 90 percent of the light photons are converted to electron-hole pairs, but in a photomultiplier tube, only about 25 percent of the photons are converted to photoelectrons at the wavelength of maximum response of its photocathode.

The net result of this sequence of steps, each with its own inefficiency, is the creation of a relatively limited number of charge carriers in the light sensor. A typical pulse will correspond to at most a few thousand charge carriers, i.e.- electrons. This figure is a small fraction of the number of electron-hole pairs that would be produced directly in a semiconductor detector by the same energy deposition. One consequence is that the energy resolution of scintillators is rather poor owing to the statistical fluctuations in the number of electrons actually obtained. For example, the best energy resolution from a scintillator for 0.662 MeV gamma rays (a common standard) is about 5 to 6 percent. You will also note that the major action of a highly energetic particle within a symmetric crystal structure is the creation of Frenkel pairs along the charged particle track. NaI:Tl$^+$ is unique in that it "self-repairs" itself at room temperature over time. Most other scintillators like ZnS:Ag^0 form irreversible vacancies and Zn$^+$ interstitials which lower its efficiency over the time it is being used.

There are many characteristics that are desirable in a scintillator, including high scintillation efficiency, short decay time, linear dependence of the amount of light generated from deposited energy, good optical quality, and availability in large sizes at modest cost. No known material meets all these criteria, and therefore many different materials are in common use, each with attributes that are best suited for certain applications.

In Table 7-10, we show the properties of some scintillators in use today. Positron emission tomography (PET) is a highly sensitive technique for diagnosing stroke and other neurological diseases such as multiple sclerosis and epilepsy. Positron-emitting radionuclides with short half-lives (Notably ^{97}Tc$_{43}$ with a half-life of 6 minutes) are used to detect cerebral blood flow, oxygen utilization, and glucose metabolism, providing both qualitative and quantitative information regarding metabolism and blood flow, such as in the heart.

The introduction of computed tomography (CT scan) in 1972 was a major advance in visualizing almost all parts of the body. Particularly useful in diagnosing tumors and other space-occupying lesions, it uses a tiny X-ray beam that traverses the body in an axial plane. Detectors record the strength of the exiting X rays.

Table 7-10

Phosphor	Peak, nm.	Decay, nsec	Yield *	Absorp**	Use
$NaI:Tl^+$	415	230	38,000	2.22	XCT,PET,Nucl.
$CsI:Tl^+$	560	680	32,000	3.24	XCT,
$LiI:Eu^{2+}$	470	1,400	11,000	2.71	neutrons
$ZnS:Ag^0$	450	200	49,000	0.84	α-ray
$CaF_2:Eu^{2+}$	435	900	24,000	0.47	β-ray
$YAlO_3:Ce^{3+}$	370	28	15,000	1.37	XCT
$CdWO_4$	470	5,000	14,000	7.68	XCT
$PbWO_4$	480	300	1,500	12.9	Nucl
$Bi_4Ge_3O_{12}$	480	300	3,800	10.8	XCT,PET,Nucl

*- Number of photons/ MeV of particle energy; **- Absorption coefficient at 150 kev particle energy (cm^{-1}); Use: XCT = x-ray computer tomography, PET = Positron emission tomography, Nucl = nuclear emission detection

This information is then processed by a computer and a cross-sectional image of the body produced. CT is the preferred examination for evaluating stroke, particularly subarachnoid hemorrhage, as well as abdominal tumors and abscesses. CT scans provide two-dimensional views of cross sections of the body, and these images must be viewed in sequence by the radiologist. Computer technology now makes it possible to construct holograms that provide three-dimensional images from digital data obtained by conventional CT scanners. These holograms can be useful in locating lesions more precisely and in mapping the exact location of coronary arteries when planning bypass surgery or angioplasty.

Scintillators used for γ-ray detection from: $^{97}Tc_{43}$ (141 kev), $^{123}I_{53}$ (159 kev) and $^{210}Tl_{81}$ (75 kev), include NaI:Tl and $Bi_4Ge_3O_{12}$. The former is available in 80 cm single-crystal plates, a form not easily achieved by other scintillator phosphor compositions.

Many new scintillators like: $La_2SiO_5:Ce^{3+}$, $Lu_2SiO_5:Ce^{3+}$, and $YAlO_3:Ce^{3+}$ are being investigated because of the known very low decay times, as short as 1 nano-second. There use would allow much shorter exposure times for patients to x-rays. The key is the ability to grow good optical grade crystals. Although a bed of phosphor particles has been tried, the packing density achieved has always been too low and the pulse output never approaches that of the single crystal NaI:Tl phosphor.

c. Long Decay Phosphors

This is the last subject we will examine. Long decay phosphors are used in phosphorescent paints as distinguished from radioluminescent paints. In the latter, radium ($^{226}Ra_{28}$) is added to a suspension of a phosphor like ZnS:Cu or ZnS:Ag. The emitted alpha particles excite the phosphor and the paint glows on a continuous basis since the half-life of Ra is 1,600 years. In contrast, phosphorescent paint glows because of the phosphors used as a pigment in its manufacture. These phosphors are prepared so that deep-traps appear in its Brilloiun bands of the solid. As a result, when they are excited by visible (mostly blue) light or near UV, they glow for hours because the deep traps are emptied very slowly at room temperature. We will not attempt to describe the exact mechanism of how these traps are depopulated. Briefly, it involves "hole-traps" near the valance band. The phosphors now in used for making such paints is shown as follows:

Table 7-8

Phosphor	Color	Peak-nm	After-Glow* (10 min)	Persistance# (min)
$CaSrS:Bi^{3+}$	blue	450	5	90
$CaAl_2O_4:Eu^{3+}:Nd^{3+}$	blue	440	35	1000
$SrAl_2O_4:Eu^{3+}$	green	520	30	2000
$SrAl_2O_4:Eu^{3+}:Dy^{3+}$	green	520	~ 100	> 2000
ZnS:Cu,Cl	y-green	530	45	200
ZnS:Cu:Co	y-green	530	40	> 500
$CaSrS:Eu^{3+}:Tm^{3+}$	red	650	12	45

* Initial After-Glow = 100 when exposed to 100 lux of 300-400 nm light
# Persistence is defined as time to decay to 0.3 mcd./m^2, the limit of light perception by the human eye.

When these phosphors are mixed with a varnish resin to form a paint, the result is used for marking stairwells and the like where a sudden loss of light is compensated by the phosphorescent paint. The output is sufficient to allow easy egress by any stranded person from the darkened vantage point.